Vehicle Design Processes, 2nd Edition

Vehicle Design Processes, 2nd Edition

Guest Editors

Ralf Stetter
Udo Pulm
Markus Till

MDPI

Basel • Beijing • Wuhan • Barcelona • Belgrade • Novi Sad • Cluj • Manchester

Guest Editors

Ralf Stetter
Mechanical Engineering
Ravensburg-Weingarten University
Weingarten
Germany

Udo Pulm
Mechanical Engineering and Production
Hamburg University of Applied Sciences
Hamburg
Germany

Markus Till
Mechanical Engineering
Ravensburg-Weingarten University
Weingarten
Germany

Editorial Office
MDPI AG
Grosspeteranlage 5
4052 Basel, Switzerland

This is a reprint of the Special Issue, published open access by the journal *Vehicles* (ISSN 2624-8921), freely accessible at: https://www.mdpi.com/journal/vehicles/special_issues/05772LR46Y.

For citation purposes, cite each article independently as indicated on the article page online and as indicated below:

Lastname, A.A.; Lastname, B.B. Article Title. *Journal Name* **Year**, *Volume Number*, Page Range.

ISBN 978-3-7258-3879-0 (Hbk)
ISBN 978-3-7258-3880-6 (PDF)
https://doi.org/10.3390/books978-3-7258-3880-6

Cover image courtesy of Ralf Stetter

Contents

Ralf Stetter, Udo Pulm and Markus Till
Vehicle Design Processes, 2nd Edition
Reprinted from: *Vehicles* **2025**, 7, 33, https://doi.org/10.3390/vehicles7020033 1

Ralf Stetter
Innovative Vehicle Design Processes Based on the Integrated Framework for Abstract Physics Modeling (IF4APM)
Reprinted from: *Vehicles* **2024**, 6, 64, https://doi.org/10.3390/vehicles6030064 5

Timo Schuchter, Markus Till, Ralf Stetter and Stephan Rudolph
Digital Integrated Design and Assembly Planning Processes for Sports Vehicles Usingthe Example of a Skateboard
Reprinted from: *Vehicles* **2025**, 7, 22, https://doi.org/10.3390/vehicles7010022 24

Raphael Mieth and Frank Gauterin
Comparative Assessment for Holistic Evaluation of Drive Systems
Reprinted from: *Vehicles* **2024**, 6, 17, https://doi.org/10.3390/vehicles6010017 42

Jannik Kexel, Jonas Müller, Ferris Herkenrath, Philipp Hermsen, Marco Günther and Stefan Pischinger
Virtual Plug-In Hybrid Concept Development and Optimization under Real-World Boundary Conditions
Reprinted from: *Vehicles* **2024**, 6, 58, https://doi.org/10.3390/vehicles6030058 54

Fabian Weitz, Niklas Leonard Ostendorff, Michael Frey and Frank Gauterin
Lifting Actuator Concept and Design Method for Modular Vehicles with Autonomous Capsule Changing Capabilities
Reprinted from: *Vehicles* **2024**, 6, 51, https://doi.org/10.3390/vehicles6030051 87

Sharath Christy Anand, Florian Mielke, Daniel Heidrich and Xiangfan Fang
Optimization, Design, and Manufacturing of New Steel-FRP Automotive Fuel Cell Medium Pressure Plate Using Compression Molding
Reprinted from: *Vehicles* **2024**, 6, 41, https://doi.org/10.3390/vehicles6020041 106

Rudolf Keil, Jan Alexander Tschorn, Johannes Tümler and Mehmet Ercan Altinsoy
Evaluation of SiL Testing Potential—Shifting from HiL by Identifying Compatible Requirements with vECUs
Reprinted from: *Vehicles* **2024**, 6, 44, https://doi.org/10.3390/vehicles6020044 130

Zhiyou Sun and Boyoung Lee
Exploring Factors Influencing Electric Vehicle Purchase Intentions through an Extended Technology Acceptance Model
Reprinted from: *Vehicles* **2024**, 6, 72, https://doi.org/10.3390/vehicles6030072 159

Mohammad Abu Tami, Huthaifa I. Ashqar, Mohammed Elhenawy, Sebastien Glaser and Andry Rakotonirainy
Using Multimodal Large Language Models (MLLMs) for Automated Detection of Traffic Safety-Critical Events
Reprinted from: *Vehicles* **2024**, 6, 74, https://doi.org/10.3390/vehicles6030074 191

Luca Vecchiato, Federico Capraro and Giovanni Meneghetti
Design, Topology Optimization, Manufacturing and Testing of a Brake Caliper MADE of Scalmalloy® for Formula SAE Race Cars
Reprinted from: *Vehicles* **2024**, *6*, 75, https://doi.org/10.3390/vehicles6030075 . . . . **211**

Sergii Panchenko, Juraj Gerlici, Alyona Lovska, Vasyl Ravlyuk and Ján Dižo
Prediction of Residual Wear Resources of Composite Brake Pads of a Modernized Brake System of Freight Wagons
Reprinted from: *Vehicles* **2024**, *6*, 97, https://doi.org/10.3390/vehicles6040097 . . . . **233**

Anis Hamza, Kamel Bousnina, Issam Dridi and Noureddine Ben Yahia
Revolutionizing Automotive Design: The Impact of Additive Manufacturing
Reprinted from: *Vehicles* **2025**, *7*, 24, https://doi.org/10.3390/vehicles7010024 . . . . **253**

vehicles

MDPI

Editorial

Vehicle Design Processes, 2nd Edition

Ralf Stetter [1,*], Udo Pulm [2] and Markus Till [1]

1 Faculty of Mechanical Engineering, Ravensburg-Weingarten University of Applied Sciences, 88250 Weingarten, Germany; till@rwu.de
2 Department of Mechanical Engineering and Production Management, Hamburg University of Applied Sciences, 20099 Hamburg, Germany; udo.pulm@haw-hamburg.de
* Correspondence: ralf.stetter@rwu.de

Received: 28 March 2025
Accepted: 8 April 2025
Published: 9 April 2025

Citation: Stetter, R.; Pulm, U.; Till, M. Vehicle Design Processes, 2nd Edition. *Vehicles* **2025**, *7*, 33. https://doi.org/10.3390/vehicles7020033

1. Introduction

This Special Issue reports on the current status of research concerning vehicle design processes. Designing vehicles is one of the most challenging tasks in engineering for several reasons. Considerable consumer expectations and intense global competition complicate vehicle design. Currently, dramatic changes are influencing vehicle design, mainly caused by the transition to electric drive systems and new vehicle application scenarios, especially in the Asian market. Cost-driven design is a necessity, and vehicles must be economical in terms of production, operation, and recycling. Sustainable design is also imperative in order to achieve ecological vehicles. The dynamics of vehicles must be considered in the design of all components, and lightweight design is of fundamental importance. Consumers expect convincing functional performance, high product quality, an appealing appearance, high reliability, interconnected functionality, and comprehensible and appealing user interfaces. Vehicles are expected to provide more and more additional services, also in connection with external service providers. These colossal requirements lead to complex multi-domain vehicle design processes because most of the important decisions are made in the design phase. Production optimization and intelligent operation are important topics, but flaws and insufficiencies in the design stage lead to considerable expenditure in later stages and suboptimal products. The design processes of vehicles involve thousands of engineers, who are spread out across the globe and need to consider multiple product versions and variants as well as multi-company product platforms. Testing necessities and legal issues frequently play an important role in these processes, and the economic and ecological quality of a product must be monitored throughout these processes. Vehicle safety and ergonomic quality must be considered even in the early stages. Naturally, only digital support makes these processes feasible. For all domains, powerful computer tools for synthesis, analysis, evaluation, and optimization have been created, and numerous attempts have been made to sensibly link the data used in all these tools. However, the multitude of domain-specific and generic data formats and the sheer size of the data still cause serious problems. Importantly, design is also connected with scheduling and project management because certain design decisions can lead to long-term testing and production preparation processes.

In recent years, research teams worldwide have proposed novel, expanded, and improved approaches, which may support, enhance, and streamline vehicle design processes. The promising approaches concentrate on the abstract levels of product modelling—especially the function level [1] and the level of abstract physics [2]. Leading companies in the field of vehicle development and production have made a shift from document-based process management to model-based systems engineering (MBSE), and several research groups have contributed to this transition [3–5]. A central scientific field in this context is the automation of design processes. Graph-based design languages (GBDLs) constitute an important example; they have led to a paradigm shift in design processes, enabling the integration of data, information, and knowledge into product and systems models and fostering design automation [6–8]. Another central scientific field emphasizes the expansion of the established design-for-X guidelines. Important examples are fault-tolerant design (FTD) [9], design for sustainability [10], and design for resilience [11]. Another promising field is the integration of control and diagnosis capabilities in vehicle components, focusing on smoother, more efficient operation and increased fault tolerance [12]. Several research groups have focused on the establishment of a full-scope digital twin of a vehicle that allows bi-directional interaction in real time and covers all the life-cycle stages of a vehicle [13]. Another set of approaches aims at the integration of artificial intelligence (AI) in design, assembly, and production processes [14,15].

The listed challenges and the multitude of approaches concerning the multi-domain design processes of vehicles have led to a prominent need for research aimed at supporting design engineers in this endeavor. This Special Issue presents a collection of research in this area.

2. A Detailed Summary of the Contents of This Special Issue

The level of product modelling concerning abstract physics is receiving increasing attention. Stetter [16] proposed an integrated framework for abstract physics modeling (IF4APM) suited for innovative vehicle design processes. In a connected study, Schuchter et al. [17] applied graph-based design languages (GBDLs) to digital integrated design and assembly-planning processes for sports vehicles. The application of the proposed framework led to two machine-executable V-models for product and assembly system development in the scope of model-based systems engineering (MBSE). It is important to note that both approaches are universal and can be used in the development and production of components for the automotive industry, their assemblies, and entire vehicles. Mieth and Gauterin [18] proposed a concise set of methods that serves as a methodological framework for the evaluation of vehicle drive systems. Their procedure involves evaluating different drive concepts based on defined criteria and comparing these evaluations with one another. To cope with the diversity of requirements in the automotive development landscape, Kexel et al. [19] developed a generic and holistic methodical systems engineering approach for virtual plug-in hybrid concept development and optimization under real-world boundary conditions. Weitz et al. [20] described the design of a lifting actuator for modular vehicles with autonomous capsule-changing capabilities and comprehensively demonstrated the dimensioning of all main components according to the boundary conditions. The development of a new plastic-intensive medium-pressure plate (MPP), forming part of a fuel-cell system, together with a steel plate meeting all mechanical and chemical requirements, was the focus of the extensive study reported by Anand et al. [21]. Elsewhere, Keil et al. [22] aimed to evaluate and optimize the effectiveness of software-in-the-loop (SiL) simulations in the vehicle software testing process. They focused on supporting the testing process by shifting specific test cases from hardware-in-the-loop (HiL) test benches to SiL-based simulations. Sun and Lee [23] aimed to explore the impact of product design dimensions on electric vehicle (EV) purchase intentions to provide a theoretical basis for companies' differentiation strategies and reflect the impact of product design on purchase intention. The emergence of multimodal large language models (MLLMs) has led to a novel approach to safety event analysis consisting of integrating textual, visual, and audio modalities. Accordingly, the framework proposed by Abu Tami et al. [24] leverages the logical and visual reasoning power of MLLMs, directing their output through object-level question–answer (QA) prompts to ensure accurate, reliable, and actionable insights for investigating safety-critical event detection and analysis. Vecchiato et al. [25] described the conceptualization, design, topology optimization, manufacturing, and validation of a hydraulic brake caliper for Formula SAE race cars made of Scalmalloy®, an innovative Al-Mg-Sc alloy that had never before been adopted to manufacture a brake caliper. Panchenko et al. [26] focused on the prediction of the residual wear resources of the composite brake pads of a modernized brake system used in freight wagons. Design for Additive Manufacturing (DfAM) encompasses two primary strategies: adapting traditional designs for 3D printing and developing designs specifically optimized for additive manufacturing. In this vein, the concise review compiled by Hamza et al. [27] focuses on the automotive sector, systematically examining DfAM's potential to redefine vehicle design, production processes, and industry standards.

Author Contributions: All authors contributed equally to this work. All authors have read and agreed to the published version of the manuscript.

Acknowledgments: This Special Issue was successfully organized with the support of the editorial team of the journal *Vehicles*. The guest editors also wish to thank all the authors whose valuable work was published in this Special Issue and the reviewers for evaluating the manuscripts and providing helpful suggestions.

Conflicts of Interest: The authors declare no conflicts of interest.

References

1. Wichmann, R.L.; Gericke, K.; Eisenbart, B. Analysing risk to function fulfilment: Applying the function integrity diagnosis and documentation method. *J. Eng. Des.* **2021**, *32*, 346–373. [CrossRef]
2. Tröster, P.M.; Schmidt, D.; Albers, A. A Designation Guide for Consistent Cross-System-Level Modeling of Embodiment Function Relations. *Procedia CIRP* **2023**, *119*, 1222–1227. [CrossRef]
3. Graessler, I.; Wiechel, D.; Oezcan, D.; Taplick, P. Tailored metrics for assessing the quality of MBSE models. *Proc. Des. Soc.* **2024**, *4*, 2545–2554. [CrossRef]
4. Henderson, K.; McDermott, T.; Salado, A. MBSE adoption experiences in organizations: Lessons learned. *Syst. Eng.* **2024**, *27*, 214–239. [CrossRef]
5. Tao, F.; Ma, X.; Liu, W.; Zhang, C. Digital Engineering: State-of-the-art and perspectives. *Digit. Eng.* **2024**, *1*, 100007. [CrossRef]
6. Voss, C.; Petzold, F.; Rudolph, S. Graph transformation in engineering design: An overview of the last decade. *AI EDAM* **2023**, *37*, e5. [CrossRef]
7. Borowski, J.; Rudolph, S. A Digital Machine-Executable V-Model for a Formula Student Racing Car. In *International Stuttgart Symposium*; Springer Fachmedien: Wiesbaden, Germany, 2024; pp. 325–335.
8. Neumaier, M.; Schopper, C.; Gundlach, T.; Gast, C.; Döring, D.; Rudolph, S. Automated packing and piping in an Airbus A320 main landing gear bay: An industrial development case study. *CEAS Aeronaut. J.* **2024**, *15*, 1201–1222. [CrossRef]
9. Stetter, R. *Fault-Tolerant Design and Control of Automated Vehicles and Processes*; Springer International Publishing: Cham, Switzerland, 2020.
10. Kumar, A.; Rajak, D.K.; Kumar, P.; Rathee, S. (Eds.) *Smart Electric and Hybrid Vehicles: Design, Modeling, and Assessment by Industry 4.0 Approaches*; CRC Press: Boca Raton, FL, USA, 2024.
11. Barrero-Arciniegas, H.A.; Bataleblu, A.A.; Don, D.; Rauch, E.; Matt, D.T. Design Parameters for Resilience in Cyber-Physical Production Systems. *Procedia Comput. Sci.* **2025**, *253*, 2316–2326. [CrossRef]
12. Witczak, M.; Pazera, M. Towards integrated tracking fault-tolerant control for Takagi-Sugeno systems. In Proceedings of the 2024 IEEE 63rd Conference on Decision and Control (CDC), Milan, Italy, 16–19 December 2024; IEEE: Piscataway, NJ, USA, 2024; pp. 2068–2073.
13. Trauer, J.B. On the Conception and Implementation of Digital Twins–Supporting Companies in the Development of Digital Twins. Ph.D. Dissertation, Technische Universität München, Munich, Germany, 2024.
14. Rudolph, S. On Some Artificial Intelligence Methods in the V-Model of Model-Based Systems Engineering. In Proceedings of the MODELSWARD, Rome, Italy, 21–23 February 2024; pp. 386–393.
15. Wang, Y.; Xu, S.; Bwar, K.H.; Eisenbart, B.; Lu, G.; Belaadi, A.; Fox, B.; Chai, B.X. Application of machine learning for composite moulding process modelling. *Compos. Commun.* **2024**, *48*, 101960. [CrossRef]
16. Stetter, R. Innovative Vehicle Design Processes Based on the Integrated Framework for Abstract Physics Modeling (IF4APM). *Vehicles* **2024**, *6*, 1345–1363. [CrossRef]

17. Schuchter, T.; Till, M.; Stetter, R.; Rudolph, S. Digital Integrated Design and Assembly Planning Processes for Sports Vehicles Using the Example of a Skateboard. *Vehicles* **2025**, *7*, 22. [CrossRef]
18. Mieth, R.; Gauterin, F. Comparative Assessment for Holistic Evaluation of Drive Systems. *Vehicles* **2024**, *6*, 403–414. [CrossRef]
19. Kexel, J.; Müller, J.; Herkenrath, F.; Hermsen, P.; Günther, M.; Pischinger, S. Virtual Plug-In Hybrid Concept Development and Optimization under Real-World Boundary Conditions. *Vehicles* **2024**, *6*, 1216–1248. [CrossRef]
20. Weitz, F.; Ostendorff, N.; Frey, M.; Gauterin, F. Lifting Actuator Concept and Design Method for Modular Vehicles with Autonomous Capsule Changing Capabilities. *Vehicles* **2024**, *6*, 1070–1088. [CrossRef]
21. Anand, S.; Mielke, F.; Heidrich, D.; Fang, X. Optimization, Design, and Manufacturing of New Steel-FRP Automotive Fuel Cell Medium Pressure Plate Using Compression Molding. *Vehicles* **2024**, *6*, 850–873. [CrossRef]
22. Keil, R.; Tschorn, J.; Tümler, J.; Altinsoy, M. Evaluation of SiL Testing Potential—Shifting from HiL by Identifying Compatible Requirements with vECUs. *Vehicles* **2024**, *6*, 920–948. [CrossRef]
23. Sun, Z.; Lee, B. Exploring Factors Influencing Electric Vehicle Purchase Intentions through an Extended Technology Acceptance Model. *Vehicles* **2024**, *6*, 1513–1544. [CrossRef]
24. Abu Tami, M.; Ashqar, H.; Elhenawy, M.; Glaser, S.; Rakotonirainy, A. Using Multimodal Large Language Models (MLLMs) for Automated Detection of Traffic Safety-Critical Events. *Vehicles* **2024**, *6*, 1571–1590. [CrossRef]
25. Vecchiato, L.; Capraro, F.; Meneghetti, G. Design, Topology Optimization, Manufacturing and Testing of a Brake Caliper MADE of Scalmalloy® for Formula SAE Race Cars. *Vehicles* **2024**, *6*, 1591–1612. [CrossRef]
26. Panchenko, S.; Gerlici, J.; Lovska, A.; Ravlyuk, V.; Dižo, J. Prediction of Residual Wear Resources of Composite Brake Pads of a Modernized Brake System of Freight Wagons. *Vehicles* **2024**, *6*, 1975–1994. [CrossRef]
27. Hamza, A.; Bousnina, K.; Dridi, I.; Ben Yahia, N. Revolutionizing Automotive Design: The Impact of Additive Manufacturing. *Vehicles* **2025**, *7*, 24. [CrossRef]

vehicles

Article

Innovative Vehicle Design Processes Based on the Integrated Framework for Abstract Physics Modeling (IF4APM)

Ralf Stetter

Department of Mechanical Engineering, Ravensburg-Weingarten University (RWU), 88250 Weingarten, Germany; ralf.stetter@rwu.de

Abstract: In industrial vehicle design processes, most companies have implemented model-based systems engineering (MBSE). As a consequence, design processes are nowadays not driven by documents, but by digital models of the vehicle to be developed and its components. These models exist on different levels of abstraction. The models on the requirements level are already well defined as well as the models of the defined product behavior and product properties. In recent years, the specification of models on the level of product functions was largely clarified, and elaborate frameworks already exist. However, this is not yet true for the level between functions and definite properties; this level can be referred to as "abstract physics". The enormous importance of this level, which, amongst others, can represent the physical effect chains which allow a vehicle component to function, is expressed by several researchers. Several research works aim at specifying models on this level, but, until now, no general consensus can be identified, and the existing model specifications are less appropriate for the early stages of vehicle design. This paper explains an Integrated Framework for Abstract Physics Modeling (IF4APM), which incorporates different perspectives of abstract physics and is suited for the early phases. The explanation is based on typical components of several kinds of vehicles. The main advantages of the proposed approach are the consistent interconnection of abstract product models, the clearness and understandability of the resulting matrices, and the aptitude to be used in the early phases of a vehicle design process.

Keywords: vehicle development; abstract physics; design models; model based design

Citation: Stetter, R. Innovative Vehicle Design Processes Based on the Integrated Framework for Abstract Physics Modeling (IF4APM). *Vehicles* **2024**, *6*, 1345–1363. https://doi.org/10.3390/vehicles6030064

Academic Editor: Patrick Gruber

Received: 4 July 2024
Revised: 26 July 2024
Accepted: 31 July 2024
Published: 3 August 2024

1. Introduction

In today's vehicle design processes, a gap can be observed between the rather clearly and concisely described functionality of the vehicle and its modules and the defined behavior and properties of them. This gap is caused by a lack of information that describes which physical phenomena realize the functions by which as well as how and to what extend these phenomena are acting. In recent years, the complexity of vehicle development is further increasing because of a growing share of intelligent components, because of increasing technical complexity of vehicles and their modules, and because of steadily increasing model ranges and product families [1]. Current advancements in production technologies, growing ecological constraints, and globally diversified customers require early design methods which allow designers to explore the available design space systematically [2]. Systematic design methods need to be based on concise and reliable products, production processes, and production system information on all levels of abstraction. Concerning the virtual models of technical systems (such as vehicles and their components), a wide consensus that four levels of abstraction can be distinguished can be observed (compare e.g., [3–6]). It is the current consensus in design science that these levels are not appropriate to be the basis for a general project planning of design, but that all levels may be used in all phases of design processes with changing intensity. The project definition and concept phases are usually characterized by a strong emphasis on requirements, functions, and abstract physics, while the detail design and production preparation phases mainly employ detailed descriptions of geometry, material, and structure. Still, changing customer expectations and unexpected testing outcomes can always lead to changed requirements and can require a revisit to other more abstract levels. The four levels of abstraction of product models are shown in Figure 1.

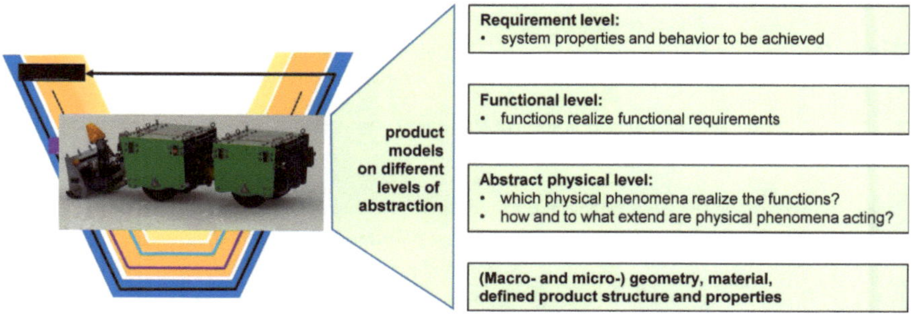

Figure 1. Vehicle product models on different levels of abstraction.

As visible in Figure 1, the most abstract level of systems models contains the requirements, i.e., the intended system properties and system behavior, in order to fulfill the customer expectations. The integration of the entities on this level of abstraction—the requirements—into engineering frameworks has been the focus of research projects for several years [7–9]; detailed specifications of model structures and contents are established. The so-called Requirements Interchange Format (ReqIF), for instance, defines a file format, which may be used for the exchange of requirements with metadata [10]. A part of these requirements will lead to functions that allow the realization of these functional requirements (second level of abstraction). For this level, detailed frameworks have been developed in the last decade [11–16]). The main focus of this paper is the third level—abstract physics. As stated above, the main content of this level is information, where physical phenomena realize the functions, as well as how and to what extend these phenomena are acting. For an analysis and understanding on the micro-level, working surfaces can also be of great interest on this level of abstraction. The least abstract level (in many publications, thhis is also described as the most "concrete" level; however, in this paper, this adjective is avoided in order to prevent confusion with a prominent building material) concerns the macro- and micro-geometry, material, product structure, and product properties. This abundant amount of information is stored in established data formats such as computer aided design (CAD) files, and the model specifications are well documented.

One may ask, why the gap between the second and the fourth level presents a real challenge for innovative vehicle design processes. Albers et al. [17] report a prominent problem for engineers to embody functional ideas in sub-systems and to localize these functions in mechanisms and their interfaces. Drave et al. [14] point out that technical systems fulfill certain functions through physical effects acting between components and emphasize a strong correlation between these physical effects and the architecture of the system. In connected research, Hoepfner et al. [18] report a lack of emphasis on this abstraction level, and that current model-based systems engineering (MBSE) frameworks do not integrate formal models for modeling the physical behavior in order to allow a seamless link to digital system models. Additionally, Zerwas et al. [19] point out that principle solutions are essential because they can efficiently describe possible solutions by specifying effects as well as active surfaces and material(s) without losing a functional orientation. The early phases are decisive because so-called breakthrough products (extremely successful products which completely replace prior products, such as fuel injection systems instead of carburetors) are usually characterized by a major change in the physical structure [20]. Especially in the early phases, it is important to note that not all aspects of a technical system can be represented with concrete parameters. In industrial companies, often so-called "engineering samples" are used to represent the intended realization of certain components, especially if their appearance is decisive for customer satisfaction [21]. In industrial vehicle development processes, mature V-model-based development architectures can be observed, especially in the field of software and control development [22]. However, in the scope of mechanical and mechatronic engineering, certain levels of abstraction are not yet covered [18]. Generally, models on all level of abstraction and during all phases of product development are desirable because the employment of models as the primary artifacts can lead to an improved semantic integration of digital assets and robust early analysis [23]. Consequently, the proposed approach aims to close the gap between the second and the fourth level visible in Figure 1. On the given abstraction level of product models, detailed information concerning material selection or manufacturing processes and necessities are consciously omitted in order to reduce complexity and to enhance transparency. It can be concluded that intensified and detailed scientific activity on this level of abstraction may lead to more efficient design processes, better knowledge reuse, and seamless digital processes in the early phases of design. This also leads to the formulation of the following research question:

How can different perspectives of abstract physics be combined in a manner that allows integration in engineering frameworks for the design of vehicles?

This research question is analyzed using several vehicle design processes focusing on remote controlled vehicles and lifting devices for vehicles (Figure 2).

Figure 2. Vehicle and vehicle component development processes.

These vehicle design processes consider remote-controlled vehicles for purposes such as snow plowing or lawn mowing [24] (Figure 2 left side) and lifting systems in the area of agricultural technology—lifting masts for tractors [25] and power lifts for universal agricultural vehicles [26] (Figure 2 right side). As stated above, the levels of abstraction of product models are not appropriate to be the basis for a general project planning of design, but explicit project schedules need to be developed. A simplified timeline of vehicle design processes is visible in Figure 3.

In Figure 3, two main process flows are represented; the project-independent technology development flow as well as the project-oriented flow, which may be sensibly divided into the concept phase and vehicle development phase. For the project-oriented flow, typical mile-stones of a vehicle development process are given (compare [1,27]). An early phase concerns the concept development. In this phase, the important vehicle characteristics are derived, employing methods such as market research and benchmarking. The important product specifications are developed and important concept decisions are made. For a given project, product models on all levels of product concretization are employed and a consideration of abstract physics is possible. However, the development of products components, which employ novel physical effect chains, very often is carried out outside the scope of a single vehicle development process, and can be understood as an independent ongoing technology development process. After the concept definition, the vehicle development process in the close sense takes place. On the basis of the product specification, a detailed system description is elaborated together with the preparation of all manufacturing and assembly processes. Still, changes of specifications are frequent, and, in some cases, the physical effect chains also need to be altered in this phase. However, the main area of interest concerning abstract physics are the technology development and the earlier phases of project oriented design. It is important to point out that the process planning and control in vehicle design processes is usually realized employing model-based system engineering (MBSE) [28–30]. In MBSE, product models on the different levels of abstraction are the primary process management objects and take over the role of other documents which are not directly product-related. The fact that models, which are generally agreed upon, are not yet available on the level of abstract physics can lead to mistakes, inefficient knowledge management, double work, and misunderstandings. Several research groups worldwide have started to address this issue; their main contributions will be described in Section 2 and form the state of the art. Based on this, the integrated framework is presented in Section 3. The vehicle design processes, which are analyzed in the ongoing research and serve for explaining the integrated frameworks, are characterized in Section 4, together with elements of the integrated framework. A final section (Section 5) concludes the paper and delivers an outlook on further research perspectives.

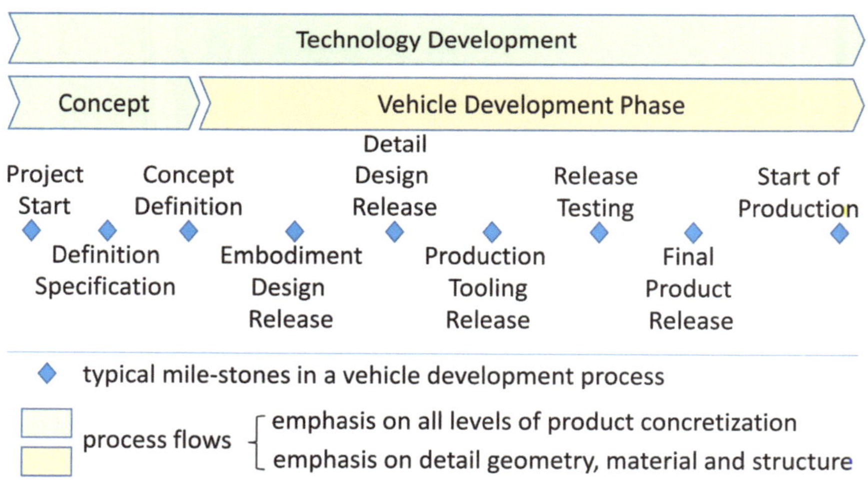

Figure 3. Timeline of vehicle design processes.

2. Perspectives of Abstract Physics

In earlier research [31], the different perspectives of abstract physics were analyzed based on the design processes of wind turbines. An overview of those perspectives is given in Figure 4.

In general, five perspectives can be distinguished as follows:

- a phenomenon-oriented perspective which consists of physical effects, effect chains, and effect networks and supports the understanding of physics and innovative system concepts;
- a control-oriented perspective which focuses on the elements of a control cycle (actor, process, and senor) and supports the synthesis of intelligent controlled systems [5];
- a behavior-oriented perspective which employs mathematical models such as differential equations in order to allow simulation and optimization;

- an interface-oriented perspective which focuses on interacting surfaces as part of a physical effect chains and allows a detailed analysis of these surfaces and their role in realizing the effect chain;
- a logic-oriented perspective which employs certain methods such as fault-tree analysis in order to analyze reliability and safety. These analyses frequently cover detailed phases of project-oriented design (compare Figure 3) and are therefore not included in the current state of the IF4APM.

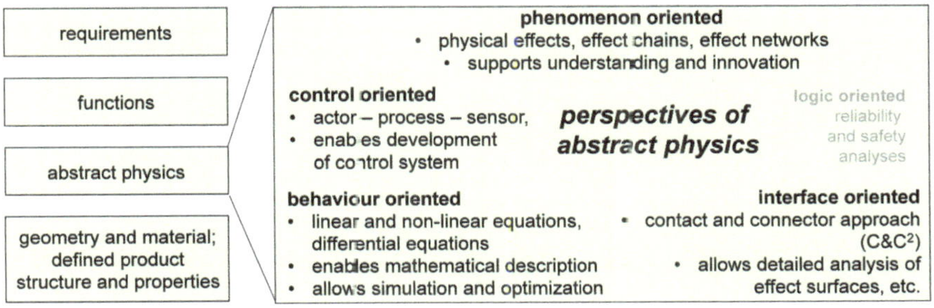

Figure 4. Perspectives of abstract physics.

2.1. Phenomenon-Oriented Perspective

The important entities of the phenomenon-oriented perspective are visible in Figure 5.

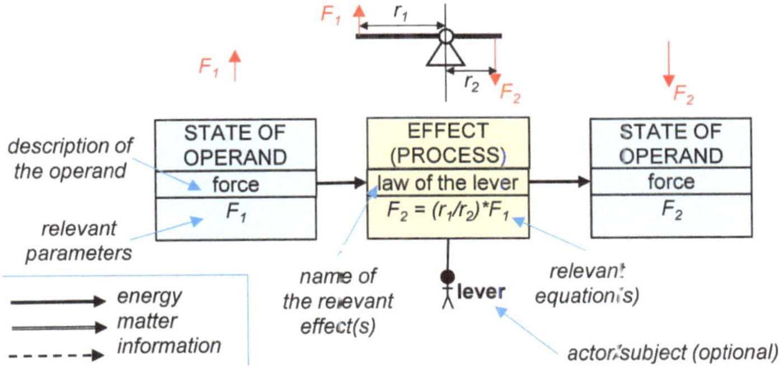

Figure 5. Entities of the phenomenon-oriented perspective.

The inputs of a physical effect (i.e., a distinct physical phenomenon which can realize an operation within a technical system [3,32]) are the states of operands (operands are the entities that are processed). The effect itself is often a transformation of energy, matter, or signal, which is based on a physical phenomenon. Common examples for physical phenomena range from statics and dynamics, over elasticity, vibration, molecular forces to thermal effects, etc. [4,33]. Additionally, interaction processes are possible (e.g., processes of actors which are within the system with actors which are outside of the system) [11]. The output(s) of the process are again states of operands. The subject—the actor—realizes the process. In the early stages of design processes, the effect might already be clear and decided, but the definite actor might still be an element of discussion.

2.2. Control-Oriented Perspective

A complementary perspective is control-oriented. In contrast to the phenomenon-oriented perspective, which is mainly focused on the actors, important aspects of this perspective are also the sensors and the application of their measurements for stabilizing the behavior of the system. The main entities of this perspective are shown in Figure 6.

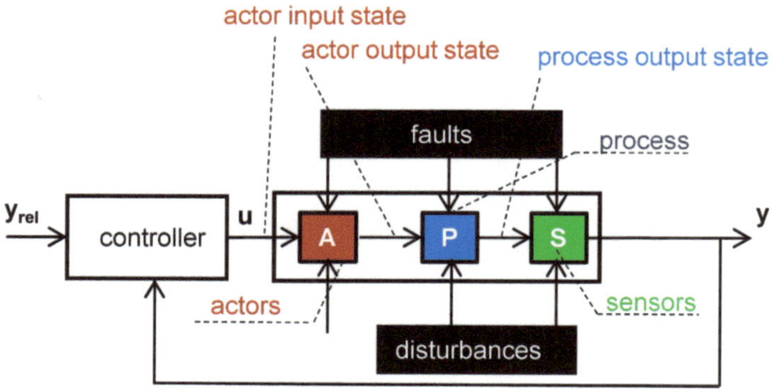

Figure 6. Entities of the control-oriented perspective.

The main element of this perspective are the control loops, consisting of actor, process, sensor, and controller. These control loops enable the use of feedback from sensor measurements for the control of the system with the goal to stabilize the behavior of the system or to follow a given trajectory [34]. Modern control systems also allow us to compensate disturbances and to detect and accommodate faults of actors, processes, and sensors [35,36]. In vehicle applications, complex control systems are present. A promising approach to tackle this enormous complexity are modularization and distribution approaches. Current research work concerning advanced vehicular mechatronic systems has led to the proposal of multi-agent systems in the control domain [37–39], including game-based control frameworks [40].

2.3. *Interface-Oriented Perspective*

Both the phenomenon-oriented and the control-oriented perspectives concentrate on the macro effects which realize the functionality and stability of a technical system. In contrast, the interface-oriented perspective focuses on the micro effects; the main entities are displayed and explained in Figure 7.

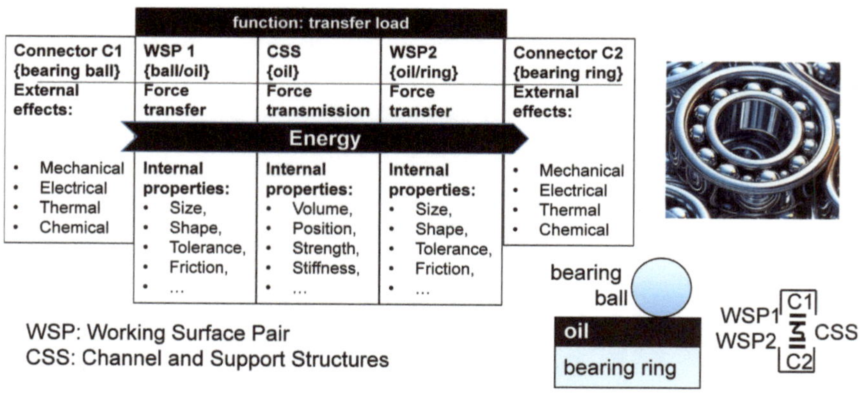

Figure 7. Entities of the interface-oriented perspective.

This perspective, which is referred to as the Contact and Channel Model (C&CM) and was introduced by Albers and Matthiesen [41], describes Working Surface Pairs (WSPs) that realize functions and Channel and Support Structures (CSSs) which connect the WSPs. WSPs are interfaces between components, and interfaces can be between solid surfaces or boundaries with the surfaces of liquids, gases, or fields, while CSSs are physical components or volumes of liquids, gases, or spaces that connect two WSPs [17,42–45].

2.4. Behavior-Oriented Perspective

Several researchers in engineering design science employ the function–behavior–structure paradigm (compare, e.g., [46–49]) and point out that functions and the concrete structure are connected by the behavior of the technical system. In some cases, this behavior can be described by means of mathematical equations. In Figure 8 an example using an electrical motor is given.

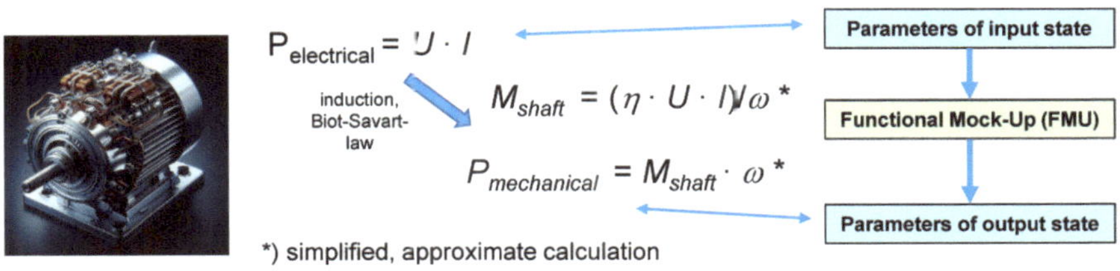

Figure 8. Entities of the behavior-oriented perspective.

Obviously, the simplified approximate calculation in Figure 8 does not represent the behavior of the electrical motor in detail. Usually, rather complex differential equations and simulations are necessary for a sensible representation and prognosis of the actual technical system. For several years, a modularization of simulation frameworks has been in development; for instance, in order to reduce complexity. A promising approach employs so-called functional mock-up (FMU) simulation components which are based on the functional mock-up interface (FMI) standard [50–53]. Today, some research groups have started to address the obvious gap between function and structure, and are employing more than one perspective of abstract physics in this endeavor. Detailed requirements concerning engineering frameworks that include abstract physics perspectives were formulated [21]. Ongoing research concerns a SysML-based modeling method that allows the mechanical domain to represent links between its requirements, functions, and principle solutions; the intended models not only contain elementary effects, but entire principle solutions with behavior models and appropriate test workflows [14,15,18,19]. Further relevant research is focusing on multi-domain matrices [54]. The IF4APM is employing the modeling intentions and solutions of earlier research and is focusing on early stages of vehicle design, where information and product structure are still unclear. It is a central intention to support design engineers with their accustomed procedure styles in vehicle development.

3. Description of the Framework

The IF4APM intends to integrate different perspectives of abstracts physics; its main entities are listed in Table 1 together with an explanation.

Table 1. Entities of the IF4APM.

Element	Explanation
Operand	The operand is the entity that is processed by the technical system, e.g., mechanical energy. Operands in a technical system appear in the general forms of matter, energy, and signal. The operand can also be a consequence or a prerequisite of energy, e.g., a force.
State	The state describes the condition of the operand e.g., the force on one side of a lever.
Process	The process describes the transformation of the operand, e.g., an enlargement of the force.

Table 1. *Cont.*

Element	Explanation
Actor	The actor is a component or sub-system of the technical system which is responsible for the realization of a process, e.g., a lever is responsible for the enlargement of the force.
Effect	The effect represents a physical phenomenon that contributes to the realization of the process, e.g., law of the lever.
Behavior	The behavior quantitatively represents the change in the elements of the technical system to realize the process. Usually, the alteration of certain parameters is the focus of the behavior.
Working Surface (WP)	The working surface is a part of an object of a system that is necessary for the realization of the process.
Working Surface Pair (WSP)	The WSP is the interface between objects that take part in the process.
Channel and Support Structure (CSS)	The CSS describes components or volumes of liquids, gases, or spaces that connect WSPs in order to realize the process.

The relations between the entities of the IF4APM and their contribution to the abstract physical structure of a technical system can be illustrated in a unified modeling language (UML)-based class diagram (Figure 9).

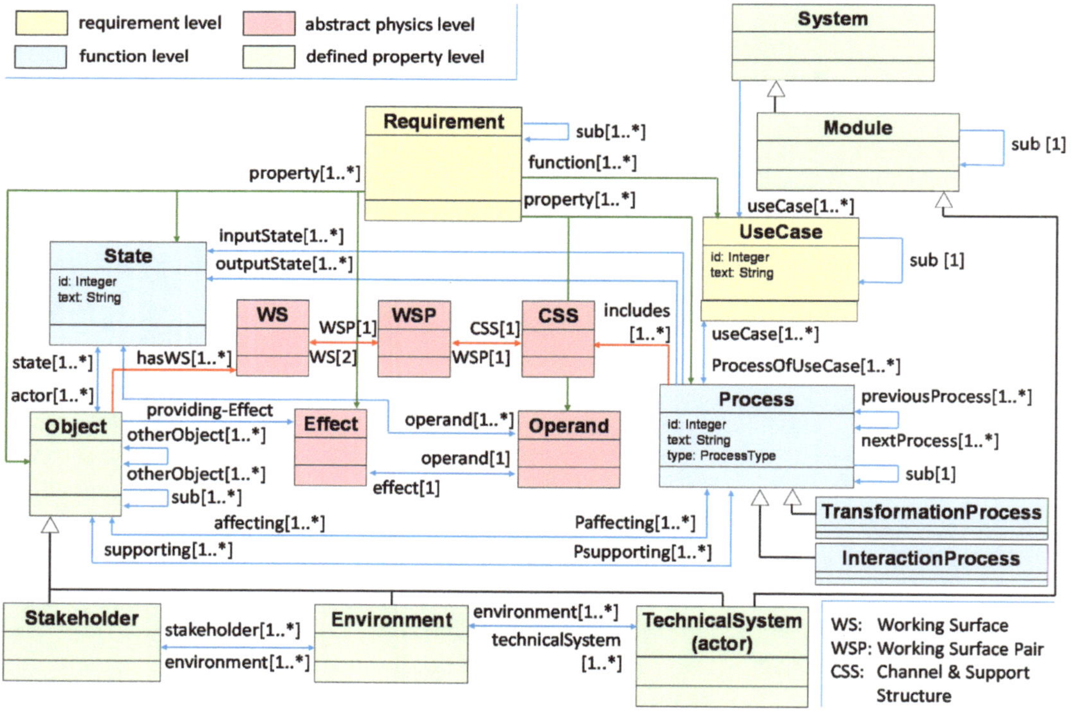

Figure 9. Relations between the entities of the IF4APM.

This class diagram was developed based on class diagrams of the framework for integrated function modeling (IFM) [13,55] and earlier research [21]. The different levels of abstraction of the respective product model are indicated with colors. The yellow entities *Requirement* (specific objectives of a system) and *Use Case* (scenarios of applying the technical system for a specific purpose) concern the most abstract level. The blue entities *State* and *Process* are located on the function level, but also play an important role on the level of abstract physics. On this level, the red entities *Effect*, *Operand*, *Working Surface*, *Working Surface Pair*, and *Channel Support Structure* are employed in the IF4APM for describing the physical effect chain. The green entities *System*, *Module*, *Object*, *Stakeholder*, *Environment*, and *Technical System* concern the level of detail geometry and material. Similar to the well-known IFM, the main representation of the IF4APM is a combination of modular matrices because they enable a concise, structured representation [11] Matrices are excellent tools for managing complex dependencies in technical systems [54]. Such matrices are a relatively intuitive means for modeling information in a structured manner. An overview of the structure of the IF4APM is given in Figure 10.

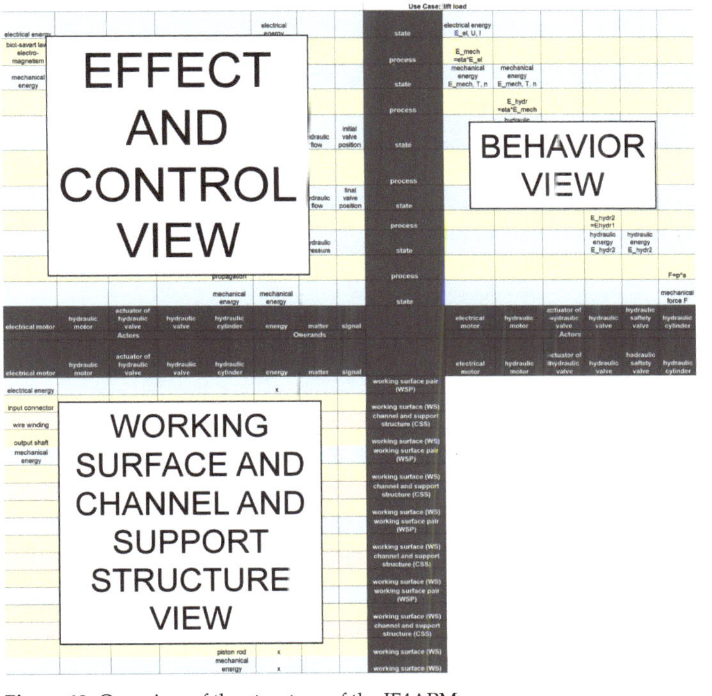

Figure 10. Overview of the structure of the IF4APM.

Three views can be distinguished as follows: the effect and control view, the behavior view, and the working surface, channel, and supports structure view. These views are explained in detail on the basis of a vehicle component example in the next section.

4. Application of the Framework

In this section, the different elements of the IF4APM are explained based on a sub-system development for an agricultural vehicle. The sub-system concerns the actuation of a lifting system (compare Figure 2), and a schematic diagram is given in Figure 11.

A hydraulic pump (which is driven by an electrical motor) is generating a certain hydraulic pressure and, under certain conditions), a hydraulic flow. A so-called 4/3 way (which is electrically actuated) valve is used for defining the motion direction of a double-acting hydraulic cylinder with a piston (up/down/none). This hydraulic cylinder creates motion which leads to the movement of devices on vehicles, e.g., lifting systems. The main physical principles are represented in Figure 12.

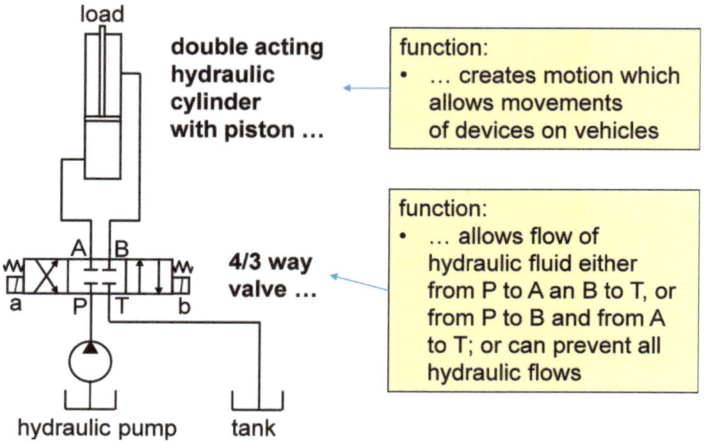

Figure 11. Sample vehicle component: lifting system.

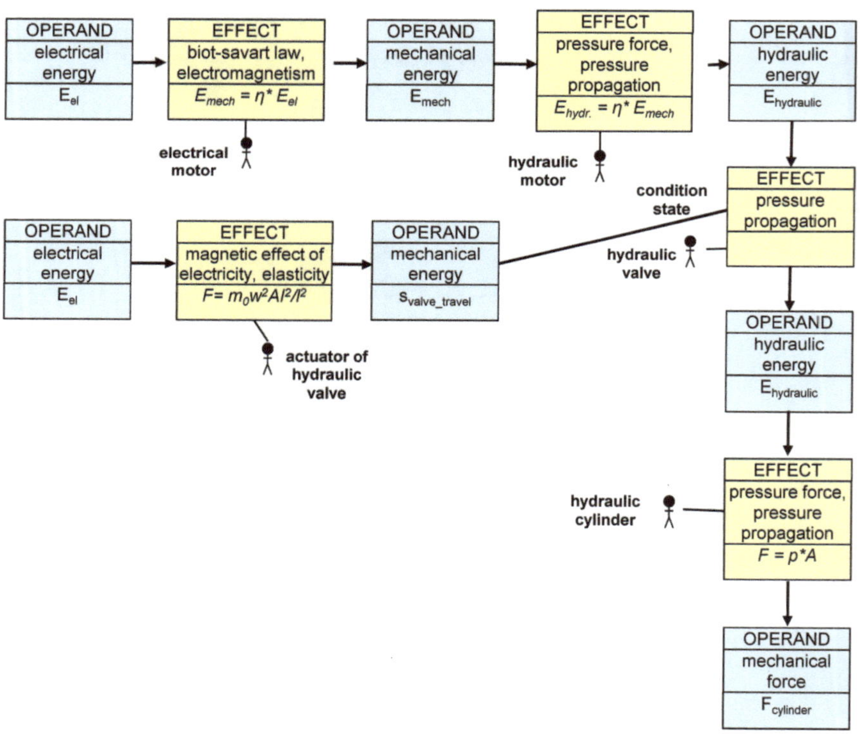

Figure 12. Main physical principles of the lifting system for vehicles.

The blue entities in Figure represent the states of operands, e.g., the electrical energy for an electrical motor. The yellow entities represent the processes and the distinct physical effects which enable these processes. For these processes, the actors, which are the function carriers and allow the physical effects to take place, are also provided in this representation. It is important to point out that it is not sensible in the early phases of design to define these actors; this should be completed during the concretization of the system design. The central view of the IF4APM is the effect and control view; an example is given in Figure 13.

electrical motor	hydraulic motor	actuator of hydraulic valve	hydraulic valve	hydraulic cylinder	energy	matter	signal		Use Case: lift load
electrical energy					electrical energy				state
biot-savart law, electro-magnetism					biot-savart law, electro-magnetism				process
mechanical energy	mechanical energy				mechanical energy				state
	pressure force, pressure propagation				pressure force, pressure propagation				process
	hydraulic energy (hydraulic pressure, hydraulic flow)				hydraulic energy	hydraulic flow			state
		electrical energy			electrical energy		initial valve position	actor input state	state (sub-process)
		magnetic effect of electricity, elasticity			magnetic effect of electricity, elasticity		magnetic effect of electricity, elasticity	actor	process
		movement of valve			electrical energy		final valve position	actor output state	condition state
			hydraulic energy		hydraulic energy	hydraulic flow			state
			pressure force, pressure propagation		pressure force, pressure propagation	pressure force, pressure propagation			process
			hydraulic energy		hydraulic energy	hydraulic flow			state
				pressure force, pressure propagation	pressure force, pressure propagation				process
				mechanical energy	mechanical energy				state
Actors					Operands			Control	

Figure 13. Effect and control view.

The effect and control view alternates between states of operands and processes. The initial state of the actor *electrical motor* is *electrical energy*; this motor transforms this energy by applying the physical effects (distinct physical phenomena) *biot-savart law* and *electro-magnetism* to *mechanical energy*, which is the final state of this first process. The operand in this process is *energy*. Additional states are necessary if sub-processes are necessary (this will be explained in detail at the end of this section). In the process lines, the physical effects that enable the process are listed, thus allowing us to describe the physical effect chain that realizes the respective functionality. The columns for actors are intended to be filled in later stages; initially, a solution-independent description of the main effects is possible in the columns for operands. The control column will be explained in detail at the end of this section. It is important to note that the IF4APM does not intend to model all states, properties, aspects, and processes of a vehicle or vehicle component, but only those which are relevant to a certain high level use case (see Figure 13, top line).

In general, a use case describes an application of a technical system for achieving a certain goal and may include different types of users and their activity in and relationship with the system [56]. Use cases can be decomposed into "sub use cases" [57] and summarized into "super use cases". For the IF4APM, a concentration on high level use cases (such as "lift load") (compare Figure 13) is sensible. Use cases allow the creation of IF4APM models as aspect models with a limited size and complexity. It is also important to point out that human beings can also act as actors in this system and can be represented in the IF4APM. They can be part of the respective use case and they can perform certain processes.

The second view of the IF4APM captures the behavior-oriented perspective (Figure 14).

Use Case: lift load					
state	electrical energy E_el, U, I				
process	E_mech =eta*E_el				
state	mechanical energy E_mech, T, n	mechanical energy E_mech, T, n			
process		E_hydr =eta*E_mech			
state		hydraulic energy E_hydr, V_dot, p			
state (sub-process)			electrical energy		
process			complex relationship (FMU)		
condition state			movement of valve		
state				hydraulic energy E_hydr1	
process				E_hydr2 =Ehydr1	
state				hydraulic energy E_hydr2	hydraulic energy E_hydr2
process					F=p*a
state					mechanical force F
Actors	electrical motor	hydraulic motor	actuator of hydraulic valve	hydraulic valve	hydraulic cylinder

Figure 14. Behavior view.

Similar to the effect and control view, the behavior view alternates between states of operands and processes. The state *electrical energy* is expressed e.g., by means of an energy quantity parameter *E_el*. The simplified calculation also used in Figure 8 allows us to calculate the output parameter *E_mech* based on the energy efficiency parameter *eta* of the motor. In the process lines, the behavior is described either by means of equations or by links to external simulations (e.g., in form of FMUs compare Section 2). It is important to note that links to other software packages, simulation platforms, and co-simulation frameworks, such as Matlab/Simulink [58], Modelica [59], and INTO-CPS [60], is also possible. The behavior entities intend to describe the abstract physical behavior that realizes the respective functionality. Only columns for actors are included because this view can usually only be sensibly filled when many properties of the product (such as the choice of actors) are already decided.

The structure of the third view of the IF4APM—the working surface and the channel and supports structure view—is different and is shown in Figure 15.

Actors					Operands			
electrical motor	hydraulic motor	actuator of hydraulic valve	hydraulic valve	hydraulic cylinder	energy	matter	signal	
electrical energy					x			working surface pair (WSP)
input connector					x			working surface (WS)
wire winding					x			channel and support
output shaft					x			working surface (WS)
mechanical energy	mechanical energy				x			working surface pair (WSP)
	input shaft				x			working surface (WS)
	rotor				x			channel and support
	output connector				x			working surface (WS)
	hydraulic energy		hydraulic energy		x	x		working surface pair (WSP)
			input connector		x	x		working surface (WS)
			cavitiy		x	x		channel and support
			output connector		x	x		working surface (WS)
			hydraulic energy	hydraulic energy	x	x		working surface pair (WSP)
				input connector	x	x		working surface (WS)
				piston	x	x		channel and support
				piston rod	x			working surface (WS)
				mechanical energy	x			working surface pair (WSP)

Figure 15. Working surface and channel and supports structure view.

The working surface and the channel and supports structure view starts with working surface pairs (WSPs) and then continues to distinct working surfaces (WSs), then listing channel and support structures (CSS) and continuing again with working surfaces (WSs) and working surface pairs. A WSP with an external source *electrical energy* brings this energy to the WS *input connector*. The CSS is the *wire winding* that represents the abstract physics of the electrical motor. The next WS is the *output shaft* which is connected to the WS *input shaft* of the hydraulic motor via the WSP *mechanical energy*. The working surface and the channel and supports structure view allow a step-wise and connected discussion of the acting physics on a detailed micro level.

As pointed out above, the control aspect of the effect and control view has not yet been discussed in detail. In order to explain the role of control functionalities in detail, an example of a controlled system will be employed. An effect chain enabling the generation of a controlled hydraulic pressure is shown in Figure 16

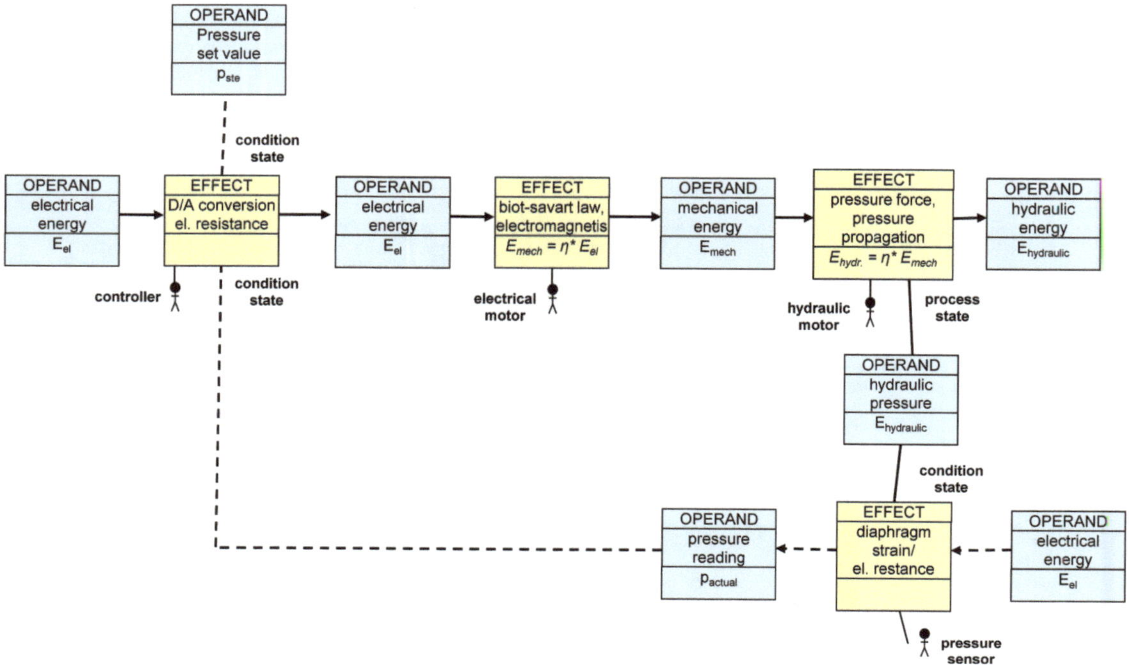

Figure 16. Physical effect chain for the generation of a controlled hydraulic pressure.

In this effect chain, the first actor is the controller of the control loop that compares the actual pressure reading with the set value and controls the flow of electric energy to an electrical motor (second actor). This motor drives an hydraulic motor which delivers the intended hydraulic pressure. For a clear and sensible representation of control functionalities, process and condition states can be employed. The connection types "condition state" (C) and "process state" (P) can be used to connect the auxiliary flows of operands to the main flow of operands [33,61]. The condition state is a prerequisite for the process to take place (e.g., the controller can only regulate the electrical flow if it has information concerning the current, actual pressure, and intended set value). A process state indicates that the state of one operand in the process is used for a second purpose; this link type allows us to assign diagnostic functions to flows of operands undergoing some kind of operation. The use of such link types allows us to clarify complex connections in functional models [61], but also in physical effect chains. The measured pressure is a process state of the main flow. This same state is used in a sub-process as a condition state in order to obtain an electrical voltage correlating with the pressure. This condition state can be used by the controller together with a condition state set value in order to realize the intended control loop. In Figure 17, how this control effect chain can be represented in the effects and control view is shown.

The set value and the measured value can be indicated in the given fields, thus representing and connecting the control effect chain. The condition states *control: measured value* and *control: set value* are a prerequisite for the controller to carry out its intended task *control output pressure*. One of the states *control: measured value* is also visible in the last line of the effect and control view; this clarifies the cyclic nature of control. The effect and control view allows an interlinked discussion of the acting physical effects and their contribution to the establishment of control loops that can guarantee a stable behavior.

Use Case: hydr. pressure

controller	electrical motor	hydraulic motor	pressure sensor	energy	matter	signal		
electrical energy				electrical energy				state
actual pressure						actual pressure	control: measured value	condition state
desired pressure						desired pressure	control: set value	condition state
D/A conversion el Resistance							control output pressure	process
electrical energy	electrical energy		mechanical energy	mechanical energy				state
	biot-savart law, electro-magnetism		pressure force, pressure propagation					process
	mechanical energy	mechanical energy	hydraulic energy	hydraulic energy	hydraulic flow			state
		pressure force, pressure propagation						process
		hydraulic energy		hydraulic energy	hydraulic flow			state
		hydraulic pressure	hydraulic pressure			hydraulic pressure		proess state
			diaphragm strain/ el. restance					process
			actual pressure			actual pressure	control: measured value	state
Actors				Operands			Control	

Figure 17. Representation of a control effect chain in the effect and control view.

5. Summary and Outlook

In the center of this publication is the presentation and explanation of the integrated framework for abstract physics modeling (IF4APM) in the scope of vehicle design. Current engineers working in vehicle design do not dispose of appropriate means to connect the function models of their vehicle components with the detailed geometry, material, and structure models. The lack of such possibilities can lead to misunderstandings, less-than-optimal knowledge management, inconsistencies, as well as isolated and incomplete simulations of the behavior of the vehicle. The presented research aims to provide engineers in vehicle design with a framework which allows the description and investigation of the physical effect chains linking functionality with detailed geometry, material, and structure. This framework intends to increase the transparency of the inner processes of a vehicle design and of the design decisions. The distinctive quality of the proposed approach is to consistently connect the product models on the levels function and detailed geometry/material in the scope of mechanical and mechatronic engineering; the specific realization of the framework leads to the aptitude being used in the early stages of the product development process, as well as to enhanced transparency and easy understandability. The knowledge that is usually captured in the minds of the engineers can be represented and stored; it can be available for further analysis, e.g., by means of big data analyses of artificial intelligence (AI). The ultimate vision concerning the application of AI for vehicle design could be that all models are available in a semantically rich representation (e.g., with all important attributes, annotations, and relationships, as well as all requirements and environmental conditions) for an AI system to learn how vehicles are developed. From this vision, the main road blocks become apparent. Future research needs to develop even richer and more interconnected models of the technical systems in a form that is completely accessible to AI. The second road block is the development of AI approaches that are powerful enough to deal with the resulting abundance of data. The third road block concerns the trustability of the results; future AI approaches not only need to deliver the design, but also a concise documentation how the design was derived and how high the levels of safety and reliability are. The basis for the presented research are design processes of remote vehicles and vehicle components; these processes also serve as a basis for the explanation of the framework. Up to now, the framework was tested using several examples; the investigation of the general applicability will be the focus of further research. One future research field is the investigation of the possibility to include the logic perspective into this framework; this could be fruitful for a better connection of reliability and safety analyses with the physical effect chains present in the vehicle. A further promising research field is a sensible integration into engineering frameworks for vehicle design, which can be based on the unified modeling language (UML), the system modeling language (SysML), graph-based design languages (GBDLs), and standards such as AutomationML.

Funding: Parts of the described research were funded in the scope of the project "Automatisierter Entwurf eines geometrischen und kinetischen digitalen Zwillings einer Rohbaufertigungsanlage für die Virtuelle Inbetriebnahme (TWIN)", which is funded by the German Federal Ministry of Education and Research. Parts of the described research were funded by the Carl Zeiss Foundation in the scope of the project AI-based digital twin (KI-basierter digitaler Zwilling—KIDZ).

Data Availability Statement: The data presented in this study are available on request from the corresponding author.

Conflicts of Interest: The author declares no conflict of interest.

Abbreviations

The following abbreviations are used in this manuscript:

AI	Artificial Intelligence
CAD	Computer-Aided Design
CSS	Channel and Support Structure
FMI	Functional Mock-up Interface
FMU	Functional Mock-up
GBDL	Graph-based Design Language
IF4APM	Integrated Framework for Abstract Physics Modeling
IFM	Integrated Function Modeling
MBSE	Model-based Systems Engineering
SysML	Systems Modeling Language
UML	Unified Modeling Language
WS	Working Surface
WSP	Working Surface Pair

References

1. Brunner, H.; Rossbacher, P.; Hirz, M. Sustainable product development: Provision of information in early automotive engineering phases. *Teh. Glas.* **2017**, *11*, 29–34.
2. Demirel, H.O.; Goldstein, M.H.; Li, X.; Sha, Z. Human-centered generative design framework: An early design framework to support concept creation and evaluation. *Int. J. Hum.-Interact.* **2024**, *40*, 933–944. [CrossRef]
3. Pahl, G.; Beitz, W. *Engineering Design: A Systematic Approach*; Springer Science & Business Media: Berlin/Heidelberg, Germany, 2013.
4. Ponn, J.; Lindemann, U. *Konzeptentwicklung und Gestaltung Technischer Produkte*; Springer: Berlin/Heidelberg, Germany, 2011.
5. Stetter, R. Fault-Tolerant Design and Control of Automated Vehicles and Processes. In *Insights for the Synthesis of Intelligent Systems*; Springer: Cham, Switzerland, 2020.
6. VDI/VDE. VDI 2221: Design of Technical Products and Systems. In *Model of Product Design*; Beuth: Berlin, Germany, 2019.
7. Holder, K.; Zech, A.; Ramsaier, M.; Stetter, R.; Niedermeier, H.P.; Rudolph, S.; Till, M. Model-Based Requirements Management in Gear Systems Design Based on Graph-Based Design Languages. *Appl. Sci.* **2017**, *7*, 1112. [CrossRef]

8. Malcher, P.; Viana, D.; Antonino, P.O.; dos Santos, R.P. Investigating Open Innovation Practices to Support Requirements Management in Software Ecosystems. In *Proceedings of the International Conference on Software Business*; Springer Nature: Cham, Switzerland, 2023; pp. 35–50.
9. Serna, A. *Requirements Engineering*; Editorial Instituto Antioqueño de Investigación: Medellín, Antioquia 2024.
10. Kosse, S.; Vogt, O.; Wolf, M.; König, M.; Gerhard, D. Requirements management for flow production of precast concrete modules. In *Proceedings of the International Conference on Computing in Civil and Building Engineering*; Springer: Cham, Switzerland, 2022; pp. 687–701.
11. Eisenbart, B.; Gericke, K.; Blessing, L.; McAloone, T. A DSM-based framework for integrated function modelling: Concept, application and evaluation. *Res. Eng. Des.* **2016**, *28*, 25–41. [CrossRef]
12. Gericke, K.; Eisenbart, B. The integrated function modeling framework and its relation to function structures. *AI EDAM* **2017**, *31*, 436–457. [CrossRef]
13. Elwert, M.; Ramsaier, M.; Eisenbart, B.; Stetter, R. Holistic Digital Function Modelling with Graph-Based Design Languages. In Proceedings of the 22nd International Conference on Engineering Design (ICED 19), Delft, The Netherlands, 5–8 August 2019; Volume 1.
14. Drave, I.; Rumpe, B.; Wortmann, A.; Berroth, J.; Hoepfner, G.; Jacobs, G.; Spuetz, K.; Zerwas, T.; Guist, C.; Kohl, J. Modeling mechanical functional architectures in SysML. In Proceedings of the 23rd ACM/IEEE International Conference on Model Driven Engineering Languages and Systems, Virtual, 16–23 October 2020; pp. 79–89.
15. Wyrwich, C.; Boelsen, K.; Jacobs, G.; Zerwas, T.; Höpfner, G.; Konrad, C.; Berroth, J. Seamless function-oriented mechanical system architectures and models. *Eng* **2024**, *5*, 301–318. [CrossRef]
16. Krüger, M.F.; Zorn, S.; Gericke, K. Combining function modelling and requirements modelling with the ifm framework. *Proc. Des. Soc.* **2023**, *3*, 987–996. [CrossRef]
17. Albers, A.; Wintergerst, E. The Contact and Channel Approach: Relating a system's physical structure to its functionality. In *An Anthology of Theories and Models of Design: Philosophy, Approaches and Empirical Explorations*; Springer: London, UK, 2014; pp. 61–72.
18. Hoepfner, G.; Nachmann, I.; Zerwas, T.; Berroth, J.K.; Kohl, J.; Guist, C.; Rumpe, B.; Jacobs, G. Towards a Holistic and Functional Model-Based Design Method for Mechatronic Cyber-Physical Systems. *J. Comput. Inf. Sci. Eng.* **2023**, *23*, 051001. [CrossRef]
19. Zerwas, T.; Jacobs, G.; Spütz, K.; Höpfner, G.; Drave, I.; Berroth, J.; Guist, C.; Konrad, C.; Rumpe, B.; Kohl, J. Mechanical concept development using principle solution models. In *Proceedings of the IOP conference Series: Materials Science and Engineering*; IOP Publishing: Bristol, UK, 2021; Volume 1097, p. 012001.
20. Stetter, R.; Niedermeier, M. Approaches towards lean products. In Proceedings of the Knowledge, Innovation and Sustainability. Proceedings of the 16th International Conference on Engineering Design, Paris, France, 28–30 August 2007; Design Society: Paris, France, 2007.
21. Stetter, R.; Till, M. A Concept for an Integrated Framework for Abstract Physics Modelling (IF4APM). *Procedia CIRP* **2024**, *120* .
22. Vitale, G.; Hollander, M. The Software-Defined Vehicle: How to Verify and Validate Software Functions. In *Proceedings of the International Stuttgart Symposium*; Springer: Wiesbaden, German, 2023; pp. 399–405.
23. Cederblahd, J.; Cicchetti, A.; Suryadevara, J. Early Validation and Verification of System Behaviour in Model-based Systems Engineering: A Systematic. *ACM Trans. Softw. Eng. Methodol.* **2024**, *33*, 1–67. [CrossRef]
24. Kraft, J. *Entwicklung Eines Universellen Schwerlast-Roboter-Fahrzeugs*; Project Report; Ravensburg-Weingarten University: Weingarten, Germany, 2020.
25. Saiger, C. *Hydraulisches Hubgerüst als Anbaugerät für Einen Kleintraktor*; Project Report; Ravensburg-Weingarten University: Weingarten, Germany, 2018.
26. Reinprecht, M. Entwicklung Eines Heckkrafthebers mit dem Schwerpunkt Fehlertoleranz. Bachelor's Thesis, Ravensburg-Weingarten University, Weingarten, Germany, 2020.
27. Mario, H.; Dietrich, W.; Gfrerrer, A.; Lang, J. *Integrated Computer-Aided Design in Automotive Development: Development Processes, Geometric Fundamentals, Methods of CAD, Knowledge-Based Engineering Data Management*; Springer: Berlin/Heidelberg, Germany, 2013.
28. Eigner, M.; Ernst, J.; Roubanov, D.; Dickopf, T. An initial approach for the application of product assembly information in the early phases of the product development process by using methods of Model Based Systems Engineering. In Proceedings of the DS 81: Proceedings of NordDesign 2014, Espoo, Finland, 27–29 August 2014.
29. Henderson, K.; Salado, A. Value and benefits of model-based systems engineering (MBSE): Evidence from the literature. *Syst. Eng.* **2021**, *24*, 51–66. [CrossRef]
30. Graessler, I.; Wiechel, D.; Oezcan, D.; Taplick, P. Tailored metrics for assessing the quality of MBSE models. *Proc. Des. Soc.* **2024**, *4*, 2545–2554. [CrossRef]
31. Stetter, R. Approaches for Modelling the Physical Behavior of Technical Systems on the Example of Wind Turbines. *Energies* **2020**, *13*, 2087. [CrossRef]
32. Koller, R.; Kastrup, N. *Prinziplösungen zur Konstruktion Technischer Produkte*; Springer: Berlin/Heidelberg, Germany, 2013.
33. Ehrlenspiel, K.; Meerkamm, H. *Integrierte Produktentwicklung: Denkabläufe, Methodeneinsatz, Zusammenarbeit, 6, Vollständig überarbeitete und Erweiterte Auflage*; Carl Hanser Verlag GmbH Co KG: Munich, Germany, 2017.
34. Lunze, J. *Regelungstechnik 1*; Springer: Berlin/Heidelberg, Germany, 2013; Volume 10.

35. Witczak, M. *Fault Diagnosis and Fault-Tolerant Control Strategies for Non-Linear Systems*; Lecture Notes in Electrical Engineering; Springer International Publishing: Berlin/Heidelberg, Germany, 2014; Volume 266, p. 229.
36. Blanke, M.; Kinnaert, M.; Lunze, J.; Staroswiecki, M. *Diagnosis and Fault-Tolerant Control*; Springer: New York, NY, USA, 2016.
37. Tang, C.; Khajepour, A. Wheel modules with distributed controllers: A multi-agent approach to vehicular control. *IEEE Trans. Veh. Technol.* **2020**, *69*, 10879–10888. [CrossRef]
38. Tang, C.; Khajepour, A. Agent-based model predictive controller (AMPC) for flexible and efficient vehicular control. *IEEE Trans. Veh. Technol.* **2021**, *70*, 9877–9885. [CrossRef]
39. Liang, J.; Lu, Y.; Yin, G.; Fang, Z.; Zhuang, W.; Ren, Y.; Xu, L.; Li, Y. A distributed integrated control architecture of AFS and DYC based on MAS for distributed drive electric vehicles. *IEEE Trans. Veh. Technol.* **2021**, *70*, 5565–5577. [CrossRef]
40. Liang, J.; Lu, Y.; Wang, F.; Yin, G.; Zhu, X.; Li, Y. A robust dynamic game-based control framework for integrated torque vectoring and active front-wheel steering system. *IEEE Trans. Intell. Transp. Syst.* **2023**, *24*, 7328–7341. [CrossRef]
41. Albers, A.; Matthiesen, S.; Lechner, G. Konstruktionsmethodisches grundmodell zum zusammenhang von gestalt und funktion technischer systeme. *Konstruktion (1981)* **2002**, *7–8*, 55–60.
42. Albers, A.; Braun, A.; Sadowski, E.; Wyatt, D.F.; Wynn, D.C.; Clarkson, P.J. Contact and Channel Modelling Using Part and Function Libraries in a Function-Based Design Approach. In Proceedings of the International Design Engineering Technical Conferences and Computers and Information in Engineering Conference, Montreal, QC, Canada, 15–18 August 2010; Volume 44137, pp. 393–404.
43. Albers, A.; Zingel, C. Extending SysML for engineering designers by integration of the contact & channel–approach (C&C2-A) for function-based modeling of technical systems. *Procedia Comput. Sci.* **2013**, *16*, 353–362.
44. Grauberger, P.; Bremer, F.; Sturm, C.; Hoelz, K.; Wessels, H.; Gwosch, T.; Wagner, R.; Lanza, G.; Albers, A.; Matthiesen, S. Qualitative Modelling in Embodiment Design-Investigating the Contact and Channel Approach through Analysis of Projects. In *Proceedings of the Design Society: DESIGN Conference*; Cambridge University Press: Cambridge, UK, 2020; Volume 1, pp. 897–906.
45. Freudenmann, T. *Ontologien zur Validierung von Produkten Basierend auf dem Contact & Channel-Ansatz (C&C^2-Ansatz)= Ontologies for the Validation of Products Based on the Contact & Channel Approach (C&C^2-Approach)*; IPEK: Karlsruhe, Germany 2014.
46. Gero, J.S.; Kannengiesser, U. The function-behaviour-structure ontology of design. In *An Anthology of Theories and Models of Design: Philosophy, Approaches and Empirical Explorations*; Springer: Berlin/Heidelberg, Germany, 2014; pp. 263–283.
47. Gero, J.; Milovanovic, J. The situated function-behavior-structure co-design model. *CoDesign* **2021**, *17*, 211–236. [CrossRef]
48. Russo, D.; Spreafico, C. Investigating the multilevel logic in design solutions: A Function Behaviour Structure (FBS) analysis. *Int. J. Interact. Des. Manuf.* **2023**, *17*, 1789–1805. [CrossRef]
49. Sanderson, D.; Chaplin, J.C.; Ratchev, S. A Function-Behaviour-Structure design methodology for adaptive production systems. *Int. J. Adv. Manuf. Technol.* **2019**, *105*, 3731–3742. [CrossRef]
50. Neema, H.; Gohl, J.; Lattmann, Z.; Sztipanovits, J.; Karsai, G.; Neema, S.; Bapty, T.; Batteh, J.; Tummescheit, H.; Sureshkumar, C. Model-based integration platform for FMI co-simulation and heterogeneous simulations of cyber-physical systems. In *Proceedings of the 10th International Modelica Conference*; Linköping University Electronic Press: Linköping, Sweden, 2014; Volume 96, pp. 235–245.
51. Larsen, P.G.; Fitzgerald, J.; Woodcock, J.; Fritzson, P.; Brauer, J.; Kleijn, C.; Lecomte, T.; Pfeil, M.; Green, O.; Basagiannis, S.; et al. Integrated tool chain for model-based design of Cyber-Physical Systems: The INTO-CPS project. In Proceedings of the 2016 2nd International Workshop on Modelling, Analysis, and Control of Complex CPS (CPS Data), Vienna, Austria, 11 April 2016; pp. 1–6.
52. Miller, M.; Pfeil, M.; Kennel, R. *Trailer Electrification—A HIL Approach for MPC Powertrain Control to Ensure Driver Safety in Micromobility*; Technical Report; SAE Technical Paper: Warrendale, PA, USA 2023.
53. Saft, P.; Pfeil, M.; Stetter, R.; Till, M.; Rudolph, S. Integration of geometry modelling and behavior simulation based on graph-based design languages and functional mockup units. *Procedia CIRP* **2024**, *120* .
54. Beernaert, T.; Verlaan, A.; Etman, P.; Giesen, P.; Van Beekum, E.; Ribeiro, M.; Bola, I.; Moser, L.; De Bock, M.; Classen, I.; et al. From Physics to Project Management: A Multi-Domain Matrix Model for the Distributed Analysis of Nuclear Fusion Systems. In Proceedings of the DS 126: Proceedings of the 25th International DSM Conference (DSM 2023), Gothenburg, Sweden, 3–5 October 2023; pp. 29–38.
55. Elwert, M.; Ramsaier, M.; Eisenbart, B.; Stetter, R.; Till, M.; Rudolph, S. Digital Function Modeling in Graph-Based Design Languages. *Appl. Sci.* **2022**, *12*, 5301. [CrossRef]
56. Scalice, R.K.; Berkenbrock, G.R.; Mendoza, Y.E.A. Use case based methodology for conceptual design of industrial mechatronic products. In Proceedings of the DS 87-4 Proceedings of the 21st International Conference on Engineering Design (ICED 17) Vol 4: Design Methods and Tools, Vancouver, BC, Canada, 21–25 August 2017; pp. 347–356.
57. Eisenbart, B.; Gericke, K.; Blessing, L. Application of the IFM framework for modelling and analysing system functionality. In Proceedings of the DS 77: Proceedings of the DESIGN 2014 13th International Design Conference, Cavtat, Croatia, 19–22 May 2014; pp. 153–162.
58. The MathWorks Inc.: Natick, MA, USA. Available online: https://mathworks.com/ (accessed on 31 July 2024).
59. The Modelica Association: Linköping, Sweden. Available online: https://modelica.org/ (accessed on 31 July 2024).

60. INTO-CPS Association: Aarhus, Denmark. Available online: https://into-cps.org/ (accessed on 31 July 2024).
61. Stetter, R.; Simundsson, A. Design for control. In Proceedings of the DS 87-4 Proceedings of the 21st International Conference on Engineering Design (ICED 17) Vol 4: Design Methods and Tools, Vancouver, BC, Canada, 21–25 August 2017; pp. 149–158.

vehicles

MDPI

Article

Digital Integrated Design and Assembly Planning Processes for Sports Vehicles Using the Example of a Skateboard

Timo Schuchter [1], Markus Till [1,*], Ralf Stetter [1] and Stephan Rudolph [2]

1 Department of Mechanical Engineering, Ravensburg-Weingarten University (RWU), 88250 Weingarten, Germany; timo.schuchter@rwu.de (T.S.); ralf.stetter@rwu.de (R.S.)
2 Institute of Aircraft Design, University of Stuttgart, 70569 Stuttgart, Germany; rudolph@ifb.uni-stuttgart.de
* Correspondence: markus.till@rwu.de

Abstract: The current product and assembly processes of system development in the vehicle industry are characterised by a multitude of different model formats, a relatively low level of data integration, and an unsatisfactory management of information. This article presents an integrated design and assembly planning process which applies several model-to-model (M2M) transformations in order to ensure a seamless transition from product requirements to an assembly system layout and design. The digital process employs a framework based on graph-based design languages (GBDLs) and achieves an integration in a model-based systems engineering (MBSE) industrial context. The underlying hypothesis that this seamless transition is possible is tested on the basis of the product and assembly system development of a sports vehicle. In this article, a skateboard is used for detailing and explaining the different modelling perspectives throughout the engineering and assembly process of this product. Due to a conscious application of GBDLs in an MBSE framework, it is possible to achieve a continuous sequence of M2M transformations which guarantees a maximum level of information integrity. These two aspects are cornerstones for a future integrated design automation of a product and its assembly system. It is important to note that the presented approach is universal and can be used in the production of components for the automotive industry, entire vehicles, and their assembly.

Keywords: graph-based design languages; assembly planning; layout planning; assembly system development

Academic Editor: Mohammed Chadli

Received: 28 January 2025
Revised: 19 February 2025
Accepted: 21 February 2025
Published: 25 February 2025

Citation: Schuchter, T.; Till, M.; Stetter, R.; Rudolph, S. Digital Integrated Design and Assembly Planning Processes for Sports Vehicles Using the Example of a Skateboard. *Vehicles* **2025**, *7*, 22. https://doi.org/10.3390/vehicles7010022

1. Introduction

The design and the production of vehicles is currently characterised by enormous challenges. Due to intensified competition and increasing customer demands a worldwide, the product variety is escalating. Globally distributed manufacturing has further led to increasingly complex supply chains where the management of heterogeneous and scattered information has become a key issue [1]. This necessitates more flexible, integrated, and intelligent product and assembly system development processes. These processes are characterised by several product and assembly system modelling attempts which are currently not connected at all or are only partly connected. Consequently, several data or process entities such as product components, assembly processes, and resources are represented many times in multiple digital models. This fact and other challenges in assembly-oriented design such as a systematic digital process chain *"from data to information to knowledge"* highlights the urgent need for a better digital interaction [2].

The main aim of this paper is to propose an integrated design and assembly planning process that enables a better digital integration as well as a higher digital consistency between several product and process models ranging from early product development to a concrete assembly system coupled with the necessary robot code. The proposed digital design and assembly planning process applies several M2M transformations in order to enable a seamless transition from initial product requirements to an assembly system design and robot code. A sports vehicle—a skateboard—serves as a concrete example for explanation. However, the proposed approach is universal and can be used in the production of components for the vehicle industry, entire vehicles, and their assembly.

In order to explain the proposed design and assembly planning process, the main information entities in the process are described here. Important results of the product and assembly system development are the EBOMs (engineering bill of materials—product parts, their quantity, and their assembly relation) from an engineering point of view) and MBOMs (manufacturing bill of materials—production process, including the manufacturing product structure) [1,3]. The bill of materials (BOMs) conveys key information that guides development at different stages and is an attempt to manage heterogeneous and scattered information [1,4]. A key step in the process of assembly planning is the generation of a sensible assembly sequence; the automation of this generation has been researched for decades but the solutions are computationally excessive, require many human operations, or require the automatic generation of constraint rules [5].

Obviously, during an integrated product and assembly system development process, an enormous amount of data in the form of different models is created and the management of the abundance of models requires a holistic approach. Over the last several decades, MBSE was found to be a versatile approach to address this challenge [6–8]. The processes described in the scope of MBSE are based on the well-known V-model (see, e.g., VDI guideline 2206 [9,10]). Figure 1 shows a V-model-based representation of the main elements of an integrated product and assembly system development process.

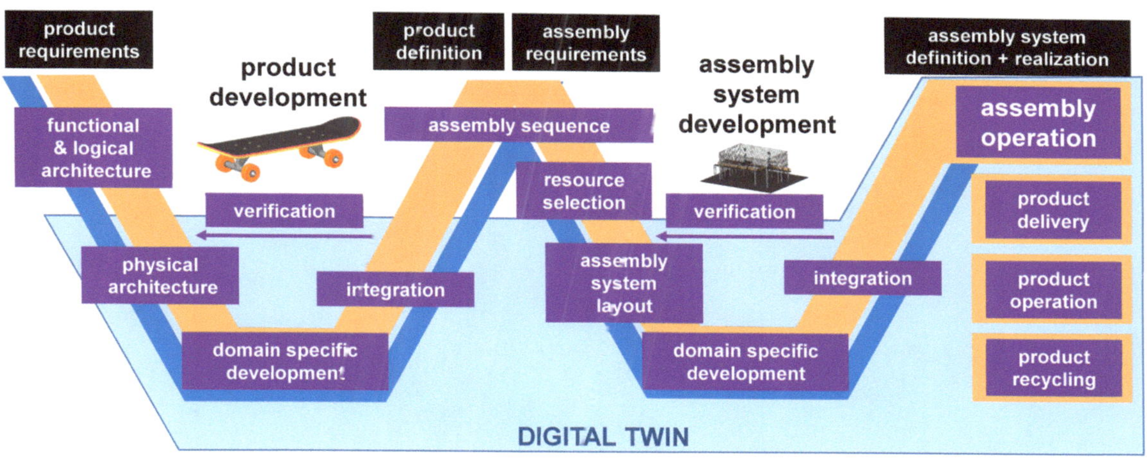

Figure 1. Life-cycle with important elements in product and assembly system development.

It is important to note that both the development cycles of the product and the assembly system are presented based on a V-model (Figure 1). The initial product model consists of the requirements concerning the functionality and decisive properties of the product to be assembled. The first cycle–product development–starts with the development of the functional and logical architecture. This architecture is decisive for a sensible realisation of the functionality and ultimately for the success of the product [11]. The next logical phase is the determination of the function carriers concerning the acting physical relationships—the physical architecture [12]. Further development is carried out in the different domains:

mechanics (including hydraulics and pneumatics), electrics/electronics, and software. It is an important quality of the V-model [9] that the integration of the domain-specific solutions is explicitly included; this decisive step enables the realisation of coordinated multi-domain systems, but requires intensive cross-domain communication. Depicted in the centre of the first cycle is the verification at different levels, which finally leads to a validation with the original product requirements. The first cycle leads to the complete set of models of the product—the product definition. Connecting the two cycles is the assembly sequence, which is the focus of this paper. In later sections of this paper, it will be shown that a model-to-model transformation of the product model to the assembly sequence is possible and that an intelligent product model can contain an implicit determination of the assembly sequence. Together with the assembly requirements, the assembly sequence enables the development cycle for the assembly system, which is essentially analogous to the first cycle. It is important to note that the verification stage may include approaches such as virtual commissioning [13] and pre-validation with virtual reality [14], with systems such as Unreal Engine, Epic Games, Cary, NC, USA [15] or Unity, Unity Technologies, San Francisco, CA, USA [16]. The second cycle culminates in the assembly system definition and realisation. On this basis, the next stages of the life-cycle, i.e., assembly operation, product delivery, product operation, and product recycling, are carried out.

As mentioned before, the two cycles are connected with the assembly sequence, which is one of the main sources of information for assembly system development, but cannot be generated without knowledge concerning the assembly system concept. It is also important to note that in this representation, the development process progresses to the recycling stage in order to be able to represent the complete scope of a digital twin (DT) (see Section 5).

Furthermore, the necessity of the continuous consideration of a circular economy is emphasized. When focusing on the models that drive the MBSE process, it becomes apparent that a shift from product information to assembly system information is just logical and inevitable. A more detailed representation of the most important models in the integrated product and assembly system development process in shown in Figure 2.

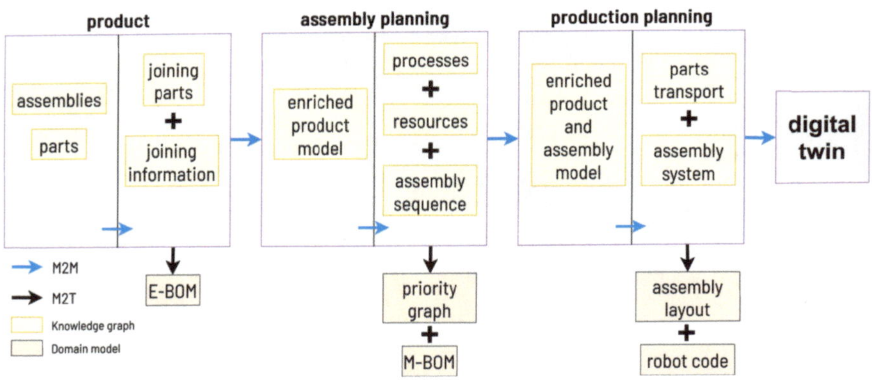

Figure 2. Model transformations in product and assembly system development.

On the very left, the items are shown, which are conventionally created and modified with computer-aided design (CAD) systems and are stored using product data management (PDM) or product life-cycle management (PLM) systems: the product parts and the assemblies are resulting from a combination of these product parts. As it will be shown in the later parts of this paper, an M2M transformation to the information entities' "joining parts" and their respective "joining information" is possible. The result of a model-to-text (M2T) transformation is the EBOM. The resulting enriched product model can be the basis of another M2M transformation determining the joining (or assembly) "processes", the necessary "resources" for the joining operations (e.g., robots), and the "assembly sequence" (see Section 3). An M2T transformation leads to a priority graph and an MBOM. The result is an even more enriched model of the product and its assembly. As it will be shown in the later parts of this paper, an M2M transformation to the information entities' "parts transport" (i.e., the manner how parts and sub-assemblies are transported within the assembly system from station to station) and a geometric and kinematic representation of the "assembly system" is possible. An M2T transformation can result in the two-dimensional assembly system layout (i.e., the assembly system floor-plan) and the robot code (i.e., the different movement commands for all manipulators within the assembly system). The stepwise enriched product and assembly model, together with the geometric and kinematic information, can be the basis for a digital twin (see Section 5).

Production planning is an anticipation of the manufacturing and assembly operations and systems for the purpose of initial development and optimisation; the production planning process in industrial companies involves various management levels and concerns different planning horizons [17]. This important industrial field is influenced by several current trends. For several years, digital twins (DTs) that provide a digital representation of a real system; enable a bidirectional exchange of parameters, states, and other data; can represent the whole system life-cycle and allow for elaborated monitoring, control, diagnosis, and optimisation processes [18–20] have been available. In this context, integrated control frameworks, both in the product itself—the vehicle—and its assembly system, are gaining importance [21,22] and their development is frequently supported by hardware in the loop (HiL) experiments [23,24], especially in later development stages. The early stages of production planning are highly influenced and altered by the widespread availability of rapid prototyping technology [17]. Based on the same technology, additive manufacturing is meanwhile applied in the production process itself—mainly for low-volume or rather small product and assembly system components [25].

In this context, important research work concerns the design for manufacturing and assembly (DFMA). A concise overview is given by Formentini et al. [26]. Bouissiere et al. proposed methods for the assessment manufacturing and assembly aspects of product architectures [27]. Conceptual design for assembly based on a mathematical framework is also a topic of current research [28]. Some current research works focus on assembly planning in the vehicle industry [29–31]. The context of sports vehicles is usually not the focus of most research work. One example is a study by Arifin et al. who address the design for manufacturing and assembly of sports equipment with the example of an aerobic walker [32]. The modular design of sports vehicles, which also considers assembly processes, is the focus of the work of Sali [33].

Obviously, an essential part of production planning is assembly planning. A complete review of the state-of-the-art automated assembly sequence generation is beyond the scope of this paper; some prominent results will be discussed below. Demoly et al. [2,34] described a product-relationship management approach that enables concurrent product design and assembly sequence planning. He et al. [1] proposed a unified bill of materials (BOMs) model based on a single source of product data A near-optimum assembly sequence search algorithm employing an objective function (scoring assembling ability as well as the efficiency of the assembly process) is presented by Enomoto et al. [5]. A related study focuses on near-optimal assembly task sequencing for multi-arm robot systems [35]. The optimisation of task allocation for human–robotic assembly processes was researched by Tram et al. [36]. The planning of optimal assembly sequences is addressed by Culbertson et al. [37]. Liu et al. [38] applied deep reinforcement learning in order to enable a physics-aware combinatorial assembly planning. The compatibility of different product models was researched by Paehler and Matthiesen [39].

Upon summarising the literature analysis concerning concurrent and integrated product and assembly system development, it can be pointed out that promising approaches for partial problems exist, but that a continuous M2M transformation-oriented approach is yet to be developed. The following sections explain elements of this kind of approach. Section 2 focuses on the product stage of the integrated process. The stages that follow concern the assembly and production planning and are addressed in Section 3 and Section 4, respectively. The dynamic simulation of the assembly system can be employed as a digital twin (DT); this is analysed and described in Section 5. This paper is concluded by a conclusion and outlook (Section 6).

2. Product Modelling in Graph-Based Design Languages

In the proposed framework, the culmination point for product modelling and the compilation of GBDLs is the so-called Design Cockpit 43® software environment [40], which is based on the fundamental research activities of Rudolph [41] and many of his co-workers (see [42,43] for two of the most recent contributions). The Design Cockpit 43® software was developed, maintained, and further extended by IILS mbH (Trochtelfingen, Germany) [40]. This software environment enables the storage of design information in the form of ontologies, the coding of product and assembly system information as vocabulary, as well the transformation of the coded knowledge into a unified central model—the design graph. The digital design process with GBDLs is shown in Figure 3 (adapted from [41,44]).

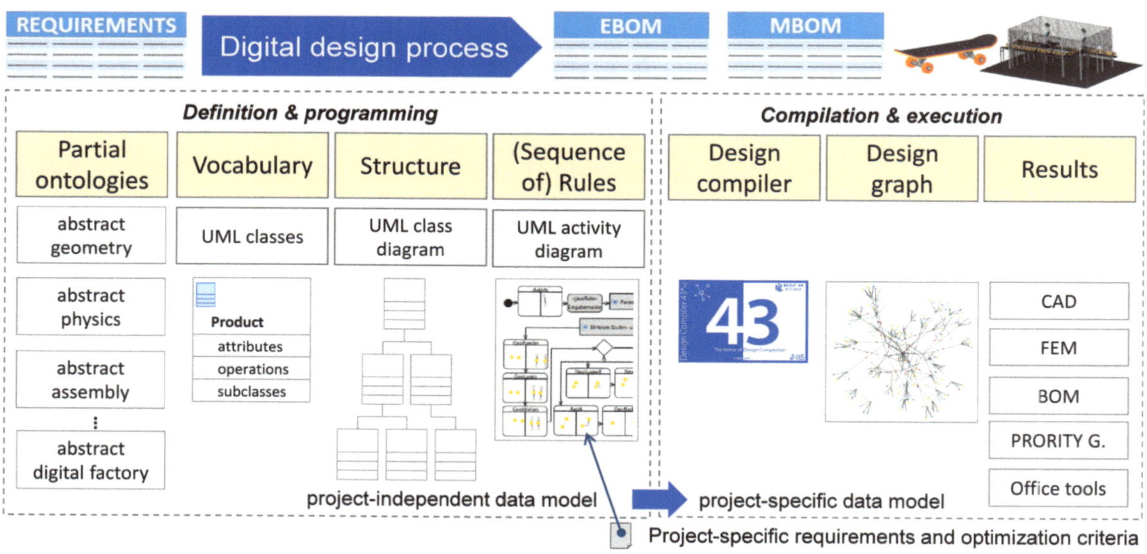

Figure 3. Digital design process with graph-based design languages (adapted from [41,44]).

In the integrated digital design process, two main phases can be distinguished. In the definition and programming phases, the main focus is on project-independent data. Based on partial ontologies, vocabulary, and structure, a sequence of rules is generated. The compilation and execution phase focuses on project-specific, instantiated data. The data from the project-independent phase are compiled leading to the design graph as a central model. From the design graph, different kinds of domain-specific data models can be compiled such as CAD files. The current research concentrates on Open Cascade, 7.8.0, Open Cascade SAS, Guyancourt, France [45], and CATIA, Dassault Systèmes, Vélizy-Villacoublay, France [46], but other CAD systems can also be integrated. For finite element analyses, the software systems Ansys, Ansys, Inc., Canonsburg, PA, USA [47], or Abaqus FEA, Dassault Systèmes, Vélizy-Villacoublay, France [48], are widely used in industry. BOMs can be represented in Microsoft Excel, Microsoft Corporation, Redmond, WA, USA [49], but professional CAD systems such as PTC Creo, 10.0, PTC Inc., Boston, MA, USA [50], can also store BOMs.

The digital design process shown in Figure 3 has already been described in detail in earlier publications [44,51]. It is important to note that the general procedure is applicable to all M2M and M2T transformations in integrated product and assembly system development. In the following, the architecture design and the underlying methodology, which is illustrated in Figure 2, will be discussed using the product model of a skateboard (see Figure 4).

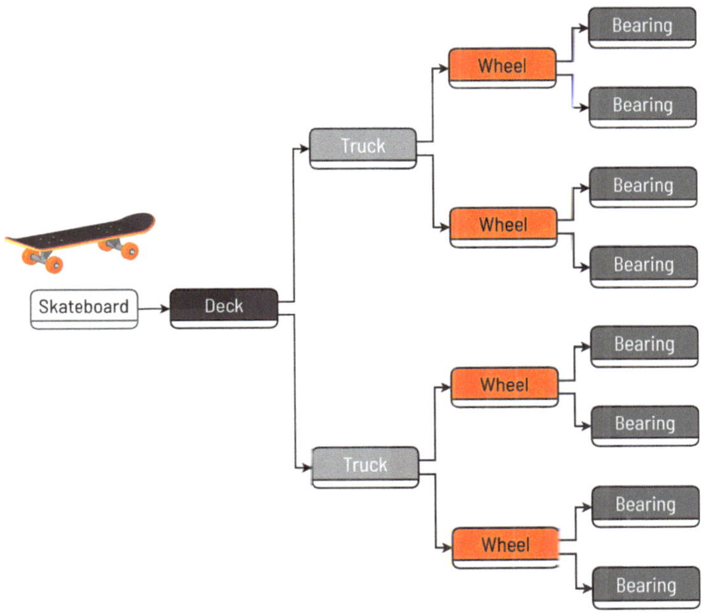

Figure 4. Product model of the skateboard as a knowledge graph.

Figure 4 represents the product model and is intended to familiarise the reader with the main structure of the product in order to ease the understanding of the assembly sequence. The model is based on the "Skateboard" node, which reflects the basic model. The "Deck" node refers to the top board of the skateboard and is therefore an essential component of the model. Two so-called "Trucks" are connected to it and represent the axles of the skateboard. Each axle is connected to two wheels, which in turn dispose of two bearings. In the following, the model described in Figure 4 is supplemented with joining information. This information describes the way in which two components are connected and contains all the information that is relevant for describing the joining process. The underlying product architecture is realised by means of three different assembly operations, which are modelled as rules. Figure 5 presents the assembly of a bearing in a wheel.

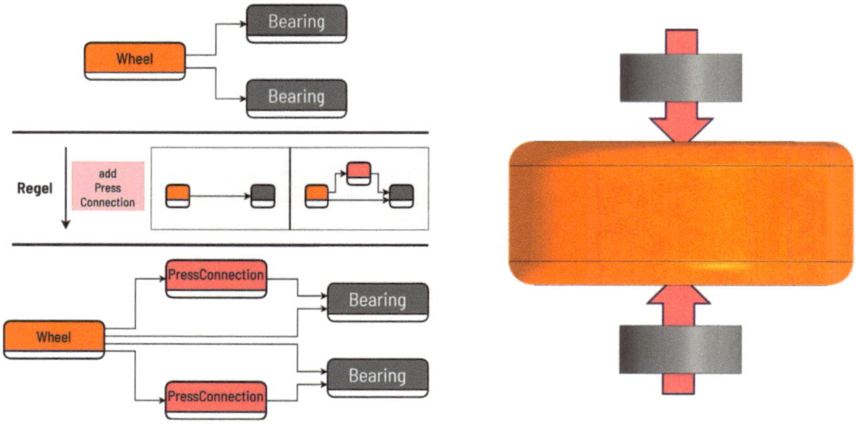

Figure 5. Assembly of bearings in wheels. (**Top left**) Knowledge graph before rule execution. (**Bottom left**) Knowledge graph after rule execution. (**Right**) A 3D CAD model.

This assembly is represented by "PressConnection" (Figure 6). The assembly operation modelled with this rule is implicitly visible on the right side of Figure 4; slightly further on the left in Figure 4 is the assembly of the wheels on an axle of a truck (Figure 6).

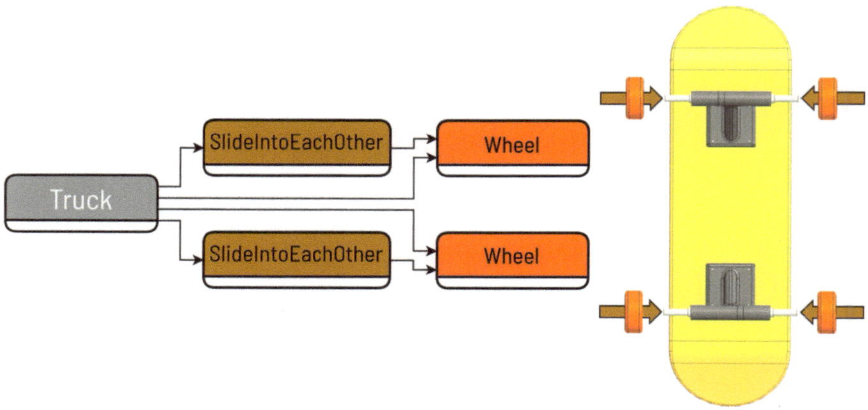

Figure 6. Assembly of wheels. (**Left**) Knowledge graph. (**Right**) A 3D CAD model.

This assembly is represented by "SlideIntoEachOther" (Figure 6). The next assembly step concerns the connection of each truck with the deck by means of screwing. This is presented in Figure 7.

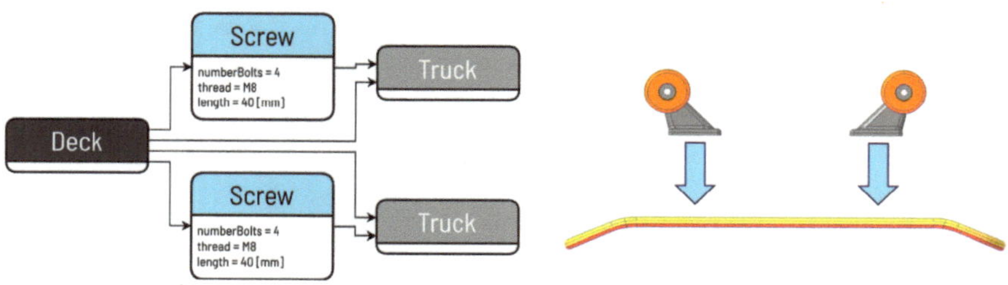

Figure 7. Assembly of trucks. (**Left**) Knowledge graph. (**Right**) A 3D CAD model.

This assembly is represented by "Screw" (Figure 7). Representing these assembly operations, the product model of the skateboard shown in Figure 4 is supplemented with joining information; the result is an enriched product model (see also Figure 2). This enriched product model is shown in Figure 8.

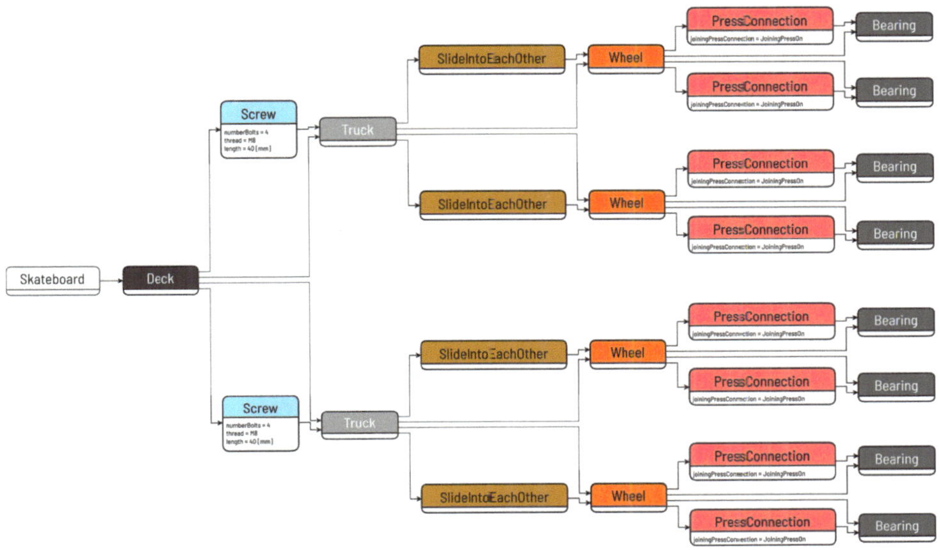

Figure 8. Enriched product model of the skateboard as a knowledge graph.

The enriched product model can also be understood as an interface to the more detailed steps of the assembly planning.

3. Assembly Planning

Assembly planning involves determining the sequence of assembly operations and the resources needed to assemble a product from its modules and components. It describes the process of organising the assembly, i.e., determining the order in which modules and parts should be assembled (the assembly sequence) and the required resources for each assembly operation. The generation of feasible assembly sequences remains a challenging research problem due to the following reasons: the number of potential assembly sequences exponentially grows, some sequences may not be physically executable due to unsatisfied physical constraints, and the geometry of objects can be highly complex [52].

The optimal sequence can be determined using a number of different methods. It is important to consider these methods because the number of possible sequences increases exponentially with each additional combination. In the case of the skateboard, which consists of $p = 15$ parts, this corresponds to a number of 1,307,674,368,000 sequences. Even if we ignore all sequences that are the inverse of another sequence, there are still 653,837,184,000 sequences. It should be noted that the majority of these sequences would not work or would only be feasible with great technical effort. Various methods have been developed to reduce this number to a minimum and to filter out the most important sequences [53–55].

In practice, the starting point is often the assembled module, which is then disassembled. A feasible disassembly sequence always corresponds to a working assembly sequence in reverse order. A common method is to first identify the mating surfaces between components. This can be performed by collision detection, the distance between components, or the structure of the model (as exemplified in Figure 8).The data may be processed to create an adjacency matrix. The resulting matrix depicts the interconnections between all components. The generated adjacency matrix may be employed as a foundation for the derivation of an interference matrix. The resulting matrix provides information regarding the components that are in contact in terms of their joining direction during the assembly process. Consequently, the assembly sequences can be derived [54]. It is likewise possible to make use of machine learning methodologies [55].

In the case of assemblies comprising a limited number of components, it is feasible to construct a Petri net that maps all sequences directly from the adjacency matrix. The corresponding sequences can be extracted from this using the enumeration algorithm, for example. Furthermore, an assembly priority graph can be derived from the adjacency matrix by enriching the boundary conditions. It is recommended that the assembly be initially analysed at the assembly level, and subsequently, the assembly sequence for the respective assemblies be determined. This is essential as the number of potential combinations increases exponentially with the level of detail.

In order to determine the number of similar assemblies, a width-based search is carried out in which the model is first simplified and then grouped. The width-based search performed on the existing graph, which functions as an enriched product model, is carried out depending on the requirements for fully and partially similar cases. A fully similar case would be defined by identical components (in terms of geometric characteristics and properties) with the same joining technique. Partially similar cases would be characterised by components that are similar in many, but not all, properties and whose joining techniques also have a large number of identical properties. As a result of this search, five assembly sets ("AssemblySet"—a collection of parts that require similar assembly operations) are identified, as shown in Figure 9.

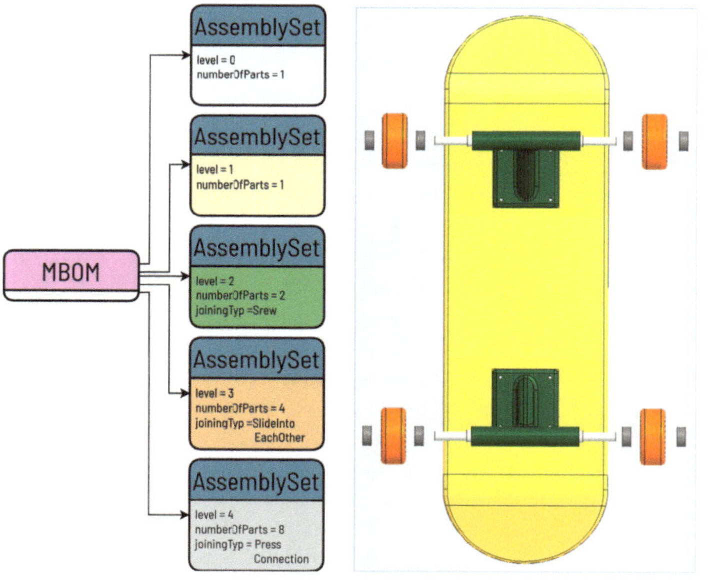

Figure 9. Assembly sets of the skateboard. (**Left**) Knowledge graph. (**Right**) A 3D CAD model.

The transformation of the product model into an assembly model is used to encapsulate the product and to transform it into a model suitable for assembly planning. Process planning is performed consistently from the previously created model, using the assembly sets as the starting point. The main processes, shown in Figure 10, are then derived from the assembly sets.

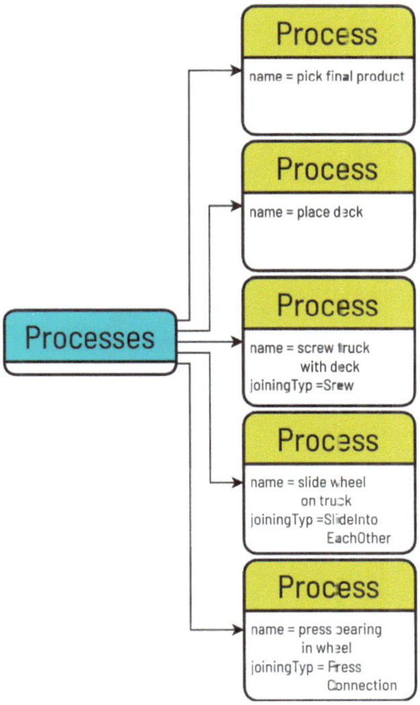

Figure 10. Process planning data model as a knowledge graph.

The main process is a set of sub-processes required to complete an assembly to the point where another assembly sequence can be started. Depending on the components used and the joining technology employed, the main processes are supplemented by sub-processes and handling steps. An example is shown in Figure 11.

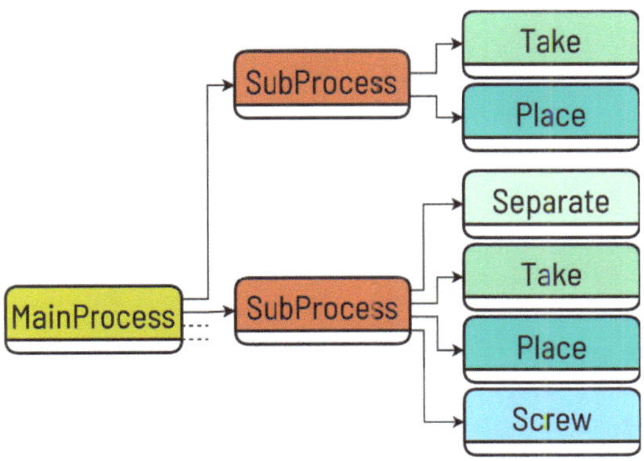

Figure 11. Partial section of the process planning for mounting the axle on the deck as knowledge graph.

The initial assembly sequence is derived from the structure of the data model, as shown in Figure 4, and the resulting assembly sets. These comprise a level that reflects the levels of the model. The assembly sequence begins with the lowest level, which, in the case of the skateboard, represents the pressing in of the layer. The correctness of the assembly sequence is validated by collision detection.

The formal description of the main and sub-processes delineates the full range of handling and joining techniques. These describe the minimum steps that must be carried out in order for the assembly step to be completed successfully. Based on the aforementioned information, an initial estimation of the cycle time can be made. Subsequently the cycle time is estimated for each individual handling step and for each joining technique. In this context, two distinct approaches are employed. In this approach, tabular values derived from the "Methods-Time-Measurement" [56] method are employed. The present analysis focuses on the temporal recording of handling steps. The analysis focuses on handling steps employed in manual assembly, with some values applicable to automated assembly. The determined value is multiplied by a correction factor, which serves as an adjustment variable. This factor is adjusted iteratively by comparing the tabular value with the subsequent actual value from the virtual commissioning and adjusting it accordingly. For all other handling steps not included in the table, a simplified approach is used in the calculation.

The calculation of t for the handling steps *take* and *place* is described in Equation (1). The product of the two factors $c_l(l_{max})$ and $c_m(m)$, the longest edge of the bounding box, l_{max}, and the average speed of the robot, v_{robot}, give the value of t:

$$t = \underbrace{c_l(l_{max})}_{\text{length heuristic}} \cdot \underbrace{c_m(m)}_{\text{mass heuristic}} \cdot l_{max} \cdot v_{robot} \tag{1}$$

It is assumed that the distance that the component has to cover can be described as a function of its dimensions (see *take* and *place*). For this purpose, the factor $c_l(l_{max})$ was defined, the value of which is calculated according to the definition given in Equation (2):

$$c_l(l_{max}) = \begin{cases} c_{l0} & \text{if } 0 \text{ mm} < l_{max} <= 50 \text{ mm}, \\ \vdots & \vdots \\ c_{ln} & \text{if } \ldots \end{cases} \qquad \text{where } n \in \mathbb{N}_0, \quad c_{ln} \in \mathbb{R} > 0 \tag{2}$$

The factor $c_m(m)$ defined in Equation (3) represents a correction factor that takes the weight of the component into account:

$$c_m(m) = \begin{cases} c_{m0} & \text{if } 0 \text{ kg} < m <= 0.1 \text{ kg}, \\ \vdots & \vdots \\ c_{mn} & \text{if } \ldots \end{cases} \qquad \text{where } n \in \mathbb{N}_0, \quad c_{mn} \in \mathbb{R} > 0 \tag{3}$$

The parameters $c_{l1}, \ldots, c_{ln}$ and $c_{m1}, \ldots, c_{mn}$ were freely selected at the start of the process. Following the method described above, a virtual commissioning is carried out. Based on the results obtained, the factors $c_l(l_{max})$ and $c_m(m)$ are adjusted iteratively. The aim of this procedure is to ensure that the result of the early estimation continuously adapts to the actual cycle time. However, verification using sufficient data is still required. If the application of the present approach does not lead to the desired results, it is possible to train a machine learning model with a sufficient database, which can then be used to predict the cycle time.

The calculation of the total t_{cycle} is described in Equation (4):

$$t_{\text{cycle}} = \sum_{i=1}^{h} t_i \quad \text{(with } h \text{ number of handle and joining steps)} \quad \textit{where } h \in \mathbb{N}_+ \tag{4}$$

Based on this modelling, an initial estimate of the cycle time is possible, as shown in Figure 12.

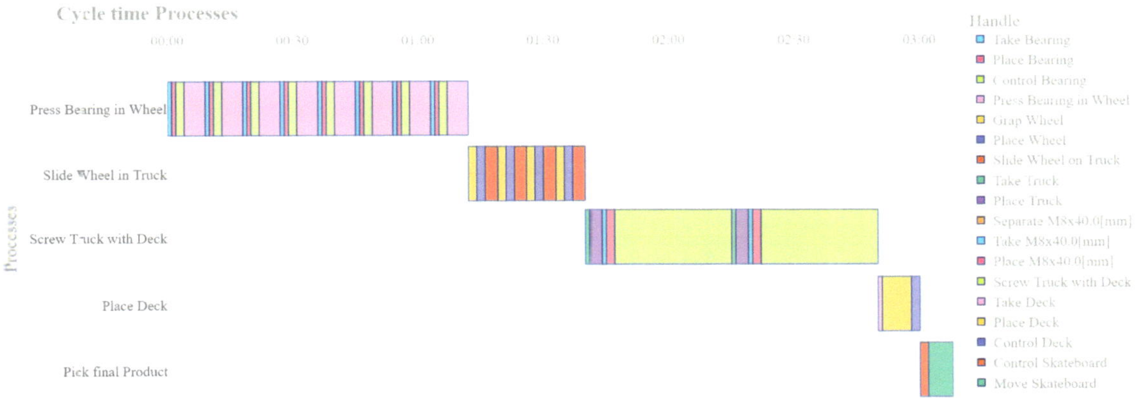

Figure 12. Cycle time diagram of the main processes.

As part of the process optimisation, the cycle time is compared with the specified requirements. If the determined cycle time fulfils the specified requirements, production planning can continue. If the specified cycle time is exceeded, the sequence is optimised. For this purpose, the main processes are split up, for example, to parallelise processes. As part of each optimisation, a check is carried out to determine whether the resulting outcome meets the specified requirements. This applies in particular to the cycle time, the available number of multi-axis robots, and the available space. The optimisation is terminated as soon as all requirements or the termination criteria are met.

Based on the assembly sets, a sensible assembly sequence could be generated via M2M transformation, which can serve as a basis for the further steps of the production planning.

4. Production Planning

In this paper, the term "production planning" is used to describe the steps following the assembly planning, i.e., planning the concrete transport of parts in the assembly system, generating the 2D assembly system layout, generating the geometry of the assembly system with all non-human resources, and generating the movement path of resources (e.g., robot code). In the given project, the descriptions of the assembly system could be generated by taking into account the cycle time and the requirements. These describe the processes that are carried out in a cell, as well as the resources required for this, such as robots, holding devices, and conveyor belts. The above descriptions enable the derivation of a 2D layout, which is shown in Figure 13.

In a final step, the 3D assembly system model is derived from the 2D layout (Figure 14). A minimum configuration can be derived by considering the main and sub-processes. The configuration encompasses all the resources indispensable for the comprehensive execution of the principal process. Such equipment may include, for instance, multi-axis robots, tables, and front conveyors. Subsequently, the principal processes are conveyed to assembly systems, wherein the cycle time has been optimised and considered in advance. In the subsequent phase, the previously identified systems are integrated with the resources of the minimum configuration. The working area of the assembly system is defined on the basis of the robot employed. It is imperative that, within the context of the current configuration, all sub-processes are conducted within the specified workspace. This entails the utilisation of the assigned resources. Moreover, a transport route must be established between the workspaces to ensure the secure and efficient transportation of parts.

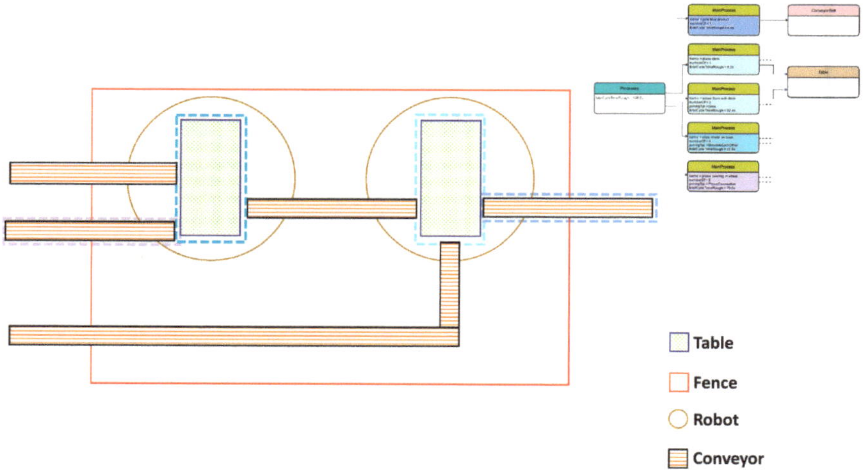

Figure 13. Two-dimensional layout of the system.

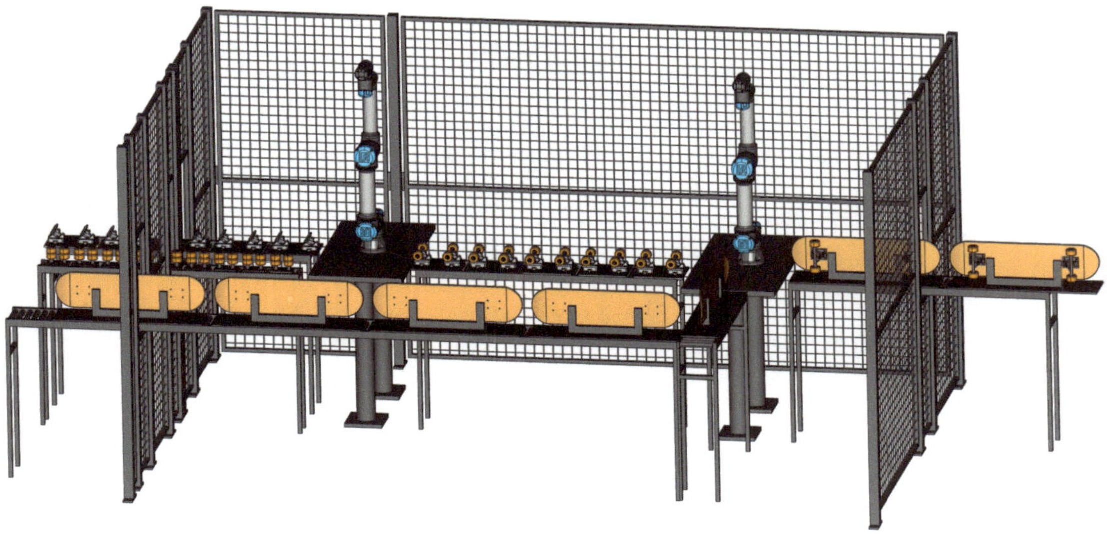

Figure 14. Three-dimensional assembly system model.

In the initial stage of the process, all resources are positioned within the designated workspace. Subsequently, an optimisation process utilising a packing algorithm is employed. Once the individual assembly systems have been generated, they are positioned globally and connected to each other using transport systems. Subsequently, a delineating structure, such as a fence, is erected around the assembly systems. The final result is a two-dimensional representation of the assembly cell, as illustrated in Figure 13. The requisite resources and layout dimensions are associated with each principal process.

In the final stage of the process, the 3D assembly system model is derived from the 2D layout. In order to achieve this, the enriched line models depicted in Figure 13 are employed, with due consideration given to the pertinent requirements. Such information was provided during the automated generation process. For example, data regarding the height of the fences were incorporated into the line model. The height of the fences is determined either by the requirements or, in the absence of a defined requirement, by the maximum height of the workspace. Subsequently, the 3D geometry is generated automatically, taking into account the aforementioned information and the 2D layout. The individual handling steps and the respective positioning are employed to derive a robot code, which is then utilised to assess the configuration. In the event that the assembly is not feasible, a return is made to the 2D layout, and the configuration is re-optimised. Conversely, if the assembly can be carried out as required, the automated design process is completed and the system can be exported to the standardised AutomationML data format.

The proposed design architecture enables the generation of simulations (see [57]) and of a digital twin of the assembly system.

5. Digital Twin

The culmination of the models created in an integrated product and assembly system development process can be employed as a dynamic simulation of the assembly system in the form of a digital twin. In general, a digital twin is understood as a virtual dynamic representation of a physical system with automatic and bidirectional exchanged data [58]. It is important to note that a full-scale digital twin (in contrast to a digital shadow) contains data of the complete product life-cycle [18]. The continuous flow of M2M transformations described in this paper leads to a unified digital twin (see Figure 2). In Figure 15, the architecture of employing this digital twin is shown.

For the given field of application, the Design Cockpit 43 [40] serves as a specialised editor to design a semantic model of a product and its assembly system. This system generates outputs in the form of ontologies and product and assembly system geometry, structure, and behaviour. The coded information is transformed into a knowledge graph. This knowledge graph can be enriched with performance data from an operating assembly system (execution—past assembly runs) and can interact with artificial intelligence (AI) and explainable artificial intelligence (XAI) [59] components in the digital twin framework [60]. The knowledge graph also allows for the execution of detailed simulations.

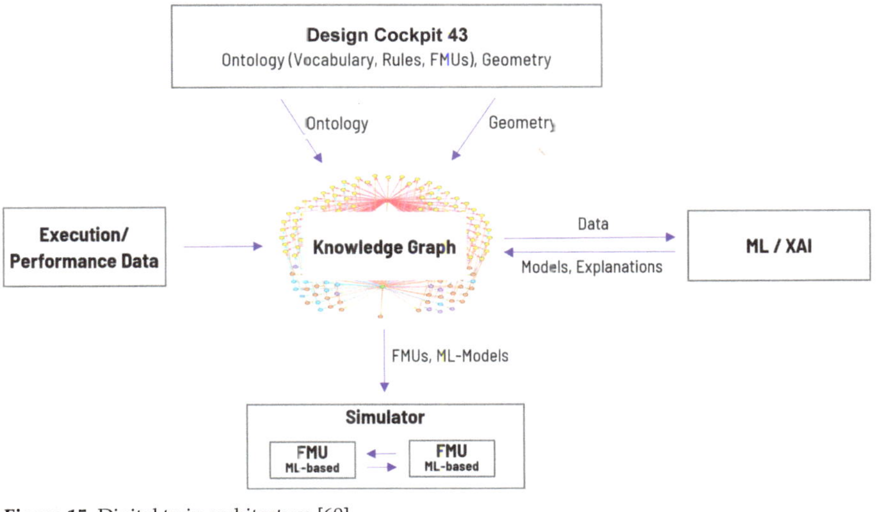

Figure 15. Digital twin architecture [60].

In order to allow for flexible simulations of the system behaviour, the behaviour is divided into so-called functional mock-up units (FMUs) [61]. In this kind of co-simulation, each FMU allows us to simulate an independent dynamic behaviour in the form of an input–output relation [62]. In order to model this relation, systems of ordinary differential equations (ODEs) are usually employed [63]. However, in the given framework, machine learning (ML)-based FMUs are also possible (see Figure 15 and also [64,65]). In this case, the input–output relation (simulating the dynamic behaviour) of the FMUs can be deduced from assembly system performance data.

As pointed out in Section 1, the step-wise enriched product and assembly model together with the geometric and kinematic information can be the basis for a digital twin; consequently, the different M2M transformations are essential steps in the digital twin creation (see Figure 2).

6. Conclusions and Outlook

The scientific core of this paper is the proposal of a consistent digital framework for the integration of product structure and geometry, assembly sequence, and assembly system structure, geometry, and behaviour. This framework is based on the application of graph-based design languages and allows for an automated design of a complete assembly system for a given product. The application of this framework leads to two machine-executable V-models for product development and assembly system development in the scope of MBSE. Unlike traditional MBSE approaches that rely on document-based process management, this framework uses abstract models, which can be automatically compiled into detailed product models and assembly system models.

The advantage of this approach is the realisation of a central, multi-domain model in the form of a knowledge graph, which can also be the basis for co-simulation based on functional mock-up units (FMUs) and for certain forms of machine learning or explainable artificial intelligence. A central element was the generation of a sensible assembly sequence; several possibilities for the challenging endeavour were explored. The resulting sequence is based on predefined assembly sets and takes into account the relevant necessities and possibilities of the assembly process. The approach is universal and can be used in the production of components for the automotive industry, their assemblies, and entire vehicles. Future research work will be aimed at even more complex products and more diverse assembly systems with human beings for certain assembly steps. Integration with AI methods is a further promising field for further research.

Author Contributions: Conceptualisation, T.S., M.T., R.S. and S.R.; methodology, T.S., M.T., R.S. and S.R.; validation, T.S.; investigation, T.S.; writing—original draft preparation, T.S. and R.S.; writing—review and editing, M.T. and S.R. All authors have read and agreed to the published version of the manuscript.

Funding: Parts of the described research were funded by the Carl Zeiss Foundation in the scope of the project AI-based digital twin (KI-basierter digitaler Zwilling, KIDZ).

Institutional Review Board Statement: Not applicable.

Data Availability Statement: The datasets presented in this article are not readily available because the data are part of an ongoing study. Requests to access the datasets should be directed to markus.till@rwu.de.

Conflicts of Interest: The authors declare no conflicts of interest.

Abbreviations

The following abbreviations are used in this manuscript:

2D	two-dimensional
3D	three-dimensional
AI	artificial intelligence
CAD	computer-aided design

DFMA	design for manufacturing and assembly
DT	digital twin
EBOMs	engineering bill of materials
FMU	functional mock-up unit
GBDL	graph-based design language
HiL	hardware in the loop
M2M	model-to-model
M2T	model-to-text
MBOMs	manufacturing bill of materials
MBSE	model-based systems engineering
ML	machine learning
ODE	ordinary differential equation
PDM	product data management
PLM	product life-cycle management
XAI	explainable artificial intelligence

References

1. He, L.; Ni, Y.; Ming, X.; Li, M.; Li, X. Integration of bill of materials with unified bill of materials model for commercial aircraft design to manufacturing. *Concurr. Eng.* **2014**, *22*, 206–217. [CrossRef]
2. Demoly, F.; Dutartre, O.; Yan, X.T.; Eynard, B.; Kiritsis, D.; Gomes, S. Product relationships management enabler for concurrent engineering and product lifecycle management. *Comput. Ind.* **2013**, *64*, 833–848. [CrossRef]
3. Layer, M.; Neubert, S.; Stelzer, R. Introducing a multipliable BOM-based automatic definition of information retrieval in plant engineering. *Proc. Des. Soc.* **2024**, *4*, 413–422. [CrossRef]
4. Cinelli, M.; Ferraro, G.; Iovanella, A.; Lucci, G.; Schiraldi, M.M. A network perspective on the visualization and analysis of bill of materials. *Int. J. Eng. Bus. Manag.* **2017**, *9*, 1847979017732638. [CrossRef]
5. Enomoto, A.; Yamamoto, N.; Yamamura, Y.; Sugawara, Y. Process knowledge integrated assembly sequence planning for control panel. *Int. J. Autom. Technol.* **2020**, *14*, 6–17. [CrossRef]
6. Gräßler, I.; Wiechel, D.; Pottebaum, J. Role model of model-based systems engineering application. In *IOP Conference Series: Materials Science and Engineering, Proceedings of the 19th Drive Train Technology Conference (ATK 2021), Aachen, Germany, 9–11 March 2021*; IOP Publishing: Bristol, UK, 2021; Volume 1097, p. 012003.
7. Pulm, U.; Stetter, R. Systemic mechatronic function development. *Proc. Des. Soc.* **2021**, *1*, 2931–2940. [CrossRef]
8. Stetter, R.; Pulm, U. Strategies and Methods for the Fault-Tolerant Function Development of Multi-Domain Systems. *Appl. Sci.* **2024**, *14*, 11646. [CrossRef]
9. VDI/VDE. *VDI/VDE 2006: Development of Cyber-Physical Mechatronic Systems (CPMS)*; Beuth: Berlin, Germany, 2020.
10. Graessler, I.; Hentze, J. The new V-Model of VDI 2206 and its validation. *At-Autom.* **2020**, *68*, 312–324. [CrossRef]
11. Eisenbart, B.; Gericke, K.; Blessing, L.T.; McAloone, T.C. A DSM-based framework for integrated function modelling: Concept, application and evaluation. *Res. Eng. Des.* **2017**, *28*, 25–51. [CrossRef]
12. Prager, F.; Tröster, P.M.; Albers, A. The C&C^2 Approach as a Thinking Tool in Mechatronic Systems. *Procedia CIRP* **2024**, *128*, 704–709.
13. Striffler, N.; Voigt, T. Concepts and trends of virtual commissioning–A comprehensive review. *J. Manuf. Syst.* **2023**, *71*, 664–680. [CrossRef]
14. Rueckert, P.; Goetsche, A.K.; Tracht, K. Assembly workshops in virtual reality as an integral part of the assembly planning process. *Procedia CIRP* **2024**, *126*, 793–798. [CrossRef]
15. Unreal Engine. Epic Games. Cary, North Carolina, USA. 2025. Available online: https://www.unrealengine.com/ (accessed on 12 February 2025).
16. Unity. Unity Technologies. San Francisco, California, USA. 2025. Available online: https://unity.com/ (accessed on 12 February 2025).
17. Fertsch, M.; Stachowiak, A.; Oleśków-Szłapka, J. Innovations-Changes in the Environment of the Production Planning Process in Enterprises. In Proceedings of the International Scientific-Technical Conference MANUFACTURING, Poznan, Poland, 14–16 May 2024; Springer: Cham, Switzerland, 2024; pp. 201–212.
18. Trauer, J.; Schweigert-Recksiek, S.; Engel, C.; Spreitzer, K.; Zimmermann, M. What is a digital twin?–definitions and insights from an industrial case study in technical product development. In Proceedings of the Design Society: DESIGN Conference, Online, 25 October 2020; Cambridge University Press: Cambridge, UK, 2020; Volume 1, pp. 757–766.
19. Javaid, M.; Haleem, A.; Suman, R. Digital twin applications toward industry 4.0: A review. *Cogn. Robot.* **2023**, *3*, 71–92. [CrossRef]

20. Pronost, G.; Mayer, F.; Camargo, M.; Dupont, L. Digital Twins along the product lifecycle: A systematic literature review of applications in manufacturing. *Digit. Twin* **2024**, *3*, 3. [CrossRef]
21. Liang, J.; Lu, Y.; Yin, G.; Fang, Z.; Zhuang, W.; Ren, Y.; Xu, L.; Li, Y. A distributed integrated control architecture of AFS and DYC based on MAS for distributed drive electric vehicles. *IEEE Trans. Veh. Technol.* **2021**, *70*, 5565–5577. [CrossRef]
22. Kargar, M.; Zhang, C.; Song, X. Integrated optimization of power management and vehicle motion control for autonomous hybrid electric vehicles. *IEEE Trans. Veh. Technol.* **2023**, *72*, 11147–11155. [CrossRef]
23. Liang, J.; Lu, Y.; Wang, F.; Yin, G.; Zhu, X.; Li, Y. A robust dynamic game-based control framework for integrated torque vectoring and active front-wheel steering system. *IEEE Trans. Intell. Transp. Syst.* **2023**, *24*, 7328–7341. [CrossRef]
24. Zhang, H.; Wang, S.; Zhao, X.; Zheng, Z. Path tracking and Handling Stability Coordinated Control of 4WS and DYC for Distributed in-Wheel Motor Drive Electric Vehicle Under Extreme Conditions. *IEEE Trans. Veh. Technol.* **2024**, *73*, 18402–18417. [CrossRef]
25. Jin, L.; Zhai, X.; Wang, K.; Zhang, K.; Wu, D.; Nazir, A.; Jiang, J.; Liao, W.H. Big data, machine learning, and digital twin assisted additive manufacturing: A review. *Mater. Des.* **2024**, *244*, 113086. [CrossRef]
26. Formentini, G.; Boix Rodríguez, N.; Favi, C. Design for manufacturing and assembly methods in the product development process of mechanical products: A systematic literature review. *Int. J. Adv. Manuf. Technol.* **2022**, *120*, 4307–4334. [CrossRef]
27. Bouissiere, F.; Cuiller, C.; Dereux, P.E.; Malchair, C.; Favi, C.; Formentini, G. Conceptual design for assembly in aerospace industry: A method to assess manufacturing and assembly aspects of product architectures. In Proceedings of the Design Society: International Conference on Engineering Design, Delft, The Netherlands, 5–8 August 2019; Cambridge University Press: Cambridge, UK, 2019; Volume 1, pp. 2961–2970.
28. Formentini, G.; Favi, C.; Cuiller, C.; Dereux, P.E.; Bouissiere, F.; Jurbert, C. Conceptual design for assembly in aerospace industry: Sensitivity analysis of mathematical framework and design parameters. *Proc. Des. Soc.* **2021**, *1*, 731–740. [CrossRef]
29. Tang, H. An integrated product-process hierarchical modeling method for development of complex assembly manufacturing systems. *Procedia CIRP* **2018**, *76*, 2–6. [CrossRef]
30. Manimuthu, A.; Venkatesh, V.; Shi, Y.; Sreedharan, V.R.; Koh, S.L. Design and development of automobile assembly model using federated artificial intelligence with smart contract. *Int. J. Prod. Res.* **2022**, *60*, 111–135. [CrossRef]
31. Kshatra, D.P.; Akhil, S.; Kiran, U.; Vineeth, Y. Process Design and system Layout for an automobile Manufacturing and assembly plant. *Int. J. Innov. Technol. Explor. Eng.* **2019**, *9*, 440–446. [CrossRef]
32. Arifin, Z.; Prasetyo, S.D.; Prabowo, A.R.; Cho, J.H. Preliminary design for assembling and manufacturing sports equipment: A study case on Aerobic Walker. *Int. J. Mech. Eng. Robot. Res.* **2021**, *10*, 107–115. [CrossRef]
33. Sali, M. Designing an Innovative Modular Platform for Sports Car Using the Generative Design Method. Doctoral Dissertation, University of Bologna, Bologna, Italy, 2024.
34. Demoly, F.; Yan, X.T.; Eynard, B.; Gomes, S.; Kiritsis, D. Integrated product relationships management: A model to enable concurrent product design and assembly sequence planning. *J. Eng. Des.* **2012**, *23*, 544–561. [CrossRef]
35. Enomoto, A.; Hayashi, N.; Inoue, R.; Tsutsumi, D.; Kajita, D.; Nakasu, N. Near-Optimal Assembly Task Sequencing and Allocation Method for Multi-Arm Robot System. In Proceedings of the 2023 IEEE 19th International Conference on Automation Science and Engineering (CASE), Auckland, New Zealand, 26–30 August 2023; pp. 1–7. [CrossRef]
36. Tram, A.V.N.; Raweewan, M. Optimal task allocation in human-robotic assembly processes. In Proceedings of the 2020 5th International Conference on Robotics and Automation Engineering (ICRAE), Singapore, 20–22 November 2020; IEEE: Piscataway, NJ, USA, 2020; pp. 158–162.
37. Culbertson, P.; Bandyopadhyay, S.; Schwager, M. Multi-Robot Assembly Sequencing via Discrete Optimization. In Proceedings of the 2019 IEEE/RSJ International Conference on Intelligent Robots and Systems (IROS), Macau, China, 3–8 November 2019; pp. 6502–6509. [CrossRef]
38. Liu, R.; Chen, A.; Zhao, W.; Liu, C. Physics-Aware Combinatorial Assembly Planning using Deep Reinforcement Learning. *arXiv* **2024**, arXiv:2408.10162.
39. Paehler, L.; Matthiesen, S. Mapping the landscape of product models in embodiment design. *Res. Eng. Des.* **2024**, *35*, 289–310. [CrossRef]
40. Design Cockpit 43. IILS Ingenieurgesellschaft für Intelligente Lösungen und Systeme mbH, Trochtelfingen, Germany. Available online: https://www.iils.de/ (accessed on 18 February 2025).
41. Rudolph, S. Übertragung von Ähnlichkeitsbegriffen. Habilitation Thesis, Fakultät Luft- und Raumfahrttechnik und Geodäsie, Universität Stuttgart, Stuttgart, Germany, 2002.
42. Neumaier, M.; Kranemann, S.; Kazmeier, B.; Rudolph, S. Automated Piping in an Airbus A320 Landing Gear Bay Using Graph-Based Design Languages. *Aerospace* **2022**, *9*, 140. [CrossRef]
43. Zech, A.; Stetter, R.; Rudolph, S.; Till, M. Capturing the Design Rationale in Model-Based Systems Engineering of Geo-Stations. *Proc. Des. Soc.* **2022**, *2*, 2015–2024. [CrossRef]

44. Elwert, M.; Ramsaier, M.; Eisenbart, B.; Stetter, R.; Till, M.; Rudolph, S. Digital Function Modeling in Graph-Based Design Languages. *Appl. Sci.* **2022**, *12*, 5301. [CrossRef]
45. Open Cascade, Open Cascade SAS, Guyancourt, France. 2025. Available online: https://www.opencascade.com (accessed on 12 February 2025).
46. CATIA, Dassault Systèmes, Vélizy-Villacoublay, France. 2025. Available online: https://www.3ds.com/products-services/catia/ (accessed on 12 February 2025).
47. Ansys, Ansys, Inc., Canonsburg, Pennsylvania, USA. 2025. Available online: https://www.ansys.com (accessed on 12 February 2025).
48. Abaqus FEA, Dassault Systèmes, Vélizy-Villacoublay, France. 2025. Available online: https://www.simulia.com (accessed on 12 February 2025).
49. Microsoft Excel, Microsoft Corporation, Redmond, Washington, USA. 2025. Available online: https://microsoft.com/en-us/microsoft-365/excel (accessed on 12 February 2025).
50. PTC Creo, PTC Inc., Boston, Massachusetts, USA. 2025. Available online: https://ptc.com/en/products/cad/creo/ (accessed on 12 February 2025).
51. Holder, K.; Zech, A.; Ramsaier, M.; Stetter, R.; Niedermeier, H.P.; Rudolph, S.; Till, M. Model-Based Requirements Management in Gear Systems Design Based On Graph-Based Design Languages. *Appl. Sci.* **2017**, *7*, 1112. [CrossRef]
52. Tian, Y.; Willis, K.D.; Al Omari, B.; Luo, J.; Ma, P.; Li, Y.; Javid, F.; Gu, E.; Jacob, J.; Sueda, S.; et al. ASAP: Automated sequence planning for complex robotic assembly with physical feasibility. In Proceedings of the 2024 IEEE International Conference on Robotics and Automation (ICRA), Yokohama, Japan, 13–17 May 2024; IEEE: Piscataway, NJ, USA, 2024; pp. 4380–4386.
53. Abdullah, M.A.; Ab Rashid, M.F.F.; Ghazalli, Z. Optimization of Assembly Sequence Planning Using Soft Computing Approaches: A Review. *Arch. Comput. Methods Eng.* **2019**, *26*, 461–474. [CrossRef]
54. Yu, J.; Yu, J.; Wang, C.; Wang, C. Method for discriminating geometric feasibility in assembly planning based on extended and turning interference matrix. *Int. J. Adv. Manuf. Technol.* **2013**, *67*, 1867–1882. [CrossRef]
55. Sınanoğlu, C.; Börklü, H.R. An assembly sequence-planning system for mechanical parts using neural network. *Assem. Autom.* **2005**, *25*, 38–52. [CrossRef]
56. Syska, A. *Produktionsmanagement: Das A–Z Wichtiger Methoden und Konzepte für die Produktion von Heute*; Gabler Verlag Wiesbaden: Wiesbaden, Germany, 2006; p. 188. [CrossRef]
57. Bendowska, K.; Zawadzki, P. Development and verification of a simulation model of an automated assembly line. *Appl. Sci.* **2023**, *13*, 10142. [CrossRef]
58. Trauer, J.; Pfingstl, S.; Finsterer, M.; Zimmermann, M. Improving Production Efficiency with a Digital Twin Based on Anomaly Detection. *Sustainability* **2021**, *13*, 10155. [CrossRef]
59. Došilović, F.K.; Brčić, M.; Hlupić, N. Explainable artificial intelligence: A survey. In Proceedings of the 2018 41st International Convention on Information and Communication Technology, Electronics and Microelectronics (MIPRO), Opatija, Croatia, 21–25 May 2018; pp. 210–215.
60. Arff, B.; Haasis, J.; Thomas, J.; Bonenberger, C.; Höpken, W.; Stetter, R. Analysis and visualization of production bottlenecks as part of a digital twin in industrial IoT. *Appl. Sci.* **2023**, *13*, 3525. [CrossRef]
61. Larsen, P.G.; Fitzgerald, J.; Woodcock, J.; Fritzson, P.; Brauer, J.; Kleijn, C.; Lecomte, T.; Pfeil, M.; Green, O.; Basagiannis, S.; et al. Integrated tool chain for model-based design of Cyber-Physical Systems: The INTO-CPS project. In Proceedings of the 2016 2nd International Workshop on Modelling, Analysis, and Control of Complex CPS (CPS Data), Vienna, Austria, 11 April 2016; IEEE: Piscataway, NJ, USA, 2016; pp. 1–6.
62. Saft, P.; Pfeil, M.; Stetter, R.; Till, M.; Rudolph, S. Integration of geometry modelling and behavior simulation based on graph-based design languages and functional mockup units. *Procedia CIRP* **2024**, *128*, 310–315. [CrossRef]
63. Elsheikh, A.; Awais, M.U.; Widl, E.; Palensky, P. Modelica-enabled rapid prototyping of cyber-physical energy systems via the functional mockup interface. In Proceedings of the 2013 Workshop on Modeling and Simulation of Cyber-Physical Energy Systems (MSCPES), Berkeley, CA, USA, 20 May 2013; IEEE: Piscataway, NJ, USA, 2013; pp. 1–6.
64. Rackauckas, C.; Gwozdz, M.; Jain, A.; Ma, Y.; Martinuzzi, F.; Rajput, U.; Saba, E.; Shah, V.B.; Anantharaman, R.; Edelman, A.; et al. Composing modeling and simulation with machine learning in Julia. In Proceedings of the 2022 Annual Modeling and Simulation Conference (ANNSIM), San Diego, CA, USA, 18–20 July 2022; IEEE: Piscataway, NJ, USA, 2022; pp. 1–17.
65. Falay, B.; Wilfling, S.; Alfalouji, Q.; Exenberger, J.; Schranz, T.; Legaard, C.M.; Leusbrock, I.; Schweiger, G. Coupling physical and machine learning models: Case study of a single-family house. In Proceedings of the Modelica Conferences, Linköping, Sweden, 20–24 September 2021; pp. 335–341.

vehicles MDPI

Article

Comparative Assessment for Holistic Evaluation of Drive Systems

Raphael Mieth [1,2] and Frank Gauterin [1,*]

[1] Institute of Vehicle System Technology, Karlsruhe Institute of Technology, 76131 Karlsruhe, Germany; raphael.mieth@mercedes-benz.com
[2] Mercedes-Benz AG, 70327 Stuttgart, Germany
* Correspondence: frank.gauterin@kit.edu

Abstract: The development of vehicle drive systems targets different goals, which are partly contradictory. While the focus is often on increasing efficiency and—depending on the type of drive system—performance, the aim is to simultaneously reduce costs, weight, and volume as much as possible. This goal generally presents a conflict of objectives; for example, a gain in efficiency usually correlates with higher costs, or an increase in performance reduces the maximum achievable efficiency. Therefore, each drive system represents a compromise among these goals, and depending on the main focus, the development can be influenced. The methods presented in this work serve as a methodological framework for the evaluation of vehicle drive systems. The procedure involves evaluating different drive concepts based on defined criteria and comparing these evaluations with one another. These criteria can be selected freely and weighted differently, depending on the individual focus. In the sense of a holistic assessment, a system evaluation factor ultimately serves as an indicator, which is composed of the rating values of the individual criteria, taking into account their specific weightings. With the help of the novel method presented in this paper, the complexity of comparing differently designed powertrains is reduced, and a holistic assessment covering relevant viewpoints is possible. Such an all-encompassing view is helpful in the early development phase and is required as an evaluation basis for further, groundbreaking decisions in concept development.

Keywords: holistic evaluation of drive systems; technology-independent assessment; cost–benefit analysis; pairwise comparison; weighted sum model

Citation: Mieth, R.; Gauterin, F. Comparative Assessment for Holistic Evaluation of Drive Systems. *Vehicles* **2024**, *6*, 403–414. https://doi.org/10.3390/vehicles6010017

Academic Editors: Ralf Stetter, Udo Pulm and Markus Till

Received: 3 January 2024
Revised: 7 February 2024
Accepted: 10 February 2024
Published: 13 February 2024

1. Introduction

The development of vehicle drive systems aims to achieve high customer benefits while maximizing profits [1]. Such high customer benefits can be represented, for example, by a high range, which depends on the size of the energy storage unit but is also influenced by the efficiency of the drive system [2,3]. Furthermore, a high usable driving performance also serves the customer benefit but is usually in conflict with high efficiency due to the characteristics of drive systems [4]. At the same time, the development process strives to minimize costs, both in terms of purchase prices from the customer and production costs for the manufacturer. In addition, a compact and lightweight design of the power unit is desirable.

To achieve these target criteria, a variety of different drive concepts can be considered. Thus, the development of vehicle drive systems constantly involves new concepts that differ from one another in terms of technology, design, execution, and more [5]. This results in an almost unlimited variety of concept variants that require a suitable approach for processing. While [6], for example, dealt with a simplified evaluation of costs and performances based on scaled and optimized drive components, this work serves the holistic evaluation with respect to initially defined criteria. The unique contributions of this study include the freely selectable evaluation criteria and independence from specific drive technology.

In order to evaluate these concepts, suitable methods are needed that reflect the individual strengths and weaknesses of the concepts. The derivation of key indicators for evaluation enables the selection of the best system based on defined evaluation criteria. The methodological approaches of this work represent a framework for the evaluation of vehicle drive systems, composed of key figures from relevant criteria. In doing so, the evaluation method is defined in a universally valid way so that every type of drive system can be compared with and evaluated holistically.

2. Methodological Approach

When considering the overall process of developing drive systems, the evaluation methods presented in this paper are placed at the end. Once the specifications and requirements are defined, the design process for the drive components begins. The drive system is then available in its entirety for a system evaluation.

As visualized in Figure 1, the methodological approach of this work begins with the definitions of specific evaluation criteria, such as efficiency or costs. These defined criteria are then weighted in relation to each other, and weighting factors are determined. With the support of experts from internal departments, these criteria can be evaluated for both a reference drive system and a sample drive system. For example, efficiency can be determined using simulated cycle efficiencies, costs are made up of material and manufacturing costs, and simulations of performance load spectra quantify the overall performance of the drives. These investigations produce concrete results, e.g., how efficient or expensive the drive systems are. In the fourth step, the calculated values, such as the cycle efficiency and determined costs, are compared with the respective values of the reference system, and rated. The final system evaluation is carried out together with the weighting factors from Step 2. Comparing it with a reference drive establishes a uniform basis for comparison.

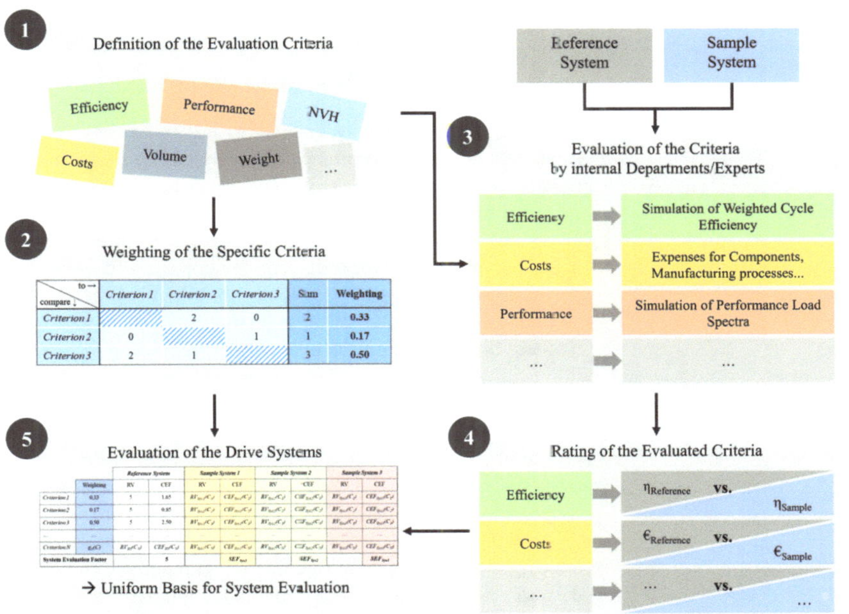

Figure 1. Scheme of the methodological approach.

3. Criteria for the Evaluation

In order to assess different drive systems comparably, concrete criteria for the evaluations must be defined. In this case, five criteria are chosen: efficiency, performance, cost, volume, and weight. However, these indicators can be freely chosen, and other technical aspects (such as thermal behavior and NVH) or even non-technical factors (such as environmental impacts or customer acceptance) could also be taken into account.

The initial aim is to deliver the greatest possible and tangible benefit for the customer in terms of efficiency and driving performance. High efficiency correlates with low consumption and, thus, a long range. The performance, on the other hand, can be measured by the maximum achievable power and torque of the powertrain.

At the same time, the weight and volume of the drive system need to be reduced as much as possible. Also, costs—including both purchase prices for the customer and production costs for the manufacturer—should be minimized. Since both high efficiency and high performance tend to correlate with increases in weight, volume, and cost, there is a conflict of objectives. This trade-off of opposing goals is visualized using a radar chart in Figure 2. An ideal drive system would combine high efficiency and great performance with the lowest possible costs, weight, and volume.

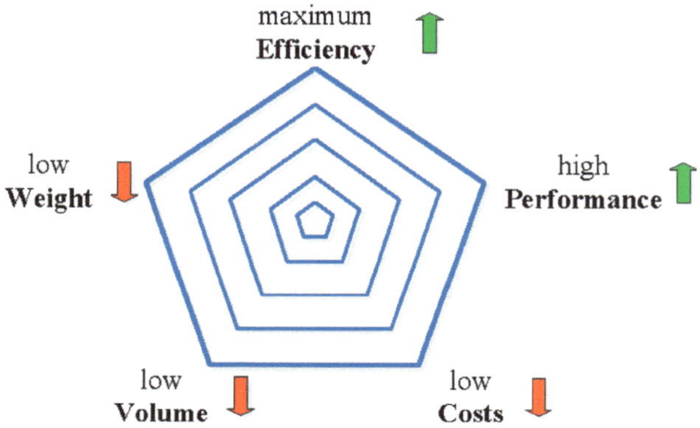

Figure 2. Evaluation criteria and the conflict of goals.

To evaluate drive systems with regard to these criteria, all the required parameters must be available in a comparable form. Comparing two drives of different maturity levels may erroneously lead to wrong conclusions. In terms of efficiency, this means that the same system boundaries and system losses must be considered. Performance, such as output power, can be measured and, therefore, evaluated relatively easily. A level playing field must also be created for costs, e.g., by ensuring that both raw material and manufacturing costs, as well as expenses for tools, are taken into account.

The determination of these evaluation indicators is not discussed further in this work; instead, the focus is on the evaluation method and its exemplary application.

4. Methods

The described conflict of opposing goals represents a multi-objective optimization problem, which is addressed by established methods in the literature, e.g., [7] and [8]. Apart from these or other well-known approaches mentioned in [9], this multi-criteria problem requires a suitable method of analysis that leads to rational decision-making.

While simultaneously optimizing multiple objects increases the complexity in decision-making, comparing only two criteria can reduce the complexity and, thus, facilitate the decision [10]. This approach of comparing only two objects at a time is the basis of pairwise comparisons. This process of mutual evaluation of the objects leads to prioritization and ultimately to a hierarchy among the criteria. This priority can be interpreted as a weighting of the specific criterion and used for a multi-criteria decision-making (MCDM) method. In this context, [11] provides a good overview of existing MCDM methods in the literature. Furthermore, [12] analyzes the characteristics and utilities of individual methodological approaches. These utility analyses, such as the weighted sum model, divide an overall problem into smaller subproblems. This type of fragmentation allows the overall problem to be captured holistically yet stated in a simplified manner, facilitating decision-making on complex problems [13].

In decision theory, pairwise comparison and utility analysis are common methods. Both are simple yet effective approaches, the combination of which is suitable for the given optimization problem, serving as a methodological basis. Furthermore, this work builds a framework for evaluating drive systems on the aforementioned methods.

4.1. Pairwise Comparison Method

A pairwise comparison (PWC) is a method of comparing entities or objects in order to judge which one has the greater amount of a quantitative property. These objects are contrasted and evaluated in pairs, with a rating score determining which of the two entities is preferred [14]. The above-mentioned evaluation criteria can serve as such objects. By juxtaposing the criteria, every criterion is assigned a priority. Based on this prioritization, a weighting factor is determined for all criteria, representing the importance of the individual criterion. Two tasks set up the approach: comparing all combinations with respect to each criterion and computing the weights of the attributes [15].

According to Formula (1), this weighting factor for each criterion C corresponds to the relative weighting $g_n(C)$, which is defined as the ratio of absolute weighting $G_n(C)$ to the sum of weighting factors for N criteria.

$$g_n(C) = \frac{G_n(C)}{\sum\limits_{n=1}^{N} G_n(C)} \quad (1)$$

Figure 3 shows an example of a tabular pairwise comparison. First, all criteria, $1 \ldots N$, are listed in both vertical and horizontal directions. This is followed by comparing vertical elements with horizontal objects and assigning a scalar between 0 and 2. For this value, 0 means less important, 2 means more important, and 1 represents equal importance.

to → / compare ↓	*Criterion 1*	*Criterion 2*	*Criterion 3*	...	*Criterion N*	Sum	**Weighting**
Criterion 1		2	0		C_1 *to* C_N	2	**0.33**
Criterion 2	0		1		C_2 *to* C_N	1	**0.17**
Criterion 3	2	1			C_3 *to* C_N	3	**0.50**
...						...	**...**
Criterion N	C_N *to* C_1	C_N *to* C_2	C_N *to* C_3			$\Sigma C_N = G_N(C)$	$g_N(C)$

Figure 3. Methodology of the pairwise comparison and weighting factor.

For example, *Criterion* 2 is less significant than *Criterion* 1, represented by the value of 2, but just as important as *Criterion* 3, which is expressed by the value of 1. The sum of these scalars gives the absolute weighting, $G_n(C)$, and the weighting factor, consequently, represents the relative weighting, $g_n(C)$. The functionality of PWC is further explained and visualized in [16].

This procedure enables the pairwise prioritization of criteria. By defining the importance of objects among themselves, this comparison method is also well suited for highlighting small differences. Subsequently, the weighting factor comes into play in the weighted sum model.

4.2. Weighted Sum Model

The weighted sum model (WSM) is a proven method for utility analyses. The goal is to compare multiple systems with respect to various criteria by evaluating individual performance in that discipline. This assessment is represented by the rating value (RV).

Moreover, each criterion can be weighted differently. Sticking with the weighting factors mentioned above, *Criterion* 3 would have the greatest impact on the evaluation among multiple systems.

Based on these weighting factors, a criterion evaluation factor (CEF) is calculated for each system as well as each criterion according to Formula (2).

$$CEF_{Sys}(C_n) = g_n(C) \cdot RV_{Sys}(C_n) \tag{2}$$

Formula (3) sums the CEF value per criterion for each system, resulting in the system evaluation factor (SEF).

$$SEF_{Sys} = \sum_{n=1}^{N} CEF_{Sys}(C_n) \tag{3}$$

This approach leads to an evaluation method that can be used, for example, to evaluate differently designed drive systems. The evaluation is represented by a single scalar. Thereby, the scale of rating values ranges from 0 (as the worst grade) to 10 (as the best grade). Figure 4 summarizes the methodical approach of the weighted sum model.

		Reference System		**Sample System 1**		**Sample System 2**		**Sample System 3**	
	Weighting	RV	CEF	RV	CEF	RV	CEF	RV	CEF
Criterion 1	0.33	5	1.65	$RV_{Sys1}(C_1)$	$CEF_{Sys1}(C_1)$	$RV_{Sys2}(C_1)$	$CEF_{Sys2}(C_1)$	$RV_{Sys3}(C_1)$	$CEF_{Sys3}(C_1)$
Criterion 2	0.17	5	0.85	$RV_{Sys1}(C_2)$	$CEF_{Sys1}(C_2)$	$RV_{Sys2}(C_2)$	$CEF_{Sys2}(C_2)$	$RV_{Sys3}(C_2)$	$CEF_{Sys3}(C_2)$
Criterion 3	0.50	5	2.50	$RV_{Sys1}(C_3)$	$CEF_{Sys1}(C_3)$	$RV_{Sys2}(C_3)$	$CEF_{Sys2}(C_3)$	$RV_{Sys3}(C_3)$	$CEF_{Sys3}(C_3)$
...	...	...	...	...	...	...	...	...	...
Criterion N	$g_N(C)$	$RV_{RS}(C_N)$	$CEF_{RS}(C_N)$	$RV_{Sys1}(C_N)$	$CEF_{Sys1}(C_N)$	$RV_{Sys2}(C_N)$	$CEF_{Sys2}(C_N)$	$RV_{Sys3}(C_N)$	$CEF_{Sys3}(C_N)$
System Evaluation Factor			5		SEF_{Sys1}		SEF_{Sys2}		SEF_{Sys3}

Figure 4. Methodology of the weighted sum model.

As shown, a reference drive system forms the basis for the evaluation of three sample systems. The rating values of this reference drive system are all equal to 5 and, thus, define the reference point. Consequently, the sum of all CEF values also results in a value of 5. Depending on the chosen rating values for the sample systems, all CEF values (as well as SEF values) can be calculated according to the above-mentioned formulas.

5. Results

In order to practically demonstrate the functionality of the presented methods of PWC and WSM, this section deals with the comparison between three drive systems and a reference drive system. First of all, all drive systems are introduced with regard to their technical specifications. In order to reveal the results, two exemplary case studies with different weightings are then considered.

5.1. Reference Drive and Sample Drive Systems

The reference drive system offers a good compromise among the criteria of efficiency, performance, cost, volume, and weight. As mentioned, these criteria influence each other. For example, efficiency and cost have an almost linear relationship, meaning that a gain in efficiency is generally accompanied by an increase in costs. Likewise, an increase in performance tends to lead to increases in costs, weight, and volume [17]. Amidst these dependencies, the reference drive system is very well balanced, which is reflected by a rating value of 5 for all criteria, according to Figure 5.

		Relative Delta				➡	Rating Values			
		Reference System	Low-Cost System	High-Efficiency System	High-Performance System	AF	Reference System	Low-Cost System	High-Efficiency System	High-Performance System
Efficiency	Weighted Cycle Efficiency	100%	−4%	+7%	−5%	10	5	3	9	3
	Mean Cycle Consumption	100%	+3%	−7%	+3%		5	4	9	4
	Mean System Losses	100%	+4%	−8%	+4%		5	3	9	3
	Mean Efficiency	100%	**−4%**	**+7%**	**−4%**		5	**3**	**9**	**3**
Performance	Performance Load Spectra	100%	−19%	−9%	+42%	2	5	3	4	9
	Drive Power	100%	−12%	−3%	+33%		5	4	5	8
	Drive Torque	100%	−16%	−6%	+34%		5	3	4	8
	Duration of Peak Performance	100%	−15%	−4%	+31%		5	4	5	8
	Mean Performance	100%	**−16%**	**−6%**	**+35%**		5	**3**	**4**	**9**
Costs	Material Costs	100%	−8%	+6%	+1%	10	5	9	2	5
	Production Costs	100%	−6%	+4%	+0%		5	8	3	5
	Mean Costs	100%	**−7%**	**+5%**	**+1%**		5	**9**	**3**	**5**
Volume	Active Volume of Drive Unit	100%	−1%	+9%	−6%	5	5	5	3	7
	Installation Space with Periphery	100%	−1%	+12%	−6%		5	5	2	7
	Mean Volume	100%	**−1%**	**+11%**	**−6%**		5	**5**	**2**	**7**
Weight	Weight of Drive Unit	100%	−6%	+6%	−3%	5	5	7	4	6
	Total Weight with Periphery	100%	−6%	+9%	−3%		5	7	3	6
	Mean Weight	100%	**−6%**	**+8%**	**−3%**		5	**7**	**3**	**6**

Figure 5. Derivation of rating values from relative deltas.

This table shows the evaluation criteria and their indicators, e.g., the expenses for material and production as cost criteria. The chart also shows all relative deltas to the reference drive system on the left-hand side, from which the rating values on the right-hand side were derived.

For this derivation, empirically determined amplification factors (AFs) are used to adjust the relative deltas to the rating value scale (0 to 10). These multipliers are necessary because relative deltas of different rating criteria are not necessarily comparable. For example, while an efficiency delta of 2% is relatively high, a performance delta of only 2% is rather negligible. For this reason, the AF for efficiency (10) is five times greater than that for performance (2). Ultimately, these amplification factors enable a comparative assessment of the criteria.

It should also be noted that the sign of the relative deltas could have an opposite effect on the rating value. For example, while the weighted cycle efficiency has a positive delta, the mean cycle consumption is higher and is, therefore, represented by a negative delta.

The resulting rating values are visualized in Figure 6, where all drive systems are compared with each other based on their already-known rating values.

This radar chart shows the comparison between all four drive systems and, thus, highlights the different emphases of the three sample drive systems. Each one has a specific purpose, which was consistently pursued during development. As the name suggests, the low-cost system is primarily aimed at low system costs, with both efficiency and performance dropping significantly compared to the reference drive system. This is represented by an $RV(Cst)$ value of 9, while both $RV(Eff)$ and $RV(Per)$ are equal to 3. On the other hand, the active volumes of both drive systems are comparable, so $RV(Vol)$ is 5 as well, and, notably, the weight is even slightly better than the reference drive system, with an $RV(Wgt)$ of 7, making the low-cost system only slightly inferior overall.

In contrast, the high-efficiency system achieves significantly higher overall efficiency with an $RV(Eff)$ of 9, but falls short of the reference drive system in all other disciplines. Since high efficiency tends to correlate with high volume, costs, and weight, these criteria are usually inversely related, which is why the ratings are so low with an $RV(Vol)$ at 2 and $RV(Cst)$ and $RV(Wgt)$ at 3 each. This is due to the strong anti-proportional dependence of the size and cost on efficiency [18]. Ultimately, the $RV(Per)$ is set to 4 but there is no clear correlation between performance and efficiency.

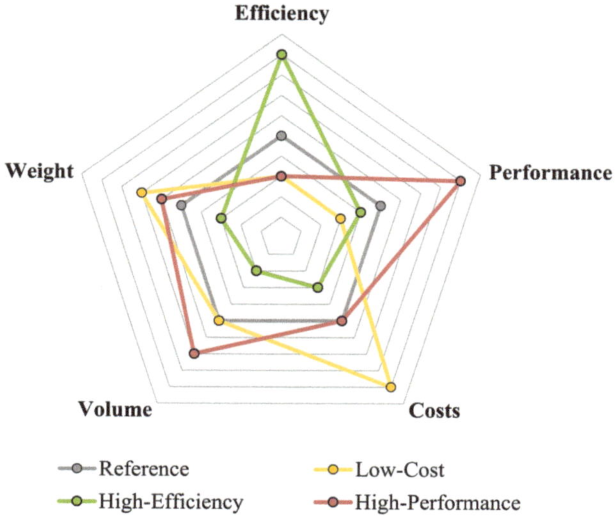

Figure 6. Comparison between the reference drive and sample drive systems in a radar chart.

Finally, the high-performance system is basically the opposite of the high-efficiency system. In particular, in the case of high performance with an $RV(Per)$ of 9, efficiency drops significantly to an $RV(Eff)$ of 3, which is why high efficiency and high performance together are two opposing targets [19]. Compared to the high-efficiency system, the focus here is on increasing performance at the expense of a reduction in efficiency. This shift in focus makes the costs with an $RV(Cst)$ of 5 comparable to those of the reference drive system. Due to the compact design of this high-performance module, gains in volume and weight can be achieved, resulting in an $RV(Vol)$ of 7 and $RV(Wgt)$ of 6.

As the method of this holistic assessment is intended to apply to conventional, electrical, and hybrid drive systems, these three drive systems serve as demonstration examples. Therefore, no further details (such as technical specifications) are provided.

5.2. Holistic Assessment

In the context of a holistic system evaluation, individual objectives play a decisive role. Therefore, the following assessment consists of two case studies (CSs) with different focal points. While the first exemplary case study focuses on efficiency as the main target, the second exemplary case study is aimed at costs.

5.2.1. Case Study 1—Focus on Efficiency

Based on the methodology of the PWC shown above, Figure 7 illustrates the weightings for CS1. As focused, efficiency is the main objective, which is represented by a weighting factor of 40%. Regardless of the targeted design of individual drives, efficiency is set as the main focus for all, followed by cost and performance. As [20] shows, efficiency is directly related to all evaluation criteria. As already mentioned, an increase in efficiency correlates with increasing costs and influences performance, which is why any gain in efficiency must always be evaluated, taking into account its cost and performance impact. For this reason, each of the two indicators is weighted relatively high at 20%. Both volume and weight are influenced by the efficiency of the drive, but with a lower priority, and, therefore, are weighted at 10%, respectively. These weighting factors ultimately express the clear gradation and prioritization of the criteria.

to → / compare ↓	*Efficiency*	*Performance*	*Costs*	*Volume*	*Weight*	Sum	**Weighting**
Efficiency		2	2	2	2	**8**	**40 %**
Performance	0		1	1	2	**4**	**20 %**
Costs	0	1		2	1	**4**	**20 %**
Volume	0	1	0		1	**2**	**10 %**
Weight	0	0	1	1		**2**	**10 %**

Figure 7. Pairwise comparison and derivation of the weighting factors according to CS1.

These derived weighting factors are now applied to the system evaluation in Figure 8 using the weighted sum model. While the weighting of the rating values in the case of the reference drive system leads to an overall *SEF* value of 5.0, the system evaluation of the sample drive system results in *SEF* values.

		Reference System		***Low-Cost System***		***High-Efficiency System***		***High-Performance System***	
	Weighting	RV	CEF	RV	CEF	RV	CEF	RV	CEF
Efficiency	0.4	5	2.0	3	1.2	9	3.6	3	1.2
Performance	0.2	5	1.0	3	0.6	4	0.8	9	1.8
Costs	0.2	5	1.0	9	1.8	3	0.6	5	1.0
Volume	0.1	5	0.5	5	0.5	2	0.2	7	0.7
Weight	0.1	5	0.5	7	0.7	3	0.3	6	0.6
System Evaluation Factor			**5.0**		**4.8**		**5.5**		**5.3**

Figure 8. Weighted sum model according to CS1.

As the table shows, the main focus on efficiency makes it clear that the high-efficiency system with a *SEF* value of 5.5 is to be preferred. At 5.3, the high-performance system also produces a higher *SEF* value than the reference drive system. Only the low-cost system is worse, with an *SEF* value of 4.8.

5.2.2. Case Study 2—Focus on Costs

The procedure for CS2 is basically analogous to the approach for CS1. In line with the focus on costs, the weighting has been selected accordingly. As costs are now weighted with a value of 40%, efficiency is weighted with 30% as the second most important criterion. The remaining criteria are determined according to the weighting factors shown in Figure 9.

to → / compare ↓	*Efficiency*	*Performance*	*Costs*	*Volume*	*Weight*	Sum	**Weighting**
Efficiency		2	0	2	2	**6**	**30 %**
Performance	0		0	1	2	**3**	**15 %**
Costs	2	2		2	2	**8**	**40 %**
Volume	0	1	0		1	**2**	**10 %**
Weight	0	0	0	1		**1**	**5 %**

Figure 9. Pairwise comparison and derivation of the weighting factors according to CS2.

In contrast to CS1, this second case study clearly demonstrates the difference caused by an alternative weighting of the criteria. While the SEF value of the reference drive system remains unchanged at 5.0, the SEF values of the sample drive systems differ from the previous ones, as shown in Figure 10.

		Reference System		***Low-Cost System***		***High-Efficiency System***		***High-Performance System***	
	Weighting	RV	CEF	RV	CEF	RV	CEF	RV	CEF
Efficiency	0.3	5	1.5	3	0.9	9	2.7	3	0.9
Performance	0.15	5	0.75	3	0.45	4	0.6	9	1.35
Costs	0.4	5	2.0	9	3.6	3	1.2	5	2.0
Volume	0.1	5	0.5	5	0.5	2	0.2	7	0.7
Weight	0.05	5	0.25	7	0.35	3	0.15	6	0.3
System Evaluation Factor			**5.0**		**5.8**		**4.85**		**5.25**

Figure 10. Weighted sum model according to CS2.

By prioritizing costs, the SEF value of the low-cost system is unsurprisingly the highest at 5.8. The SEF value of the high-efficiency system, on the other hand, drops significantly to just 4.85. Interestingly, the high-performance system remains more or less the same at 5.25.

6. Discussion

To determine whether a benefit, such as an increase in efficiency, outweighs its cost, a cost–benefit analysis (CBA) is a common method for making decisions. Additional benefits often correlate with an increase in costs, so the purpose of this evaluation method is to quantify and weigh the benefits against the costs using a cost–benefit ratio. The goal is to assess the economic efficiency and determine whether it is worth the investment [21].

Benefits and costs also correlate in the case of drive systems, as an increase in energy efficiency is accompanied by an increase in system costs. Using the example of CS1, according to which efficiency is weighted most heavily (see Figure 8), the sample systems are evaluated with a cost-efficiency analysis. The diagram in Figure 11 shows the rating scales of efficiency over cost and marks the drive systems mentioned.

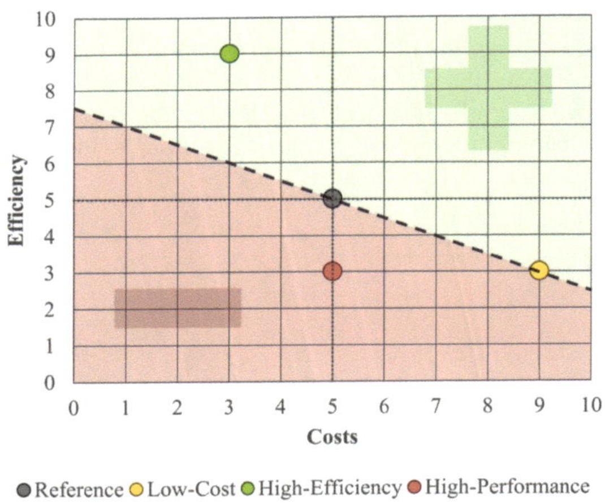

Figure 11. Cost-efficiency analysis and evaluation of the sample systems.

Since the efficiency weighting $g(Eff)$ of 40% is twice as high as the cost weighting $g(Cst)$ of 20%, the ratio between these weightings is 2:1. In terms of cost and efficiency, this ratio allows systems to be considered equivalent, where, for example, a cost advantage and an associated efficiency disadvantage balance each other out. Based on this ratio, the system cost rating value, $RV(Cst)$, must be two points better to compensate for one point lower efficiency. This dependency is expressed by the equation in Formula (4) and is represented by the dashed straight line in Figure 11.

$$f(RV(Cst)) = 7.5 - \frac{g(Cst)}{g(Eff)} \cdot RV(Cst) \quad (4)$$

As shown, the low-cost system lies on the dashed line, which means that the cost-efficiency ratio is equal to that of the reference drive system. The lower efficiency is compensated by lower system costs, so the ratio remains the same. The high-efficiency system, on the other hand, exceeds this ratio and performs better than the reference. This can be interpreted to mean that the higher system costs are justifiable relative to the efficiency gain. In addition, the increase in efficiency for this system may allow for a reduction in battery size or capacity, which in turn reduces the cost and weight. In contrast, the high-performance system has a worse ratio due to the lower efficiency at the same costs. The focus on performance brings unfavorable conditions with regard to the cost-efficiency ratio.

7. Conclusions and Extension

The methods presented in this work serve as a framework for the evaluation of drive systems regarding various criteria. Thereby, the selection and weighting of these criteria are adjustable and can be adapted depending on the priorities in the development. The functionality of the methods is explained using the example of the three sample systems presented. Despite completely different emphases, these systems are brought to a comparable basis and compared under the same aspects. The exemplarily chosen weightings remain constant. As a result, the respective focal points of the three drive systems are reflected in the system evaluation factors, demonstrating and validating the applicability of the methodology.

The rating values are currently determined by the user and are, therefore, liable to subjective assessment. Through enhancements to the method, the determination of rating values will be automated by using concrete examination results as a basis. For example, a simulation-based objectification of the system efficiency and performance is carried out, as well as a breakdown of the costs, weight, and volume of the drive components.

A system evaluation based on a tabular utility analysis requires a certain level of complexity and a good imagination. Alternatively, a system evaluation based on the surface area of the individual pentagonal mesh shapes could serve as a better illustration. The larger the surface area, the better the overall system. However, this would not take into account the different weighting factors. For example, the area of the high-performance system is greater than that of the high-efficiency system despite a poorer *SEF* value. This challenge could be countered by a three-dimensional (3D) radar chart, where the height of each corner corresponds to the weighting factor of the criterion. In this case, the mesh would be in 3D space, with the individual pentagons placed on this plane. The area as an indicator would then become the volume.

Author Contributions: Conceptualization, methodology, software, validation, formal analysis, investigation, resources, data curation, writing—original draft preparation, writing—review and editing, visualization, project administration, funding acquisition, R.M.; supervision, F.G. All authors have read and agreed to the published version of the manuscript.

Funding: This research is funded by the Publication Fund of the Karlsruhe Institute of Technology; funding number 02034100761.

Institutional Review Board Statement: Not applicable.

Informed Consent Statement: Not applicable.

Data Availability Statement: No publicly archived datasets were analyzed.

Conflicts of Interest: Author Raphael Mieth was employed by the company Mercedes-Benz AG. The remaining authors declare that the research was conducted in the absence of any commercial or financial relationships that could be construed as a potential conflict of interestThe authors declare no conflicts of interest.

Abbreviations

The following abbreviations are used in this manuscript:

AF	amplification factor
C	criterion
CS	case study
Cst	cost
CBA	cost–benefit analysis
CEF	criterion evaluation factor
Eff	efficiency
MCDM	multi-criteria decision-making
NVH	noise vibration harshness
Per	performance
PWC	pairwise comparison
RV	rating value
SEF	system evaluation factor
Sys	system
Vol	volume
Wgt	weight
WSM	weighted sum model

References

1. Buhmann, K.M.; Criado, J.R. Consumers' preferences for electric vehicles: The role of status and reputation. *Transp. Res. Part D Transp. Environ.* **2023**, *114*, 103530. [CrossRef]
2. Chakraborty, P.; Parker, R.; Hoque, T.; Cruz, J.; Du, L.; Wang, S.; Bhunia, S. Addressing the range anxiety of battery electric vehicles with charging en route. *Sci. Rep.* **2022**, *12*, 5588. [CrossRef] [PubMed]
3. Shrestha, S.; Shah, B.B.M.; Chitrakar, S.; Shrestha, B.P. Measures to resolve range anxiety in electric vehicle users. *Int. J. Low-Carbon Technol.* **2022**, *17*, 1186–1206. [CrossRef]
4. Grunditz, E.A.; Thiringer, T. Electric Vehicle Acceleration Performance and Motor Drive Cycle Energy Efficiency Trade-Off. In Proceedings of the 2018 XIII International Conference on Electrical Machines (ICEM), Alexandroupoli, Greece, 3–6 September 2018; pp. 717–723. [CrossRef]
5. Sanguesa, J.; Torres-Sanz, V.; Garrido, P.; Martinez, F.J.; Marquez-Barja, J.M. A Review on Electric Vehicles: Technologies and Challenges. *Smart Cities* **2021**, *4*, 372–404. [CrossRef]
6. Weiß, F. Methodik zur Optimalen Konzeptauslegung Elektrifizierter Fahrzeugantriebsstränge. Ph.D. Thesis, Technische Universität Chemnitz, Chemnitz, Germany, 2017. [CrossRef]
7. Li, C.; Wu, R.; Yang, W. Optimization and selection of the multi-objective conceptual design scheme for considering product assembly, manufacturing and cost. *SN Appl. Sci.* **2022**, *4*, 91. [CrossRef]
8. Yue, Z. An extended TOPSIS for determining weights of decision makers with interval numbers. *Knowl.-Based Syst.* **2011**, *24*, 146–153. [CrossRef]
9. Ma, J.; Kremer, G.E.O. A systematic literature review of modular product design (MPD) from the perspective of sustainability. *Int. J. Adv. Manuf. Technol.* **2016**, *86*, 1509–1539. [CrossRef]
10. Krejčí, J. *Pairwise Comparison Matrices and Their Fuzzy Extension—Multi-Criteria Decision Making with a New Fuzzy Approach*; Springer Nature: Berlin/Heidelberg, Germany, 2018; Volume 366. [CrossRef]
11. Baumann, M.; Weil, M.; Peters, J.F.; Chibeles-Martins, N.; Moniz, A.B. A review of multi-criteria decision making approaches for evaluating energy storage systems for grid applications. *Renew. Sustain. Energy Rev.* **2019**, *107*, 516–534. [CrossRef]
12. Taherdoost, H.; Madanchian, M. Multi-Criteria Decision Making (MCDM) Methods and Concepts. *Encyclopedia* **2023**, *3*, 77–87. [CrossRef]
13. Kühnapfel, J.B. *Nutzwertanalysen in Marketing und Vertrieb*; Springer Gabler: Wiesbaden, Germany, 2019; Volume 2. [CrossRef]

14. Ramík, J. *Pairwise Comparisons Method—Theory and Applications in Decision Making*; Springer Nature: Berlin/Heidelberg, Germany, 2020; Volume 690. [CrossRef]
15. Bozóki, S.; Dezső, L.; Poesz, A.; Temesi, J. Analysis of pairwise comparison matrices: An empirical research. *Ann. Oper. Res.* **2013**, *211*, 511–528. [CrossRef]
16. Dean, M. *A Practical Guide to Multi-Criteria Analysis*; Technical Report; University College London: London, UK, 2022. [CrossRef]
17. Kalt, S. Automatisierte Auslegung Elektrischer Antriebsmaschinen zur Anwendungsspezifischen Optimierung. Ph.D. Thesis, Technische Universität München, Munich, Germany, 2021.
18. Hadraoui, H.E.; Zegrari, M.; Chebak, A.; Laayati, O.; Guennouni, N. A Multi-Criteria Analysis and Trends of Electric Motors for Electric Vehicles. *World Electr. Veh.* **2022**, *13*, 65. [CrossRef]
19. Wang, A. Economic efficiency of high-performance electric vehicle operation based on neural network algorithm. *Comput. Eltrical Eng.* **2023**, *112*, 109026. [CrossRef]
20. Weiss, M.; Cloos, K.C.; Helmers, E. Energy efciency trade-ofs in small to large electric vehicles. *Environ. Sci. Eur.* **2020**, *32*, 46. [CrossRef]
21. Koopmans, C.; Mouter, N. Cost-Benefit Analysis. In *New Methods, Reflections and Application Domains in Transport Appraisal*; Elsevier Inc.: Amsterdam, The Netherlands, 2021; Volume 7, Chapter 1, pp. 173–183. [CrossRef]

vehicles

MDPI

Article

Virtual Plug-In Hybrid Concept Development and Optimization under Real-World Boundary Conditions

Jannik Kexel *, Jonas Müller , Ferris Herkenrath, Philipp Hermsen , Marco Günther and Stefan Pischinger

Chair of Thermodynamics of Mobile Energy Conversion Systems (TME), RWTH Aachen University, 52074 Aachen, Germany

* Correspondence: kexel@tme.rwth-aachen.de; Tel.: +49-241-80-24209

Abstract: The automotive industry faces development challenges due to emerging technologies, regulatory demands, societal trends, and evolving customer mobility needs. These factors contribute to a wide range of vehicle variants and increasingly complex powertrains. The layout of a vehicle is usually based on standardized driving cycles such as WLTC, gradeability, acceleration test cases, and many more. In real-world driving cycles, however, this can lead to limitations under certain boundary conditions. To ensure that all customer requirements are met, vehicle testing is conducted under extreme environmental conditions, e.g., in Sweden or Spain. One way to reduce the development time while ensuring high product quality and cost-effectiveness is to use model-based methods for the comprehensive design of powertrains. This study presents a layout methodology using a top-down approach. Initially, powertrain-relevant requirements for an exemplary target customer are translated into a specification sheet with specific test cases. An overall vehicle model with detailed thermal sub-models is developed to evaluate the different requirements. A baseline design for a C-segment plug-in hybrid vehicle was developed as part of the FVV research project HyFlex-ICE using standardized test cases, highlighting the influence of customer profiles on the design outcome through varying weighting factors. The target customer's design is analyzed in four real driving scenarios, considering variations in parameters such as the ambient temperature, traffic, driver type, trailer pulling, and battery state-of-charge, to assess their influence on the target variables. In the next step, the potential of hardware technologies and predictive driving functions is examined in selected driving scenarios based on the identified constraints of the baseline design. As a result, four application-specific technology packages (Cost neutral, Cold country, Hot country, and Premium) for different customer requirements and sales markets are defined, which, finally, demonstrates the applicability of the holistic methodology.

Keywords: mobile propulsion systems; model-based system design; RDE; emissions; thermal management; PHEV; technology package; predictive controls

Citation: Kexel, J.; Müller, J.; Herkenrath, F.; Hermsen, P.; Günther, M.; Pischinger, S. Virtual Plug-In Hybrid Concept Development and Optimization under Real-World Boundary Conditions. *Vehicles* **2024**, 6, 1216–1248. https://doi.org/10.3390/vehicles6030058

Academic Editors: Ralf Stetter, Udo Pulm and Markus Till

Received: 4 June 2024
Revised: 8 July 2024
Accepted: 9 July 2024
Published: 15 July 2024

1. Introduction

The development challenges in the automotive industry are constantly increasing due to the large number of vehicle variants, the growing complexity of powertrains, and future regulatory requirements [1,2]. In addition, there is a growing demand for vehicles that do not have a negative impact on the environment over their entire life cycle. To further reduce development times while maintaining high product quality and cost-effectiveness, the use of new model-based methods in early development phases is necessary [3–6].

For organizational reasons, the optimization of the derived target variables is usually carried out in isolation between individual disciplines, so that possible interactions cannot be considered holistically [3]. The state of the art in powertrain design is described below. In [7], both an operating strategy and a component design of various hybrid drives in the NEDC were carried out using a statistical design of experiments (DoE). Based on this, the approach was extended in [8] to include more realistic driving cycles, taking into account driving performance requirements. In [9], the pollutant emissions, under the Euro 6 RDE boundary conditions, are also considered. Different optimization algorithms for the energy management of hybrid vehicles are discussed in [10–13], with a focus on energy consumption. DANZER and TEUSCHL also evaluate the differential costs of designing different powertrains compared to a reference vehicle [14,15]. In addition, [14] evaluates other customer requirements such as the acoustic driving comfort based on the noise in the engine compartment. The driving scenarios in which this evaluation was performed are not explained in detail. A rudimentary approach to evaluating noise-vibration-harshness (NVH) is used in [16], and is based on the number of engine starts and gear changes. For the optimal concept design of electrified powertrains, WEISS considers the driving performance, consumption, and economy as target variables while taking into account different boundary conditions, such as a minimum electric range or the reproducibility of acceleration in different driving cycles [17].

Temperatures are often considered only for single components (e.g., in [9,18] for the exhaust system or in [19] for the vehicle interior) or as simplified model (e.g., as a thermal network of combustion engine, transmission, and electric motor in [8,17] or as a time-dependent control function of the electric motor in [20]). Cooling systems have been designed with the help of simulations for decades. However, the focus was initially only on the dimensioning of the water pump and the radiator [21,22]. EILEMANN shows that a holistically designed thermal management system can contribute to the reduction of CO_2 emissions [23]. Therefore, several publications deal with the investigation of thermal management technologies such as electric water pumps or control valves [24–29]. They are limited to conventional combustion engine drives. However, according to [19,30–32], the thermal management requirements for electrified powertrains are higher due to the different temperature levels and vehicle operating modes.

In [19,31], a simulation approach is used that can consider system relationships between multiple target variables. Genender et al., investigate the influence of different heating measures on the thermal comfort in hybrid and electric vehicles [19]. Shutty et al., show the results of different case studies on the influence of temperatures on component losses, the thermal safety of electric motors, and the thermal coupling between a battery and the vehicle interior [31]. BESTE limits her study to the high-voltage system [30].

From the state of the art, it can be deduced that, although the hybrid vehicle system design is continuously considered with an increasing number of different target variables, this has not yet been done completely or with a sufficient level of detail and consideration of interactions. As a result, unexpected deviations from defined specification targets can occur, especially in real driving scenarios of full vehicle tests, requiring repeated vehicle tests or mechanical adjustments. This can lead to an increase in development costs and a delay in the start of production [33]. This means, in detail, that it is necessary to take into account the interactions within powertrain systems and all their components at the earliest possible stage of development, including their mechanical as well as thermal boundary conditions. Consequently, the full potential for the targeted product over the whole field of application and range of use can be utilized. However, the early development stages have recently been characterized by CO_2 emission optimization with simplified approaches and a focus on basic driving cycles such as the Worldwide harmonized Light vehicles Test Procedure (WLTP).

Against this background, we developed a holistic methodical approach in the FVV project *HyFlex-ICE* for the holistic designing and optimization of vehicles and powertrains [34]. This methodology was developed using an exemplary hybrid powertrain (combined hybrid) but is generally applicable and transferable to other transport applications. As a baseline for our in-depth investigation, an optimal propulsion system configuration and operating strategy was developed for this reference vehicle.

Building on that, this research investigates the impact of real-world driving scenarios on the optimal operation of a plug-in hybrid electric vehicle (PHEV) by addressing the following questions:

Which real-world driving scenarios can limit the optimal operation of a PHEV in terms of CO_2, NVH, drivability, and pollutant emissions?

1. What is the impact of predictive controls on the powertrain limitations?
2. What is the impact of hardware technologies and rule-based control adaptations on the powertrain limitations?
3. By addressing these questions, this work aims to contribute to the development of future vehicles with optimized performance across real-world driving conditions.

2. Vehicle Requirements Management and Optimal Hybrid Powertrain System Design

For the identification of suitable technical solutions which enable the best implementation of operating modes, all boundary conditions and limitations regarding system operation must be considered. To address this objective, a systems-engineering framework was formulated and devised within the context of the Vehicle Requirements Management guidelines, ensuring that the relevant interests of all stakeholders are being taken into account.

Starting from a generic or specific product idea, such as a vehicle, a top-down approach as outlined in [35,36] is employed to gather all stakeholder requirements in an abstract and solution-neutral manner. These requirements are subsequently decomposed for subsystems, gradually transitioning into more specific technical descriptions [35,37]. Utilizing stakeholder- or customer-specific weighting factors, target values and constraints are assigned to the neutral attributes and, finally, merged into a comprehensive requirements catalogue [35,36].

For this research, corresponding features (functional requirements) and targets (non-functional requirements) are determined for all aspects of the vehicle and filtered with respect to powertrain-related requirements. On the basis of a pre-study analyzing powertrain limitations, the electrification level was defined as plug-in hybrid (PHEV), cf. [34,36]. In alignment with current market trends, the propulsion system was integrated into a c-segment sports utility vehicle (SUV), cf. [34,36].

This methodology offers flexibility in its initiation point and applicability across various propulsion systems and means of transportation, rendering it generic, scalable, and universally applicable. Furthermore, it extends to encompass multi-market, multi-derivative, and multi-platform strategies, thereby increasing its complexity with expanded steps [36].

2.1. Pre-Dimensioning and Modeling of Vehicle and Operating Strategy

A preliminary powertrain design was established based on a comprehensive requirements catalog. This catalog prioritized vehicle performance and efficiency, while incorporating weighting factors specific to the target customer segment. To achieve this balance, a multi-objective optimization approach was implemented. This approach considered not only efficiency targets but also their trade-off with the system cost and customer-desired NVH characteristics.

To reduce the design space for subsequent statistical design of experiments (DoE) optimization, an analytical pre-dimensioning and conceptual design methodology was developed. This methodology employs a combined backward-forward simulation approach, utilizing analytical vehicle resistance equations and simplified component characteristics. The primary objective is to estimate the minimum, maximum, and overall orders of magnitude of the system components. This approach aligns with the principles of model-based systems engineering, as described in [38,39]. For pre-dimensioning, a dedicated standalone MATLAB application was developed, aiming for fast initial propulsion system integration. This application uses an analytical, iterative process to estimate component properties for a range of different propulsion system configurations, enabling the identification of effects of parameter variations on critical scenarios and component properties [34]. These results are subsequently used as the preliminary base design.

The analytical sizing and conceptual design process is illustrated schematically in Figure 1. It demonstrates the integration of requirements management with the analytical sizing of component dimensions. Unlike the approach presented in [40], our methodology enables the preliminary design of various powertrain types and configurations tailored to specific customer requirements. The technical details of the algorithms in the background and the extended application possibilities will be discussed in detail in a follow-up publication.

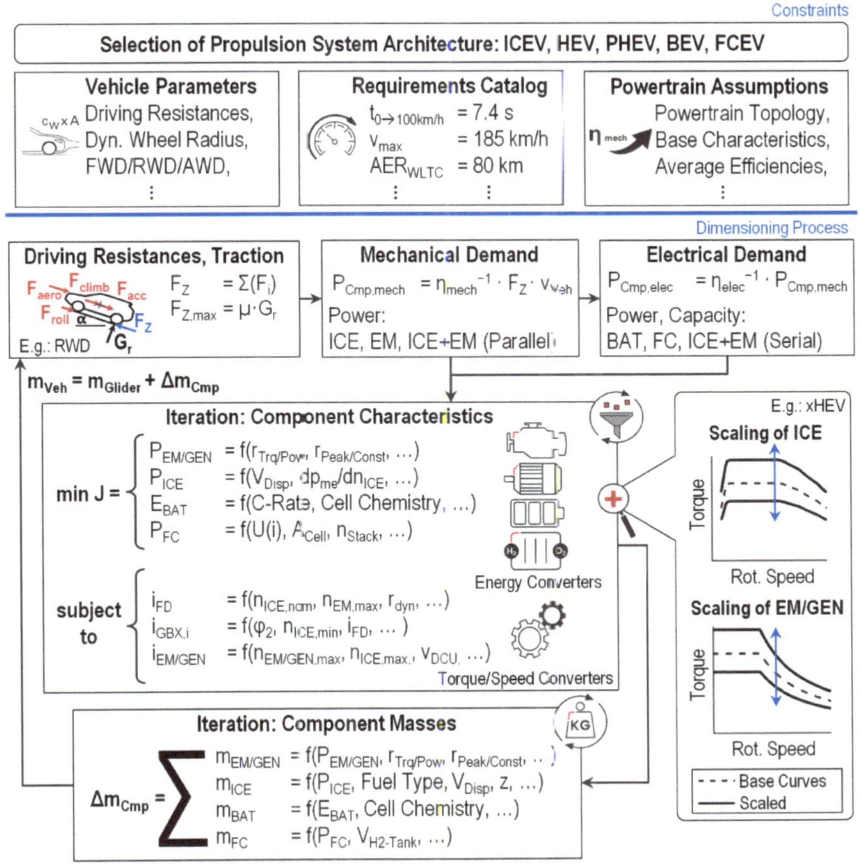

Figure 1. Pre-dimensioning tool for analytical propulsion system conceptualization and model-based vehicle requirements management.

While the sizing of the components in hybrid propulsion systems is primarily defined by performance criteria, a comprehensive design approach must consider the intricate interactions between the operating strategy and system layout. These interactions encompass the delicate balance between the performance, efficiency, emissions, and NVH. Additionally, thermal system limitations and interactions between mechanical, thermal, and electrical boundary conditions must be carefully addressed. For this purpose, a physical-empirical simulation model based on a closed-loop forward simulation approach is used for the system optimization of the propulsion concepts with respect to the various target variables of the requirements catalog. This vehicle model and this operating strategy model are based on SEIBEL [7], BALAZS [3], and BÖHMER [9]. In the *HyFlex-ICE* [34,36] and *ZITE* [41] research projects, this model was further developed, especially regarding thermal, emission, and NVH modeling. The corresponding schematic illustration of the overall model is shown in Figure 2.

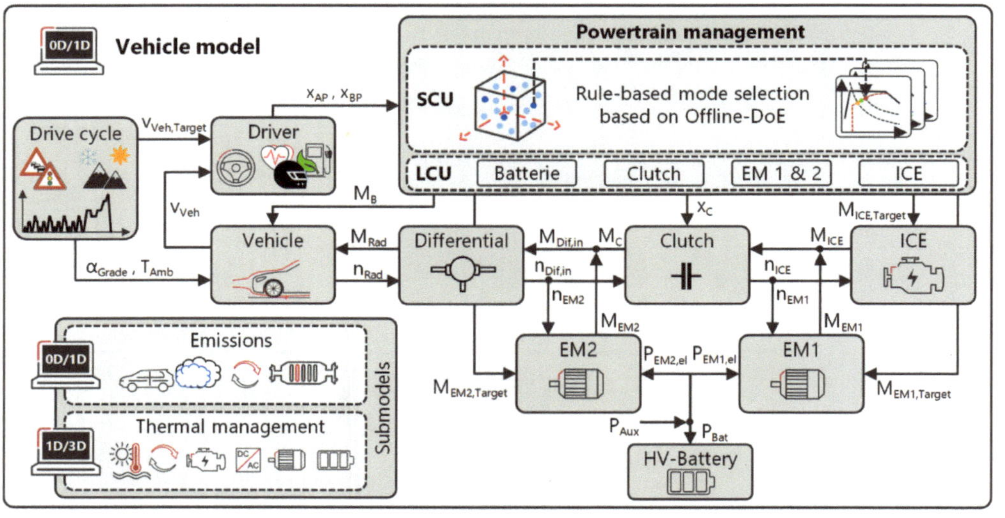

Figure 2. Schematic illustration of the closed-loop forward simulation model for a combined hybrid electric vehicle with no transmission.

The integrated system simulation considers all powertrain components on the basis of data from benchmark validated models. Furthermore, different levels of complexity and detail of the subsystems can be selected depending on the specific development phase [34,36]. For the optimal base layout, simplified thermal and emission models were used to reduce the simulation time. Optimization of the powertrain involves scaling methods for all components, cf. [7,8]. Additionally, the heuristic operating strategy incorporates a parameter set for managing driving mode changes and battery recharging, and was developed by [42,43].

2.2. Optimized Combined Plug-In Hybrid Powertrain Design

Based on the defined catalog of requirements, the system components were designed and optimized with the help of the simulation models presented herein, considering the vehicle characteristics. For this purpose, extensive simulation studies were performed to verify that all requirements were at least met. Utilizing the described modeling approach, combining high accuracy with a low computational demand, multi-objective optimization via the DoE is conducted [44,45]. The left side of Figure 3 shows the schematic structure of the powertrain. The different colors represent the mechanical, electrical, and chemical energy flows. The modeling approaches used for the baseline design and the investigation of real driving scenarios are shown on the right side with the check mark. For the electrical components, the base design uses a time-dependent function for thermal derating to achieve realistic results in the performance test cases. Simplified thermal networks were also used for the combustion engine and high-voltage battery to provide temperatures as input variables for other sub-functions. The catalyst was not included in the base design as the concept had already been validated with a similar vehicle class in the WLTC [18]. In the real driving scenarios, detailed thermal models were used for all components, as the thermal effects increase significantly compared to the standardized tests.

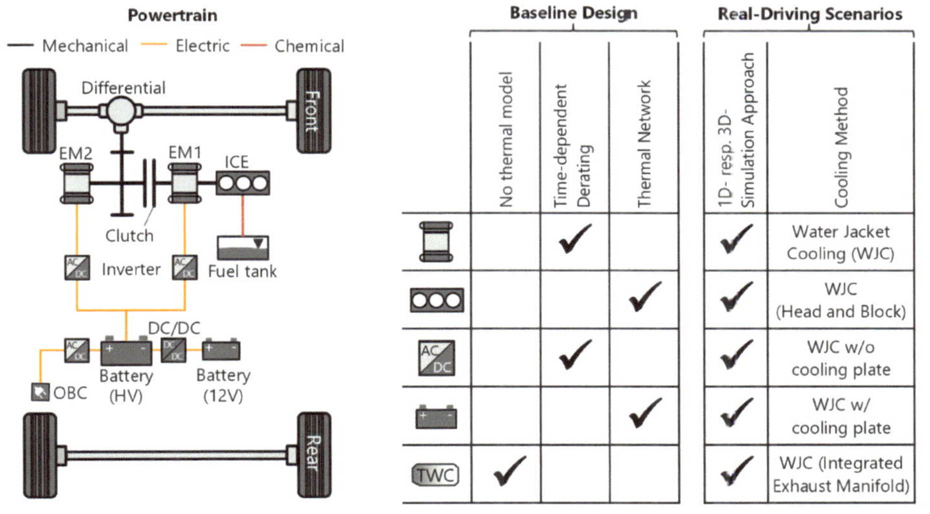

Figure 3. Schematic powertrain architecture of the combined PHEV and thermal simulation approach for the baseline design and the real-driving scenarios.

For the DoE approach, a test plan of variable hardware and operating strategy parameters is generated within xCAL [45,46]. This includes a flexible combination of all parameters to quantify the effects of parameter changes and can be optimized by the results of the analytical pre-dimensioning approach. Subsequently, leveraging longitudinal dynamic simulation outcomes, a regression model is trained in xCAL utilizing a Gaussian process model [45,46].

The optimization process considers all requirements, such as performance, CO_2 emissions, NVH, and costs, weighted in accordance with the target customer preferences. The state of charge (SOC) balance of the battery is a secondary condition for optimization. Finally, the mathematical optimization is validated using optimal parameter sets within the simulation model [34,36].

Table 1 briefly summarizes the resulting optimal base layout of the propulsion system. The main hardware design is primarily defined by the performance maneuvers. The detailed multidimensional, polynomial model and specifics regarding these critical maneuvers are elaborated in [34,36]. As a result, the hardware sizes and operating strategy parameters provide the best compromise with regard to a *CO_2-* and *cost-optimal* powertrain layout. This configuration leads to the following optimal baseline results: the WLTC charge sustaining CO_2 emissions of $m_{Base,WLTC,CS}$ = 109.1 gCO_2/km, NVH-masking noise exceedance of $N_{Base,WLTC}$ = 0.028 mJ/m^2, and delta costs compared to a reference powertrain of C_{Base} = 1074€, cf. [34].

In general, it is essential to consider that the specific component characteristics are heavily influenced by the detailed maneuver definition. Therefore, a particularly high priority must be given to the definition process for these test cases.

These defined and optimized propulsion system characteristics, including the rule-based operating strategy, serve as a basis for the following in-depth analysis to identify possible limitations and violations of all of the defined requirements in *Real-World Driving Scenarios*. Based on this, both the optimization of the operating strategy, as well as hardware adaptions, and the implementation of technology packages is carried out.

Table 1. Vehicle and propulsion system characteristic for the optimized C-segment SUV PHEV (combined hybrid).

Parameter		Value	Unit
Vehicle, C-Segment SUV	Mass Drag coefficient Rolling resistance	1.84 0.75 8.1	t m^2 kg/t
ICE, 3-Cyl. DI TC	Displacement Rated power	1.4 105	L kW
Generator (EM2), PMSM	Max. power Torque-to-Power ratio	115 2.0	kW Nm/kW
E-Motor (EM2), PMSM	Max. power Torque-to-Power ratio	246 3.8	kW Nm/kW
Battery	Energy (NMC cell)	14.7	kWh
Final Drive	Gear ratio	3.7	1

2.3. *Advanced Modeling Approach*

In general, many propulsion system limitations are caused by emission limitations, thermal boundaries, and environmental conditions. For this reason, a detailed and advanced modeling approach for the NVH, determined mainly by the thermal and emission system, is introduced in the following to consider this in the early development stage. All relevant aspects were brought together and integrated into a uniform and holistic simulation toolchain, cf. [36].

2.3.1. NVH Modeling Approach

The acoustic passenger comfort is assessed using a map-based approach that can be used in the early concept phase. The masking noise generated by rolling and aerodynamic drag is compared with the ICE noise, including airborne and structure-borne noise emissions at the position of the driver's headrest. To determine a target value, a scatter band for the masking noise including a transfer function to the described position from the TME database and from the FVV project "Noise in the vehicle interior with electrified drives" [47,48] is used (see the left diagram in Figure 4).

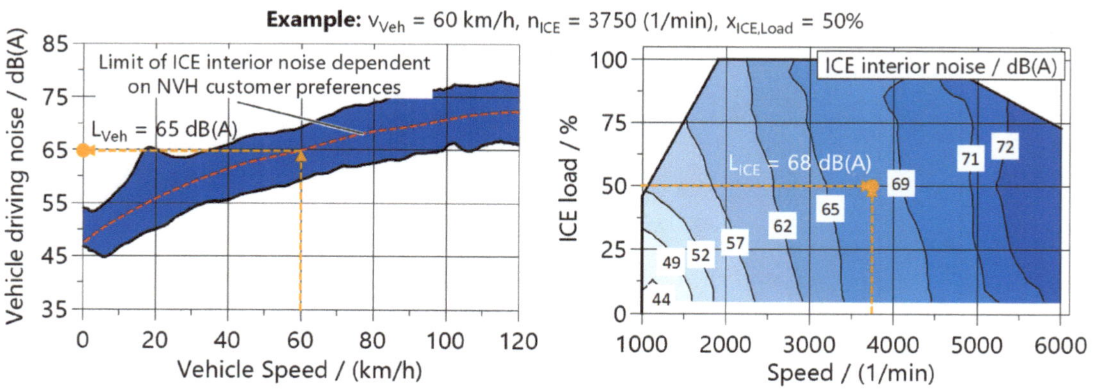

Figure 4. Overview of the rolling and wind noise as a function of the vehicle speed (**left**) and the ICE interior noise as a function of the engine operating point (**right**).

Depending on the customer weighting of the NVH criterion, a corresponding curve can be inserted into this scatter band (cf. red line on the left side of Figure 4). The interior noise of an exemplary ICE is shown in the right diagram in Figure 4. In the example shown in the figure (see orange lines in Figure 4), a rolling and wind interior noise of L_{Veh} = 65 dB(A) is generated at v_{Veh} = 60 km/h. At the ICE operating point n_{ICE} = 3750 1/min and $x_{ICE,L}$ = 50%, the interior noise is L_{ICE} = 68 dB(A). In a single timestep, the excess masking noise ΔL_M can therefore be calculated according to Equation (1).

$$\Delta L_M = L_{Veh} - L_{ICE} \tag{1}$$

This approach is extended to obtain a characteristic evaluation measure for the noise emissions over the entire driving cycle. For this purpose, the logarithmic sound intensity levels L_{Veh} and L_{ICE}, with the reference sound intensity $I_0 = 10^{-12}$ W/m^2, are first converted into intensities according to Equation (2).

$$I = I_0 \cdot 10^{\frac{L}{10}} \tag{2}$$

From this, the integral of the deviation between I_{Veh} and I_{ICE} is calculated according to Equation (3), if $(I_{Veh} - I_{ICE}) > 0$. The ratio $N_{Veh\text{-}ICE}$ in J/m^2 considers both the duration and the intensity of the masking noise overshoot.

$$N_{Veh-ICE} = \int_{t_0}^{t_{Trip}} (I_{Veh} - I_{ICE})\,dt \tag{3}$$

2.3.2. Thermal Modeling Approach

Limitations of propulsion systems can be caused by thermal constraints of the ICE, by the emission system, or by electrical components such as the HVB or EM. In addition, the interaction with the cabin climatization and passenger comfort must also be considered. Therefore, detailed thermal subsystems and determining their interactions are necessary for holistic powertrain and control strategy development [32,49,50]. Consequently, the vehicle model is coupled with advanced thermal management submodules based on a 1D–3D approach considering the powertrain's components and its respective cooling systems. A physical-empirical calculation approach is used for the heat transfer calculations of the component structure, coolant, oil, and air. In addition, an FVV research project has investigated thermal cabin comfort based on the predicted mean vote (PMV) [34,51,52].

Compared to conventional powertrains, hybrid vehicles have additional low-temperature circuits due to the different optimum temperature levels of the electrical components. Lowering the fluid temperature while maintaining the same heat flow results in a lower component temperature. The schematic structure of the thermal management system is shown in Figure 5. The colors symbolize the different temperature levels of the high temperature circuit (HTC), low temperature circuits I and II (LTC), and the cooling circuit (RC). All cooling circuits have electric water pumps (see HTC, LTC I, and II) or an electric compressor (see RC) to regulate the heat flows on demand.

A variety of cooling approaches and media are available for the thermal management of individual components. Indirect cooling by means of a water jacket has proven successful for ICE as well as for electric powertrain components [53]. Therefore, water jacket cooling is used for the ICE, inverter, high-voltage battery (HVB), and electric motor (EM) in the baseline simulations. However, the trend towards more compact electric powertrain components with high power densities requires increasingly efficient cooling methods and higher cooling capacities. Direct cooling approaches are investigated in the technology assessment (see Section 5).

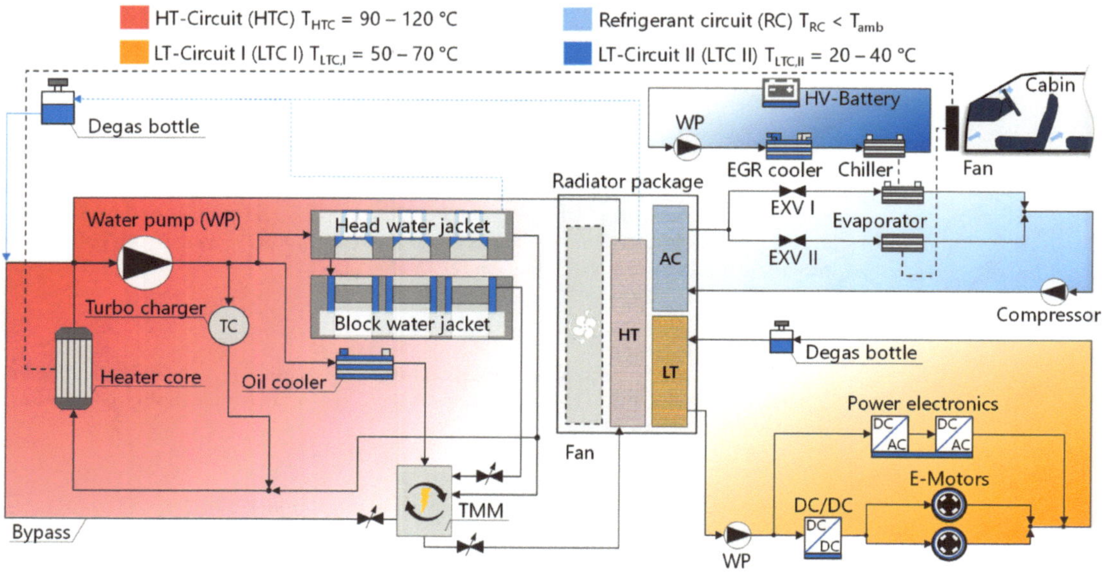

Figure 5. Schematic layout of the different cooling circuits in a hybrid electric vehicle.

2.3.3. Emissions Modeling Approach

The emission modeling approach is based on [54] and has been continuously developed in the literature [9,41,55,56]. The raw emissions model includes stationary engine maps and empirical-physical correction functions for transient operating conditions. The thermal behaviors of the exhaust gas and the catalyst are represented by advanced thermal networks which allow the simulation of the local structure and gas temperatures. Based on that, the conversion efficiency of the catalyst can be calculated. Further corrections regarding the air–fuel ratio and the oxygen storage capacity are considered. A detailed description can be found in the above-mentioned publications.

Figure 6 shows the baseline setup of the exhaust aftertreatment system (EATS), which is designed to comply with the existing EU6 emissions standard. The basis for this is the ICE, which is operated by the *Miller* principle, with stochiometric operation ($\lambda = 1$) in the entire engine map. The EATS combines a close-coupled three-way catalyst and an uncoated underfloor gasoline particulate filter (GPF) of the first generation. In the new condition, the GPF has an average filtration efficiency of η_{GPF} = 70%, which is considered as the worst-case condition.

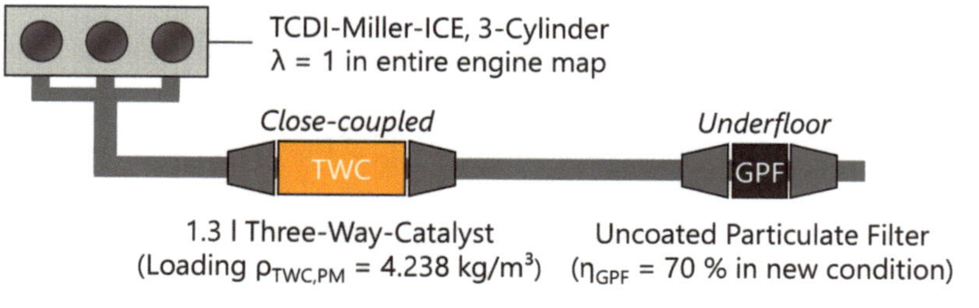

Figure 6. Setup of EU6 base exhaust aftertreatment system used for the simulation study.

3. Results of the Most Challenging Real-World Driving Scenarios

The precise definition of the individual test cycles has a major influence on the development of a hybrid vehicle. In addition to the standardized test procedures, it is necessary to investigate real driving conditions, such as those in cold or hot countries. The selected scenarios cover a wide range of different requirements and, by using the tool chain, correspond to real route and gradient profiles as well as realistic speed profiles [34]. The following driving cycles are examined in this work:

- Urban driving cycle (UDC) at low and high ambient temperatures [57];
- Cross-country driving cycle (CCDC) at low and high ambient temperatures [57];
- German highway driving cycle (GHDC) at moderate ambient temperatures [57];
- Mountain driving cycle (MDC) at moderate ambient temperatures [57].

Since the first three driving cycles occur frequently in everyday life, they are particularly relevant to the design. Mountain driving, such as on the Großglockner High Alpine Pass [58], is rather rare but still important for the powertrain component layout. In addition to the climatic boundary conditions, such as the temperature, there are also other variation parameters such as the number of occupants, the traffic situation, the trailer load, and the battery charge level at the start of the trip. The weight of an additional passenger corresponds to the average weight ($m_{Passenger}$ = 76 kg) of the population in Germany [59]. This results in the test matrix in Table 2, which was simulated using the simulation model from Section 2. The vehicle control strategy is based on the target customer design for the WLTC. To identify the limitations, a statistical analysis is performed in Section 3.1.

Table 2. Vehicle and propulsion system characteristics for the optimized C-segment SUV PHEV (combined hybrid). (UDC: urban driving cycle, CCDC: cross-country driving cycle, GHDC: German highway driving cycle, MDC: mountain driving cycle).

Parameter	UDC	CCDC	GHDC	MDC
Ambient temperature	−10 °C 40 °C	−10 °C 40 °C	0 °C	0 °C
State-of-charge	17% 50% 95%	17% 50% 95%	17% 50% 95%	17% 50% 95%
Vehicle payload	76 kg 349 kg	76 kg 349 kg	76 kg 349 kg	76 kg 349 kg
Trailer	750 kg	750 kg	750 kg	750 kg
Traffic	No traffic Low traffic High traffic	No traffic Low traffic High traffic	No traffic Low traffic High traffic	No traffic Low traffic High traffic
Driver type	Defensive Dynamic	Defensive Dynamic	Defensive Dynamic	Defensive Dynamic

3.1. Simulation Results of Real-World Driving Scenarios

3.1.1. Statistical Evaluation of Component Temperatures

In the following, the simulation results of the real driving scenarios are statistically evaluated and presented in box plots. First, the temperatures of the individual components are analyzed to identify the thermal influence on the drivetrain. Figure 7 shows the maximum temperatures of the HVB and the EM in the respective driving cycles. The power electronics do not exceed the maximum allowable temperature in any scenario and are not considered in the following analysis. In general, it should be noted that the median and the 25% or 75% quartile can be identical. This is especially true when driving in the pure electric operating mode.

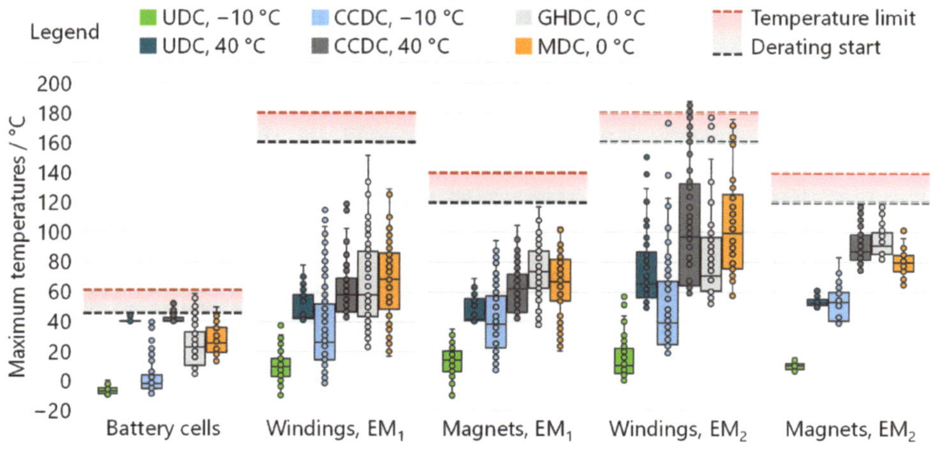

Figure 7. Statistical evaluation of the maximum temperatures occurring in the battery cells, the EM windings, and the EM magnets (UDC: urban driving cycle, CCDC: cross-country driving cycle, GHDC: German highway driving cycle, MDC: mountain driving cycle).

The cells of the HVB as well as the windings and magnets of the EM only heat up slightly in the UDC at T_{amb} = −10 °C due to the reduced power availability [60,61] and the low vehicle power demand in the urban driving cycle. This means that all components remain within the permissible temperature range even at T_{amb} = 40 °C. In the CCDC, GHDC, and MDC, the HVB reaches higher temperatures due to self-heating by cell ohmic losses [62]. Demanding driving scenarios with a sporty driving style lead to a thermal derating of the HVB so that the maximum cell temperature of $T_{HVB,cell,max}$ = 60 °C is not exceeded. Similarly, the maximum power of EM1 and EM2 is also limited by reaching critical temperature ranges. The magnet temperatures of EM1 have a wider range because, in the serial mode, different power outputs are required depending on the state-of-charge and the HVB temperature. At high EM speeds, the iron losses increase, resulting in higher magnet temperatures for EM1 than for EM2. As the EM2 can use the electrical energy of the battery and the generator (EM1) in the serial operating mode, the highest temperatures occur at the winding heads of EM2.

3.1.2. Statistical Evaluation of the Operating Modes

The resulting influences on the operating strategy are shown in Figure 8 based on the driving shares of the individual operating modes. In the UDC, due to the low vehicle speed, the vehicle is only operated in the electric and serial modes to increase the overall efficiency of the drive. The proportion of electric driving increases at higher ambient temperatures due to the higher battery power. The same trend can also be observed in OBVs. As the speed level on cross-country roads and highways increases, the proportion of driving in the parallel operating mode increases.

Due to the thermal derating of the HVB or EM during sporty driving and high traffic volumes in the CCDC at T_{amb} = 40 °C, the ICE is used for boosting, so that the minimum only changes slightly compared to T_{amb} = −10 °C. In the mountain driving cycle, the driving performance requirements increase due to the additional gradient. For this reason, the accelerations at low vehicle speeds are performed in the serial operating mode. Due to recuperation during downhill driving, the median proportion of electric driving is still over 50%. The effects of the large variance in component temperatures and driving modes on the individual target variables are explained in the following section.

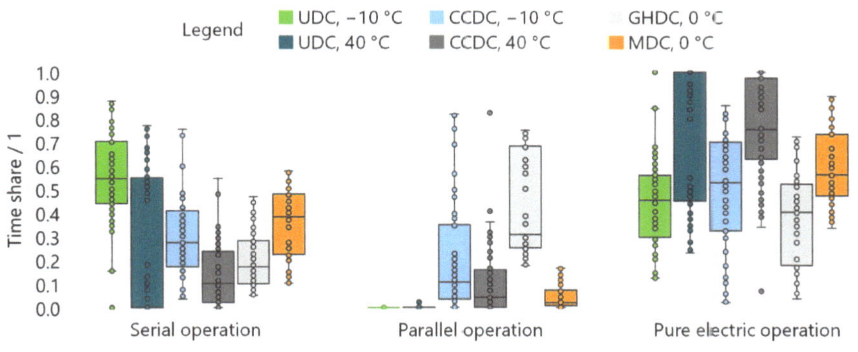

Figure 8. Statistical evaluation of the time-based operating strategy driving shares (UDC: urban driving cycle, CCDC: cross-country driving cycle, GHDC: German highway driving cycle, MDC: mountain driving cycle).

3.1.3. Statistical Evaluation of CO_2 and ICE Noise Emissions

The distribution of CO_2 emissions on the left side of Figure 9 shows that the median for trips at high ambient temperatures is below the fleet limit value and the baseline reference due to the high proportion of electric trips (see Section 2: 109 g_{CO2}/km). However, up to 421 g_{CO2}/km are emitted on short trips in the UDC under demanding conditions, such as driving with a trailer, a fully loaded vehicle, and a sporty driving style. At low ambient temperatures, both the 25% quartile and the median are above the fleet limit. This is due to the higher proportion of parallel and series operation and the higher losses of the ICE. For the OBV, the median decreases at low ambient temperatures due to the longer driving distance compared to the UDC. At high ambient temperatures, the CO_2 emissions are below the limit. Nevertheless, there is an increase compared to the UDC due to the higher performance requirements in the CCDC. In the CCDC, the engine speed is directly linked to the wheel speed at high speeds in parallel operation, which can lead to a less favorable operating point with higher engine friction and fuel consumption. In the MDC, the gradient leads to an additional increased power demand compared to the other driving cycles, so that the median is at a higher level.

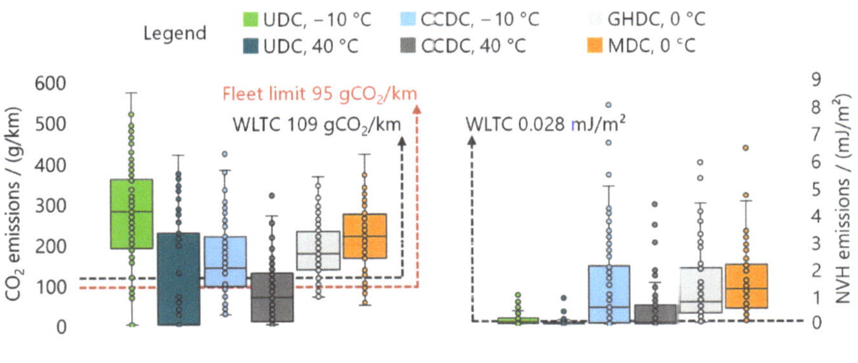

Figure 9. Statistical evaluation of CO_2 and ICE noise emissions (UDC: urban driving cycle, CCDC: cross-country driving cycle, GHDC: German highway driving cycle, MDC: mountain driving cycle).

Higher electric driving shares are also advantageous for low noise emissions (see right side of Figure 9), as high ICE speeds can be avoided. In contrast, the overruns are significantly higher at low ambient temperatures due to the low electrical battery power and the correspondingly decreased electrical driving share. In this case, the missing electrical power is usually provided in the serial operating mode, so that the noise emissions increase due to a high engine speed level at high driving performance requirements. Noise emissions in the GHDC are rather low at high vehicle speeds due to the sufficient masking noise, while in the MDC, high exceedances occur at lower vehicle speeds and higher power requirements.

3.1.4. Statistical Evaluation of Pollutant and Particulate Emissions

Figure 10 shows the statistical evaluation of the pollutant and particulate tailpipe emissions according to $d_{EU7,Clove}$ = 16 km. It should be noted that the distance of the urban driving cycle is only d_{UDC} = 7.5 km. This means that longer distances can also lead to higher emissions. The emissions depend, among other things, on the combustion chamber wall temperature [49] and the catalyst temperature [63].

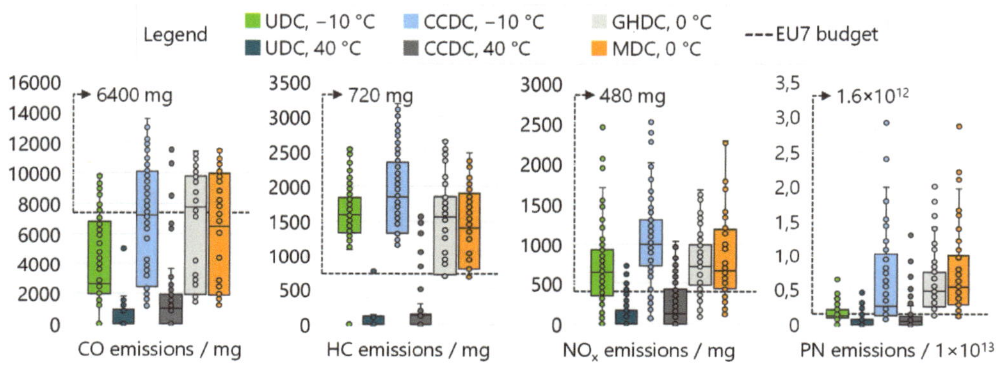

Figure 10. Statistical evaluation of pollutant and particulate tailpipe emissions according to $d_{EU7,Clove}$ = 16 km. (UDC: urban driving cycle, CCDC: cross-country driving cycle, GHDC: German highway driving cycle, MDC: mountain driving cycle).

To initially minimize raw emissions during the TWC heating process, an ICE operating point with increased speed in the medium load range is suitable (cf. Section 5.1.1, Figure 11). When the battery state-of-charge is low, the TWC is preheated at the beginning of the driving cycle. If the driver has high performance requirements in heavy traffic with repeated hard accelerations, the engine may start unexpectedly even if the HVB is fully charged. This shifts the operating point to higher loads and possibly also to higher engine speeds, resulting in emission breakthrough if the TWC is not fully heated. This effect is reflected in the exceeding of the available emission budget in all driving cycles with cold ambient temperatures, as the HVB cannot be used for boosting.

In addition, low starting temperatures and high-load operating points lead to significantly higher particulate raw and tailpipe emissions. This is particularly evident in the MDC due to the high driving power in the serial operating mode. In addition to temperature, the filtration efficiency of the GPF ($\eta_{GPF,new}$ = 70%) is crucial. Apart from the scenarios at T_{amb} = 40 °C, the 25% quartile of the remaining driving cycles already exceeds the available budget of PN = 1.6×10^{12}. Similar to the PN, the HC emissions also increase at low starting temperatures [64]. In addition, the control of the air–fuel ratio during catalyst heating has a major influence on the resulting raw emissions, so that scenario-specific modifications offer potential for optimizing tailpipe emissions.

3.1.5. Selection of Challenging Driving Scenarios

Statistical evaluation allowed the identification of powertrain limitations in real driving scenarios. The CCDC is suitable for demonstrating the optimization, as thermally relevant restrictions are caused by both poor heating behavior and the thermal derating of components at high ambient temperatures. These are much less severe in the UDC due to the lower driving performance requirements. The driving scenarios selected as examples are shown in Table 3. At cold ambient temperatures, low state-of-charge scenarios with a high traffic volume and dynamic driving behavior are particularly critical, as the ICE must provide the driving power for the EM. For high ambient temperatures, the focus is on high HVB charging levels, as the electric powertrain components are then subjected to high thermal loads.

Table 3. Overview of selected challenging real-world driving scenarios for further investigations in Section 4.

Driving Scenario	**CO_2 g/km**	**SOC %**	**Duration s**	**NVH mJ/m^2**	**CO [1] mg**	**HC [1] mg**	**NO_x [1] mg**	**PN [1] 1×10^{12}**
Baseline WLTC/Legal limits	109/95	-	-	0.028/-	-/6400	-/720	-/480	-/1.60
Cold CCDC I [2]	298.5	15.5	2828	2.679	3140	1825	925	10.6
Cold CCDC II [3]	138.8	24.4	4161	0.368	6553	1872	730	1.72
Hot UDC I [4]	128.8	28.2	754	0.336	1789	108	179	2.30
Hot CCDC I [5]	54.8	34.8	2789	0.630	8494	1328	287	7.42

[1] Evaluation after $d_{Emi,Clove}$ = 16 km. [2] $T_{ambient} = -10$ °C, SOC_{inital} = 17%, $m_{Payload}$ = 76 kg, heavy traffic and dynamic driver. [3] $T_{ambient} = -10$ °C, SOC_{inital} = 50%, $m_{Payload}$ = 76 kg, heavy traffic and defensive driver. [4] $T_{ambient}$ = 40 °C, SOC_{inital} = 50%, $m_{Payload}$ = 826 kg, heavy traffic and dynamic driver. [5] $T_{ambient}$ = 40 °C, SOC_{inital} = 95%, $m_{Payload}$ = 349 kg, heavy traffic and dynamic driver.

Based on these elaborated limitations, with regard to several requirements in *Real-World Driving Scenarios*, in the following, the developed control and hardware measures are explained, aiming for a reduction in and optimization of these limitations. In detail, this comprises adjustments of the rule-based operating strategy, model predictive control (MPC) strategies, sophisticated hardware adaptations, and innovative technologies. Firstly, the technical background and algorithms are explained. Secondly, the results of the challenging scenarios including interdependencies and trade-offs are investigated.

4. Optimization-Based Control Approach

The investigated optimization-based control approach consists of two stages, the trip planning and the energy management stages [34]. The trip planning stage is used to provide an optimal target trajectory of the vehicle speed and the target state-of-charge (SoC) as inputs for the energy management. [65]. The energy management is referred to as predictive powertrain management (PPM) in the following.

4.1. Predictive Powertrain Management

The PPM was implemented in MATLAB/Simulink using a model predictive control approach. For the prediction trajectory, a horizon of 5 s was used for the vehicle velocity. The SoC trajectory was not considered in this work. Instead, a charge depletion charge sustaining (CS) strategy is pursued for this implementation, just how it is implemented in the deterministic approach that the PPM is compared to. Furthermore, a theoretical certification is possible. As part of the multi-objective optimization, the common objective of low CO_2 is extended by the objective of positive NVH performance. This is done by the formulation of the cost function according to Equation (4):

$$
\begin{aligned}
J_{Total} &= w_{\Delta SOC} \cdot J_{\Delta SOC}(N_{HS}) \\
&+ \frac{\sum_{k=0}^{N_{HS}-1} w_{\eta System} \cdot J_{\eta System}(k) + w_{\eta ICE} \cdot J_{\eta ICE}(k) - w_{\Delta NVH} \cdot J_{\Delta NVH}(k)}{N_{HS}} \\
&+ \frac{\sum_{k=-1}^{N_{HS}-1} w_{\Delta nICE} \cdot J_{\Delta nICE}(k) + w_{\Delta T-ICE} \cdot J_{\Delta T-ICE}(k) + w_{\Delta NVH} \cdot J_{\Delta NVH}(k)}{N_{HS}+1}
\end{aligned} \tag{4}
$$

where J_{Total} is the total cost for the entire horizon, $J_{\Delta SOC}$, $J_{\eta System}$, $J_{\eta ICE}$, $J_{\Delta NVH}$, $J_{\Delta nICE}$, and $J_{\Delta T\text{-}ICE}$ are the individual cost terms, and $w_{\Delta SOC}$, $w_{\eta System}$, $w_{\eta ICE}$, $w_{\Delta NVH}$, $w_{\Delta nICE}$, and $w_{\Delta T\text{-}ICE}$ are the corresponding weights. N_{HS} is the number of horizon steps.

To evaluate the NVH criteria, the ICE interior noise was compared to the masking noise (cf. [34]). A genetic algorithm (GA) was used [66] as the optimization algorithm, and allowed the design of a flexible setup of the control, as well as parallelization and early termination, which can be beneficial for the computational time [67]. As mentioned above, the prediction horizon of $t_{hor} = 5$ s was used. This horizon was discretized by a step count of n = 10. This control parameterization was selected as a compromise between computational time on the one hand and result quality on the other hand. With this setup, a DoE approach was performed to estimate the optimal parameterization of the cost function weights. Fast-running reduced order models (ROMs) were built by using 0D map approaches for the battery, the combustion engine, and the electric machines. Due to the short prediction horizon and their thermal inertia, the component temperatures were simplified as constant. As a baseline comparison, the results of the rule-based (RB) energy management were compared to the PPM.

4.2. Results of Predictive Powertrain Management in Real-World Driving Scenarios

The comparison between the RB energy management and PPM was conducted in the UDC and CCDC real-world driving scenarios. Due to the low temperatures in the CCDC and the resulting low possible charge and discharge power rates, a thermally managed battery scenario is also considered. In addition to the CO_2 and NVH behavior, the results of the NO_x, CO, HC, and PN were evaluated. The RB results are used as a reference for each respective cycle and are shown in Table 4.

Table 4. Simulation results of (1) UDC (T_{amb} = 40 °C, T_{Cabin} = 50 °C, $m_{Trailer}$ = 750 kg, SOC_{init} = 50%, dynamic driver and high traffic); (2) CCDC (T_{amb} = −10 °C, $T_{HVB,init}$ = −10 °C, T_{Cabin} = −10 °C, $m_{Trailer}$ = 0 kg, SOC_{init} = 17%, dynamic driver and high traffic); (3) CCDC, warm battery (T_{amb} = −10 °C, $T_{HVB,init}$ = 30 °C, T_{Cabin} = −10 °C, $m_{Trailer}$ = 0 kg, SOC_{init} = 17%, dynamic driver and high traffic). Potential analysis of the MPC approach compared to a rule-based EMS.

Relative Difference MPC–RB EMS	**Hot UDC I** **T_{amb} = 40 °C** **$T_{HVB,init}$ = 40 °C**	**Cold CCDC I** **T_{amb} = −10 °C** **$T_{HVB,init}$ = −10 °C**	**Cold CCDC I** **T_{amb} = −10 °C** **$T_{HVB,init}$ = 30 °C**
CO_2 emissions	−12.9%	−3.0%	−4.6%
Electric energy consumption	4.8%	−103.6%	131.4%
Total energy consumption	−1.6%	−3.7%	−0.6%
HVB state-of-charge	−1.1%	2.3%	−3.6%
NO_x tailpipe emissions [1]	−14.0%	−28.4%	−31.5%
HC tailpipe emissions [1]	25.0%	−5.8%	−2.9%
CO tailpipe emissions [1]	−9.3%	−2.2%	0.6%
PN tailpipe emissions [1]	−64.1%	−12.3%	−17.7%
Time share electric mode	−3.0%	−5.0%	−2.5%
Time share serial mode	4.6%	3.6%	6.6%
Time share parallel mode	−1.6%	1.4%	4.1%
NVH—exceedance of masking noise	−96.2%	−17.1%	−29.0%

[1] Evaluation after $d_{Emi,Clove}$ = 16 km.

Overall, the MPC approach emitted fewer CO_2 emissions compared to the rule-based strategy. Also, the total energy consumption in all three cycles is −1.6% in the UDC, −3.7% in the CCDC without thermal preconditioning of the battery, and −0.6% in the CCDC with thermal preconditioning of the battery. This is also explained by a system efficiency increase.

In the UDC cycle, the CO_2 emissions are −2.9% less, while the electrical consumption is +4.8% higher. This indicates that the use of electrical energy is favored more by the MPC in the case that the boundaries (UDC: high SOC) allow the usage of the battery in a wide operation range. This can also be derived from the comparison of the CCDCs. In the CCDC without thermal preconditioning, the increase in system efficiency is achieved through the choice of operating point in serial and parallel operation, whose time share increases by 3.6% and 1.4%, respectively; in the CCDC without thermal preconditioning of the battery, the efficiency gain is also achieved using electrical energy, which results in a delta SoC of −3.6%.

In terms of the reduction of the NVH exceedances a significant reduction can be detected in all three investigated cycles. While there is a reduction by −96% in the UDC, which is driven mostly in electric mode, the reductions in the CCDC cycles are −17.1% in case of low initial battery temperature and −29% in case of the CCDC with preconditioned battery compared to RB.

In UDC NO_x (−14%), CO (−9.3%) and particle number (−64.3%) were reduced, while an increase of +25% can be seen for the HC emissions. For the CCDC with at least eight-times higher engine runtimes compared to the UDC, the NO_X reductions are even more significant, with −28.4% and −31.5%, while the HC can be reduced by −5.8% and −2.9% and the particulate number by − 12.3% and −7.7%. In the case of the cold battery cycle, the CO is reduced by −2.2%, and in the case of the preconditioned battery, the CO emission is marginally increased by 0.6%. Even though the emissions are mostly reduced by the MPC approach, neither pollutant emissions or particle emissions are optimization objectives of the current implementation. A more detailed analysis of emission reduction using MPC control approaches will be investigated in a following publication by the authors. However, in the significant NVH exceedance reduction, the potential of the multi-objective setup of the PPM is showcased.

5. Hardware Adjustments and Hardware Optimization

The simulation results indicate a significant discrepancy between standardized test procedures, such as the WLTC, and real-world driving conditions, particularly under extreme boundary conditions. To mitigate the adverse effects of personal transportation on various stakeholders, including drivers, the general population, and the environment, the following section examines technical solutions and deterministic control enhancements aimed at improving the performance of PHEVs.

5.1. Technical Measures and Rule-Based Controls Enhancement Approach

The investigated technologies are derived from an engineering assessment that evaluates the most challenging scenarios and aligns these with current development trends. This section first describes these technologies and their impacts on the propulsion system. The technologies were incorporated into the existing simulation model, which was used for the baseline simulations. Within the simulation study, technologies are individually analyzed for their potential to optimize the target parameters, see Section 5.2. Based on this, an initial technology assessment is performed, and appropriate packages are assembled, see Section 5.3. Finally, the technology packages are validated in the respective *Real-World Driving Scenarios*, see Section 5.4.

5.1.1. Engine Torque Limitation

An effective way to reduce cold start emissions is to limit the engine torque or engine power. This intervention in the engine control system is explained in Figure 11.

The maximum allowable engine power is limited until the TWC operating temperature is reached. Once the TWC has reached its light-off temperature in the range of approximately $T_{TWC,LO}$ = 350 °C, the maximum engine power can be released again. This may result in reduced drivability during heavy accelerations if the electrical system is unable to boost due to a low battery state-of-charge or thermal derating of the HVB. Depending on the application of the limitation, the impact on driving performance can range from barely noticeable restriction for the driver to significant.

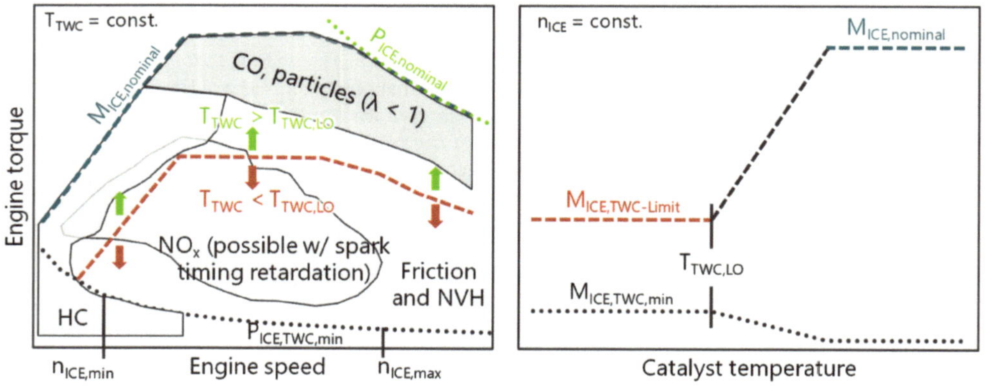

Figure 11. Exemplary engine torque limitation control dependent on the catalyst temperature.

5.1.2. Load Point Shifting

Another deterministic extension of the control strategy is load-point shifting (LPS) of the ICE. Notably, in the serial mode during the acceleration phases, LPS is well suited to reduce masking noise overshoots. The case example in Figure 12 shows the influence of two different LPS variants on the interior noise of the ICE and the resulting engine power.

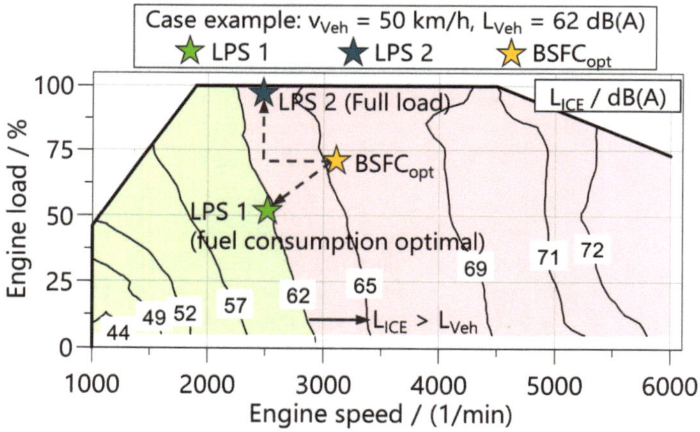

Figure 12. Influence of different variants of load-point shifting on the interior noise of the ICE.

LPS 1 shows the shift on the fuel-consumption-optimized curve, while LPS 2 shows a shift of the operating points on the full-load curve. Both variants reduce noise emissions but have different side effects on the battery state-of-charge, CO_2, and pollutant emissions. In this case example, LPS 1 does not exceed the vehicle noise of 62 dB(A) (see green marked area in Figure 12). However, LPS1 is limited in power, which would require a shift into the red marked area for high vehicle power demands. Since only the operating strategy needs to be adjusted, this measure is cost neutral.

5.1.3. Coolant Heat Storage

When the ICE operating temperature has been reached, the coolant heat storage (CHS) stores hot coolant from the high-temperature cooling circuit (HTC). At the next vehicle start, the thermal energy in the CHS can be used for indirect heating of the ICE by convective heat transfer from the coolant to the combustion chamber walls. For thermal conditioning prior to engine start, an electric water pump in the HTC is required to pump high coolant flow rates of the CHS (V_{CHS} = 5 L) through the engine water jackets for combustion chamber heating. A thermal simulation model, which considers the heat exchange with the environment, was used to analyze the cool-down behavior of the CHS in the parked state. The schematic structure of the model is shown in Figure 13 on the right with the characteristic values of the different insulation materials, air and polyurethane (PU). A heat transfer coefficient of $\alpha_{WWS,amb}$ = 7 W/(m²K) is assumed for the convective heat transfer from the CHS to the environment. The simulation results on the left side of Figure 13 show the influence of different insulation layers on the cool-down of the coolant in the CHS at T_{amb} = −10 °C and T_{amb} = 40 °C.

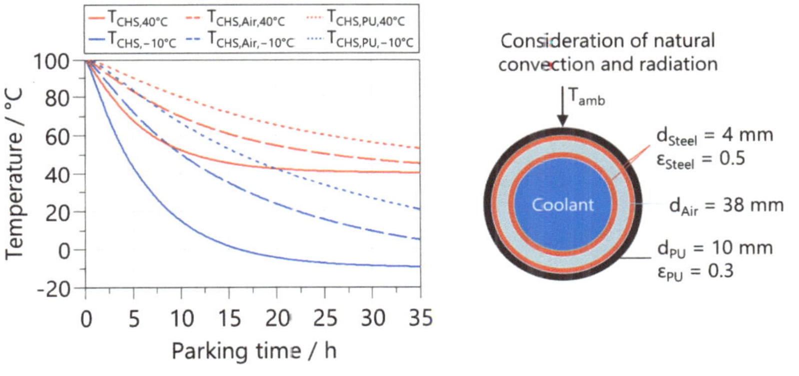

Figure 13. Influence of different insulation concepts on the cool-down of the coolant in the heat storage at T_{amb} = −10 °C and T_{amb} = 40 °C (**left**), and schematic layout of the thermal simulation model (**right**).

According to [68], the average parking duration in Germany is t_{Park} = 19 h/day. In this work, it is assumed that the vehicle is parked in the evening and used again the next morning, so that the flow temperatures are read for a parking period of t_{Park} = 12 h. The HT incurs additional costs C_{HT} = 56€ for the valves, pipes, and housing. Latent heat storage systems can additionally improve the cooling behavior and increase the amount of energy stored [69,70].

5.1.4. Engine Encapsulation

Underhood and underfloor insulation has long been state of the art for NVH reduction [3]. Engine encapsulation (EE) represents an extension which can be implemented as a system close to the engine (e.g., injector insulation) or as an engine-enclosing housing [71]. RÖCHLING uses the composite material Al-Seeberlit for simultaneous thermal and engine noise insulation. The thermal and acoustic influence of the engine encapsulation was derived from vehicle measurements. The left side of Figure 14 shows the cool-down of the ICE at different ambient temperatures for the baseline variant and with EE during parking. The results at T_{amb} = 20 °C are from a measurement by RÖCHLING and were used to calibrate a thermal model for the simulation of other ambient temperatures.

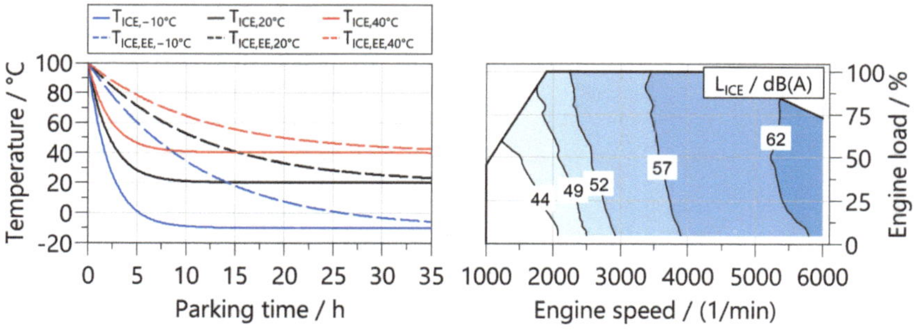

Figure 14. Influence of the engine encapsulation on the ICE cool-down at T_{amb} = −10 °C, T_{amb} = 20 °C, and T_{amb} = 40 °C (**left**) and on the interior noise of the ICE (**right**) based on the measurement results from [72].

Analogous to the previous section, a parking duration of t_{Park} = 12 h is also assumed for the definition of the coolant and oil temperatures. Röchling's measurements also show that the interior noise of the ICE can be reduced by $\Delta L_{ICE,EE}$ = 10 dB(A) compared to the baseline (see right diagram in Figure 14) [72]. The additional costs amount to C_{EE} = 70€ [73].

5.1.5. High-Voltage Battery Heater

To extend the application range of HVBs at low temperatures, there are several options for actively heating the cells. Some manufacturers take advantage of the poor cell efficiency at low temperatures to self-heat the cells (see GHDC and MDC driving scenarios at T_{amb} = 0 °C in Figure 7). However, for battery systems with high thermal inertia, this can require long heating times. For this reason, two technologies that generate heat via an electrical resistance, so-called PTC elements, are investigated. The heating performance is many times higher than that with self-heating of the cells. Indirect heating of the battery is possible with self-regulating PTC heating elements in the battery cooling circuit (see HV heater in Figure 15). When the electric water pump is activated, the heat from the PTC heater is dissipated convectively and transferred to the cells via the cooling plate. HV systems with heating capacities up to $P_{HVB,PTC1}$ = 9 kW are available on the market [74].

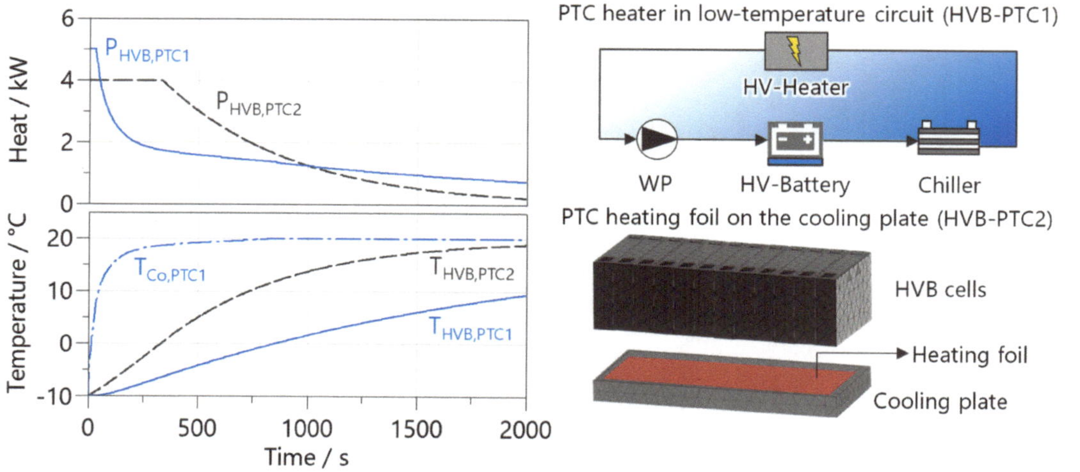

Figure 15. Heating performance (**left**) and structure (**right**) of the two battery heating concepts.

The intelligent control considers a continuous reduction in heating power as the battery or coolant temperature rises. The HV heater is integrated into the simulation model of the cooling system and considers the thermal inertia and the additional hydraulic pressure loss. Costs of $C_{HVB,PTC1}$ = 78€ are incurred for the PTC heater and the HV controller.

It is also possible to heat the cells directly. This is done by attaching a heating foil to the side or bottom of the cell. For ease of integration, the variant on the cooling plate or the underside of the cell is used here. Depending on the manufacturer, the heating elements enable a maximum heat flux of 20 kW/m^2 < $q_{HVB,PTC2,max}$ < 350 kW/m^2 [75–77]. As the battery power is limited during a cold start and excessive temperature differences within the cell must be avoided, the maximum heating power for the entire system is limited to $P_{HVB,PTC2,max}$ = 4 kW. This results in a heating power of $P_{HVB,PTC2,cell}$ = 41.7 W/cell for a single cell, corresponding to the number of cells $N_{HVB,cell}$ = 96. The costs of the heating foils and the controls are summarized as $C_{HVB,PTC2}$ = 120€.

The diagram in Figure 15, on the left, shows the average cell temperature and heat dissipation over time for the two heating concepts. In the first concept, the coolant temperature initially rises sharply to allow heat transfer to the cells. After reaching $T_{Co,HVB}$ = 20 °C, the heating power is reduced to avoid wasting unnecessary energy on cooling later. Due to the lower thermal resistance, the cell heats up faster in the second concept. The power is adjusted according to the temperature difference within the cells. This results in higher electrical energy consumption for the second concept.

5.1.6. Exhaust Aftertreatment System Adaptions

An electrically heated catalyst (EHC) enables faster heating of the TWC, minimizing the drivability limitations associated with conventional TWC heating strategies. In combination with a secondary air pump (SAP), thermal conditioning is possible prior to engine start. Figure 16 shows a possible exhaust aftertreatment concept with SAP, EHC, and coated GPF. The current development trend promises filtration efficiencies of $\eta_{GPF,new}$ > 90% in a new condition [78]. Despite the lack of thermal influence on the GPF filtration efficiency, this technology is helpful in reducing particulate emissions in cold regions.

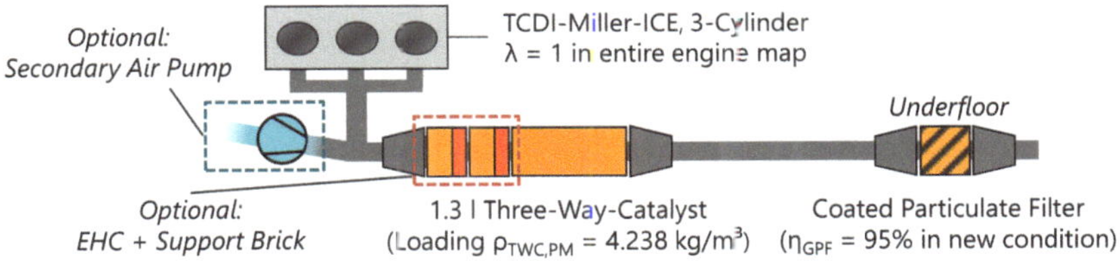

Figure 16. Exhaust aftertreatment concept with optional secondary air pump, electric heating disks, and coated gasoline particulate filter.

For the the propulsion system in this study, a heating disk with a support catalyst and a preheating time of $t_{EHC,VH}$ = 30 s is implemented [79]. This ensures rapid conversion of the pollutant emissions due to the low thermal inertia of the support catalyst. Two heating disks can also be connected in series to increase the heating power (see Figure 16). The necessity of a 48 V on-board power supply requires at least an additional DCDC to achieve heating powers of $P_{EHC,max}$ = 4 kW per heating disk. This results in an additional cost of C_{EHC} = 139€ for an EHC (4 kW) with SAP and a 48 V on-board power supply.

5.1.7. Electric Combustion Chamber Heater

The heat input through indirect heating measures such as the heat exchanger is limited by the heat transfer coefficient and the coolant temperature. Therefore, an electric combustion chamber heater (ECCH), in which heat is transferred directly from a heating wire to the liner, is investigated in this paper. Figure 17 shows the concept with cast-in round wire in the cylinder on the right.

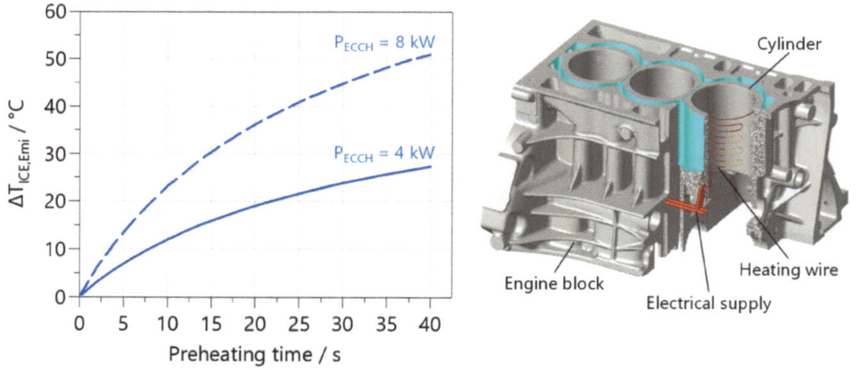

Figure 17. Electric combustion chamber heater and its heating behavior with different preheating times and heating outputs.

Various concepts with different materials were investigated as part of a design phase. Detailed results will be published by the authors in the future with a focus on tailpipe emissions. The effects on the strength of the engine and the manufacturing process were not considered in this work. To analyze the heating behavior, the concept was implemented in the thermal simulation model of the ICE. The electrical equivalent model of the heating coils calculates the resulting heat as a function of the electrical resistance. For comparability with the EHC, the heating power is limited to $P_{ECCH,max}$ = 8 kW.

The temperature differences for different preheating times and heating powers are shown in the diagram in Figure 17. The heating process can be accelerated by using higher heating powers. As with the EHC, a 48 V on-board power supply is used to minimize the cost of HV protection measures [80]. The cost for the integrated heater, 48 V system, and ECU function adaptations is estimated at C_{ECCH} = 125€.

5.1.8. EM and HVB Direct Cooling Approaches

The results of the real driving scenarios underline the limitations of indirect cooling approaches. To quantify the influence of direct cooling approaches, the existing jacket cooling of the electric motor (see Figure 5) is extended by oil spray cooling of the winding heads.

Figure 18 shows the concept on the right side. Direct cooling of the winding heads requires the use of a dielectric fluid, such as transmission oil, to prevent a short circuit. An oil pump supplies the distribution gallery and the cooling nozzles on top of the EM. Fundamental research by Bennion shows that the heat transfer coefficient (HTC) depends mainly on the flow velocity [81]. The oil temperature and the surface topology of the impact surface have only a minor influence and are not considered below. The left diagram in Figure 18 shows that the increase in HTC is degressive with an increasing flow velocity. In addition, under real conditions, not all surfaces of the winding head are reached directly by the cooling nozzles, so an HTC reduction is to be expected.

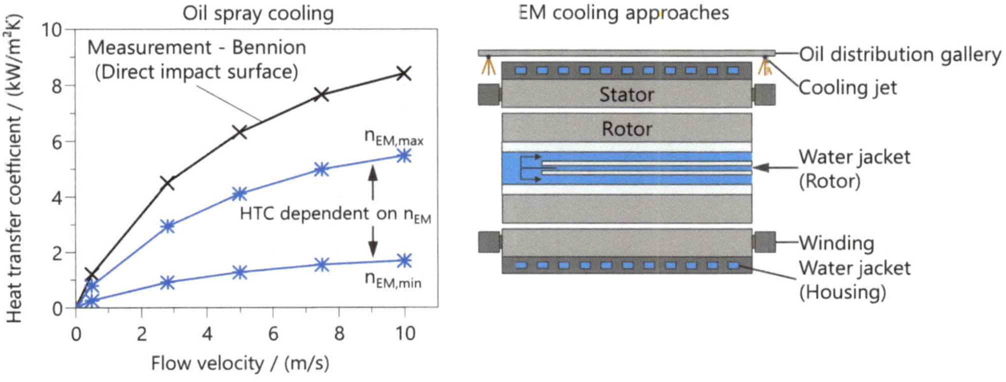

Figure 18. Heat transfer coefficients of the oil spray cooling according to [81] (**left**) and schematic overview of the oil spray cooling of the winding head (**right**).

At low speeds, the proportion of wetted surfaces decreases further due to the lower centrifugal forces [82]. The energy consumption of the electric oil pump and the heat exchange of the oil with the low-temperature cooling circuit (see Figure 5) are considered in the simulation model. The cost of extending the EM cooling system for the additional components, such as the oil pump and oil cooler, is $C_{EM,OC}$ = 78€.

In battery systems, immersion cooling is used as a direct cooling method where the cells are in direct contact with the coolant. The concept used in this work is shown in Figure 19 on the right and uses a dielectric fluid analogous to EM direct cooling.

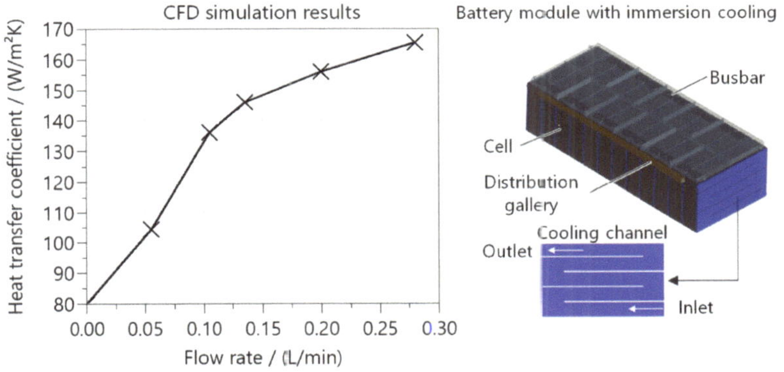

Figure 19. CFD simulation results for determining the water-side heat transfer coefficients (**left**) and structure of the HVB immersion cooling system (**right**).

Heat is transferred through the side surfaces of the battery cells, increasing the heat flow significantly due to the surface area, which is more than five times larger than the bottom surface of the cell. Another advantage is the lower temperature gradient compared to the cooling plate in the baseline design, which can reduce aging effects. Deflections in the cooling channels are used to achieve a uniform heat transfer. The heat transfer coefficients were derived from CFD simulations and are shown in Figure 19 on the left. As expected, the heat transfer coefficient increases with an increasing volume flow. The change to immersion cooling is assumed to be cost neutral. The cost of integrating the cooling channels into the module housing is offset by the elimination of the cooling plate. Minor modifications to the existing cooling system result in costs of $C_{HVB,IC}$ = 30€.

5.1.9. Sodium-Ion Battery Technology

This study investigates sodium-ion (SiB) battery technology as a potential alternative to nickel–manganese–cobalt (NMC) cells. A key advantage of SiB battery technology lies in its significantly lower internal resistance at low temperatures compared to NMC cells. This translates to superior discharge capabilities at cold temperatures, as illustrated in Figure 20, which depicts the scaled maximum power achievable by a battery system employing SiB technology relative to a baseline NMC system. However, the implementation of SiB technology necessitates a significant revision of the existing hybrid control strategy. To fully exploit the potential of SiB technology, the control algorithms were specifically optimized for its unique characteristics. Derived from [83], the costs are estimated using $f_{cost,SiB} = 1.33$ for a given battery size compared to NMC technology. In general, SiB battery technology exhibits a downward cost trend, with projections indicating a continued decrease in coming years. This cost reduction enhances the economic viability of SiB technology for diverse applications. A disadvantage is the additional mass of about $m_{add,Bat,SiB} = 80$ kg compared to the base system, due to the low gravimetric energy density of $w_{Bat,SiB} = 102$ Wh/kg compared to $w_{Bat,NMC} = 132$ Wh/kg for the NMC cell technology [84].

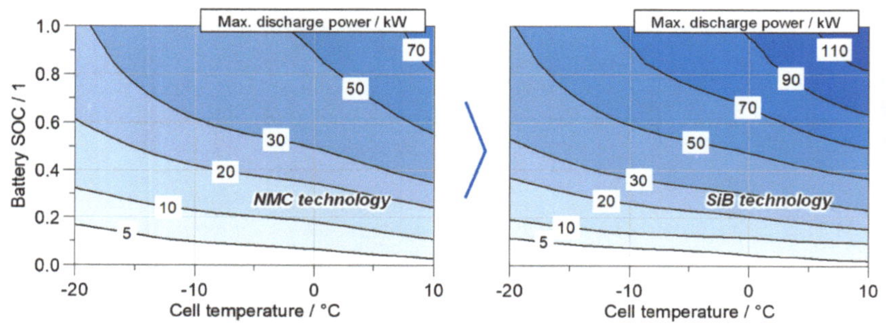

Figure 20. Comparison of maximum available battery system discharging power between SiB and NMC battery cell technology at low ambient and thus cell temperatures.

5.2. Results and Conclusions for Single Technologies

Table 5 shows the results of the single hardware and control strategy adaptations in the selected real-world driving scenarios (see Section 3.1.5). The optimization potentials differ due to the different load profiles of the specific driving scenarios, but generally show the same tendencies. The effects are analyzed and explained using the results of the cross-country driving cycle in cold and hot conditions.

Limiting the maximum engine torque during the TWC heat-up phase reduces pollutant emissions. However, the HC limit is still not met due to the low starting temperature. The lower engine load slows down the engine warm-up, resulting in a slight increase in particulate emissions compared to the baseline. The driving time increases by only $\Delta t_{T,DB} = 11$ s, as the warm-up phase is a small part of the total trip. The impact on CO_2 and noise emissions is correspondingly small.

By shifting the load point on the fuel consumption optimized curve (LPS1) to lower engine speeds and loads, the ICE completely prevents the masking noise from being exceeded. The disadvantage is an increase in the travel time by $\Delta t_{T,LPS1} = 211$ s. The second variant (LPS2) shifts the operating point to the full load curve to continue to meet higher vehicle power requirements. This increases the driving time by only $\Delta t_{T,LPS2} = 17$ s, but also reduces the NVH optimization potential. The higher CO emissions due to the higher engine load are a further drawback. Otherwise, all emissions can be reduced with both variants. However, this is accompanied by a lower SOC at the end of the trip, as less electrical energy is recharged into the HVB in series operation. Accordingly, this adaptation of the operating strategy must be suspended at a very low battery state-of-charge to avoid a complete discharge of the HVB

Table 5. Simulation results of the individual technologies in the overland driving scenario Cold CCDC I (red: worst value of the target variable, green: best value of the target variable).

Driving Scenario		CO_2 g/km	SOC %	Duration s	NVH mJ/m^2	CO [1] mg	HC [1] mg	NO_x [1] mg	PN [1] 1×10^{12}
Baseline WLTC/Legal limits		109/95	-	-	0.028/-	-/6400	-/720	-/480	-/1.60
Cold CCDC I		293.6	15.5	2833	3.791	3463	1608	1037	14.3
ICE torque limitation		292.9	15.5	2844	3.798	2199	1050	439	14.5
Load point shifting	Variant 1	244.3	12.8	3044	0	1647	1171	760	3.44
	Variant 2	298.8	15.3	2850	3.619	6048	1339	792	14.1
Coolant heat storage	w/o pre-heating	291.4	15.5	2833	3.781	2824	1071	1010	13.7
	with pre-heating	291.3	15.5	2857	3.779	3216	907	950	13.6
ICE encapsulation		288.0	15.5	2833	0.004	1987	132	915	12.9
GPF 2nd generation		293.6	15.6	2833	3.790	2426	1329	772	3.16
HVB PTC heater	Variant 1	274.1	18.3	2833	3.471	3361	1645	1034	14.3
	Variant 2	252.0	19.6	2833	3.088	3708	1645	1007	13.6
Electric heated catalyst (EHC)	4 kW w/o pre-heating	294.1	15.6	2833	3.784	3770	1525	683	14.3
	8 kW w/o pre-heating	294.6	15.5	2833	3.790	3439	1472	584	14.4
EHC & secondary air pump	4 kW w/pre-heating	293.8	15.6	2833	3.791	3288	817	550	14.3
	8 kW w/pre-heating	293.9	15.6	2833	3.789	1936	336	158	14.4
Electric combustion chamber heater		294.6	15.0	2833	3.791	3051	368	887	13.5
Sodium-ion battery technology		296.4	14.2	2854	2.779	2480	1447	1249	9.88

[1] Evaluation after $d_{Emi,Clove} = 16$ km.

The starting temperature of the engine with engine encapsulation (EE) is $T_{ICE,Start} = 25$ °C (see Figure 14). The fuel consumption is reduced due to lower engine friction and the earlier use of exhaust gas recirculation. In addition, the thermal energy of the coolant can be used for interior air conditioning, so that the electrical energy consumption of the auxiliary heater is reduced. Furthermore, particulate and HC emissions are reduced due to improved fuel mixture formation [64,85,86]. Since the heat storage must first transfer the thermal energy from the coolant to the combustion chamber and oil by convection, the potential is somewhat lower than with the EE. Pre-conditioning with the electric water pump can increase the potential of the heat storage.

The second-generation coated gasoline particulate filter enables significantly lower particulate emissions thanks to its higher filtration efficiency. The additional coating of the GPF makes it possible to reduce pollutant emissions without compromising the drivability, CO_2, and noise emissions.

The use of battery heating increases the proportion of electric driving, which can reduce the CO_2 and noise emissions of the ICE. In addition, the state-of-charge of the HVB increases at the end of the driving cycle, as the maximum charging power also increases due to the higher cell temperatures. The pollutant emissions decrease only slightly, as the HVB heating process is not yet completed during the cold start. The optimization potential of direct heating (variant 2) is higher, as the heating process can be carried out faster and the benefits of the higher battery power can be used earlier.

Without preheating, the electric heated catalyst shows only a slight improvement in pollutant emissions, with simultaneously higher CO_2 emissions due to the electrical energy required. Pre-conditioning with the secondary air pump prior to the first engine start can significantly increase the potential to achieve the lowest pollutant emissions compared to the other technologies when using two heating disks. Compared to the EHC, the electric combustion chamber heating (ECCH) reduces HC and particulate emissions as soon as they are generated in the combustion chamber. The advantage is that the electrical energ. can be partially compensated by reduced engine friction.

The SiB cell technology enables a strong reduction in tailpipe emissions due to reduced engine loads during the cold start and the catalyst heating phase. In addition, the increased battery power output can reduce the NVH emissions. A disadvantage is the increased CO_2 emissions due to the increased battery system mass. This also results in longer travel times.

To quantify the influence of the HVB charging status and the driver type, the individual technologies are analyzed in the CCDC scenario 'cold CCDC II.' Only those technologies are considered that are also influenced by the above-mentioned variation parameters. The simulation results for defensive driving and a medium state-of-charge are shown in Table 6.

Table 6. Simulation results of the individual technologies in the overland driving scenario Cold CCDC II (red: worst value of the target variable, green: best value of the target variable).

Driving Scenario		CO_2 g/km	SOC %	Duration s	NVH mJ/m^2	CO [1] mg	HC [1] mg	NO_x [1] mg	PN [1] 1×10^{12}
Baseline WLTC/Legal limits		109/95	-	-	0.028/-	-/6400	-/720	-/480	-/1.60
Cold CCDC II		136.8	24.5	4160	0.529	7318	1775	829	2.07
ICE torque limitation		135.3	24.2	4160	0.525	1629	915	232	2.00
Load point shifting	Variant 1	125.6	20.7	4168	0	4417	1557	670	1.61
	Variant 2	151.2	25.7	4160	0.062	9746	1380	449	5.26
TWC heating strategy		139.1	24.5	4160	0.368	2248	1521	780	1.66
HVB heater	Variant 1	106.8	15.7	4160	0.447	7354	1820	1019	1.91
	Variant 2	99.5	15.4	4160	0.376	7643	1819	896	1.82
Sodium-ion cell technology		134.1	17.4	4158	0.471	2386	1325	124	2.32

[1] Evaluation after $d_{Emi,Clove}$ = 16 km.

The high battery state-of-charge also increases the proportion of electric driving. In combination with the lower driving performance requirements due to the defensive driver, this leads to improved results for the target variables compared to the Cold CCDC I scenario. The potential for battery heating also increases as more electrical energy is available due to the higher state-of-charge.

In this scenario, the cold start of the ICE during the driving cycle leads to high raw emissions or to emission breakthrough in the cold TWC. Accordingly, pollutant emissions can be significantly reduced by using torque limitation, as in the previous scenario. In addition, a reduction in particulate emissions can also be observed here, underlining the high sensitivity of the boundary conditions to the results. A reduction in drivability can be avoided due to the higher battery charge level. This can also be seen from the slight increase in driving time ($\Delta t_{T,LPS1}$ = 8 s) for the first LPS variant.

The adaptation of the TWC heating strategy enables thermal preconditioning of the TWC directly at the beginning of the trip. This results in a significant reduction in pollutant emissions, as the TWC has already reached the light-off temperature when the engine starts unexpectedly. However, this is accompanied by an increase in CO_2 emissions. Predictive control strategies offer a way to further optimize this interaction.

Thanks to the higher HVB state-of-charge, the SiB cell technology reaches its full optimization potential without an increase in travel time. A reduction in CO_2 emissions is possible because more electrical energy is available to propel the vehicle. This also leads to a slight optimization of the NVH emissions. Except for the PN, an even higher reduction in tailpipe emissions is observed. The PN emissions increase due to non-optimal engine operating points, lower engine temperatures, and, thus, increased PN raw emissions.

In the previous chapter, the optimization measures for cold country driving scenarios were evaluated. This section analyzes individual technologies for the CCDC at T_{amb} = 40 °C. The focus is on the investigation of adjustments to the control-based operating strategy, as well as high-performance cooling methods to avoid thermal derating of the electrical components (HVB and EM). The results for the "hot country" CCDC scenario are shown in Table 7.

Table 7. Simulation results of the individual technologies in the cross-country driving scenario Hot Country (red: worst value of the target variable, green: best value of the target variable).

Driving Scenario	CO_2 g/km	SOC %	Duration s	NVH mJ/m^2	CO [1] mg	HC [1] mg	NO_x [1] mg	PN [1] 1×10^{12}
Baseline WLTC/Legal limits	109/95	-	-	0.028/-	-/6400	-/720	-/480	-/1.60
Hot Country II	54.8	34.8	2789	0.630	8494	1328	287	7.42
ICE torque limitation	56.4	35.5	2792	0.531	1602	163	100	5.44
TWC heating strategy	47.4	33.2	2793	0	7217	917	246	5.94
HVB direct cooling	47.4	28.6	2784	0.303	7749	1309	280	4.61
EM direct cooling	54.8	35.5	2788	0.614	8618	1329	286	7.38
HVB and EM direct cooling	44.6	28.5	2784	0.363	7749	1309	280	4.61

[1] Evaluation after $d_{Emi,Clove}$ = 16 km.

Similar to the Cold CCDC II scenario, unexpected engine starts, in this case triggered by thermal derating of the EM or HVB, lead to high tailpipe emissions. Therefore, as with torque limitation, an appropriate operating strategy adjustment is required to reduce fuel and particulate emissions. The first variant of the LPS also shows the same trend as the previous scenarios. The very low drivability restrictions of $\Delta t_{T,DB}$ = 3 s and $\Delta t_{T,LPS1}$ = 4 s should be highlighted positively.

In this driving scenario, the HVB output power is thermally regulated so that direct cooling of the cells can reduce both CO_2 and noise emissions. The reduction in pollutants is small, so the TWC heating strategy must be adjusted. In addition to the HVB, the EM also reaches limit temperatures, although the optimization potential of oil spray cooling as a single measure is low. A further reduction in CO_2 emissions is possible by combining the two direct cooling methods.

5.3. Developed Technology Packages Based on Single Technology Evaluation

The previous analysis of the individual technologies shows that a combination of different technologies is necessary for an optimal overall vehicle design. Table 8 provides an overview of the selected technology packages and the associated total costs.

In addition to the legal requirements, the evaluation of these methods can also consider other constraints such as the manufacturer's image, different customer segments, and the country of sale. The cost-neutral package aims to fulfill the minimum legal requirements. It is therefore suitable for low-end users (see target customer classification [87]), as only small additional costs are to be expected. According to Brokate et al. [88], the target customer defined here has a higher willingness to pay due to his social status and value orientation, which means that he is particularly open to new drive technologies. For this reason, the hot water tank, the electric heating disk with a secondary air pump, and the battery heating element (HVB-PTC 1) are also used in the cold country package. To avoid thermal shutdown of the electrical components, direct cooling of the battery and electric motor is part of the hot country package. Compared to the target customer, the willingness to pay increases again in the premium customer segment. This makes it possible to use the more expensive battery heater (HVB-PTC 2) and engine encapsulation. The SiB cell technology is not part of this package, due to its low benefit–cost ratio (delta costs of 2887€). Nevertheless, the use of SiB battery cells may be suitable for high-performance sports cars that do not allow any drivability restrictions.

Table 8. Overview of technology packages and the differential costs compared to the baseline powertrain for different customer preferences.

Cost Neutral	Cold Country	Hot Country	Premium
• GPF 2nd generation • Load point shifting • ICE torque limitation	• GPF 2nd generation • Load point shifting • ICE torque limitation • Coolant heat storage • Secondary air pump & electric heated catalyst (4 kW) • HVB heater (PTC 1)	• GPF 2nd generation • Load point shifting • ICE torque limitation • EM and HVB direct cooling	• GPF 2nd generation • Load point shifting • ICE torque limitation • Engine encapsulation • Secondary air pump & electric heated catalyst (4 kW) • HVB heater (PTC 2) • EM and HVB direct cooling
0€	273€	108€	437€

5.4. Results and Conclusions for Technology Packages

The corresponding simulation results of the individual technology packages for the Cold CCDC I driving scenario are shown in Table 9. Together with an increase in travel time of $\Delta t_{T,CNP}$ = 200 s, the cost-neutral package results in optimized CO_2, noise, and pollutant emissions. Except for the HC emissions, the regulatory limits are met. It should be noted that no conformity factors for extended driving conditions, e.g., extreme ambient temperatures, were considered in this paper. In combination with the heat storage or engine encapsulation, the electric heating disk enables a further reduction in exhaust emissions. For the premium package, it should be emphasized that the torque limitation during cold starts can be reduced by a thermal preconditioning of the TWC. Furthermore, the additional use of the LPS leads to an optimization of the ICE operating range, whereby the drivability restrictions (see cost-neutral package) can be avoided by the battery heater.

Table 9. Simulation results of the technology packages for the overland driving scenarios Cold CCDC I & II (red: worst value of the target variable, green: best value of the target variable).

Driving Scenario	CO_2 g/km	SOC %	Duration s	NVH mJ/m^2	CO [1] mg	HC [1] mg	NOx [1] mg	PN [1] 1×10^{12}
Baseline WLTC/Legal limits	109/95	-	-	0.028/-	-/6400	-/720	-/480	-/1.60
Cold CCDC I	293.6	15.5	2833	3.791	3463	1608	1037	14.33
Cost neutral	243.4	12.8	3053	0	1163	960	187	0.96
Cold country with 30 s pre-heating	229.4	13.8	2833	0	1082	448	169	0.91
Premium with 30 s pre-heating	210.1	15.3	2833	0	957	134	158	0.81
Cold CCDC II	136.8	24.5	4160	0.529	7318	1775	829	2.07
Cost neutral	125.2	20.6	4168	0	1130	897	109	0.40
Cold country with 30 s pre-heating	100.8	13.7	4161	0	754	270	33	0.34
Premium with 30 s pre-heating	93.3	13.9	4161	0	745	52	22	0.22

[1] Evaluation after $d_{Emi,Clove}$ = 16 km.

Analogous to the results of the individual technologies, a higher optimization potential can be observed for the Cold CCDC II scenario due to the higher state-of-charge and the defensive driving style. The significant drivability restriction from the first scenario is only $\Delta t_{T,CNP}$ = 6 s for the cost-neutral package. Otherwise, the packages show the same trend, as already explained for the Cold CCDC I scenario.

The results of the hot country driving scenario are shown in Table 10. The cost-neutral package shows a similar behavior to that in the Cold CCDC II scenario, as the ICE performance restriction can be compensated by the HVB if the state-of-charge is sufficiently high. Due to the already high starting temperatures of T_{amb} = 40 °C, all emissions are below the available budgets. A further improvement in CO_2 emissions is possible through the combination of improved cooling and LPS compared to the consideration of individual technologies. The electric heating disk in the premium package is only activated when the engine is started for the first time. Preheating the TWC with a secondary air pump is only possible in this scenario if the engine start can be predicted. Although the pollutant emissions can be further reduced compared to the other packages, the general level was already very low beforehand. Therefore, it can be concluded that the benefit of the electric heating disk is not very high in hot regions.

Table 10. Simulation results of the technology packages for the overland driving scenario Hot Country (red: worst value of the target variable, green: best value of the target variable).

Driving Scenario	CO_2 g/km	SOC %	Duration s	NVH mJ/m^2	CO [1] mg	HC [1] mg	NO_x [1] mg	PN [1] 1×10^{12}
Baseline WLTC/Legal limits	109/95	-	-	0.028/-	-/6400	-/720	-/480	-/1.60
Hot Country	54.8	34.8	2789	0.630	8494	1328	287	7.42
Cost neutral	48.7	33.5	2793	0	1825	146	97	1.07
Hot country with 30 s pre-heating	38.1	26.5	2785	0	1070	135	91	0.41
Premium with 30 s pre-heating	37.0	26.1	2785	0	604	44	3	0.27

[1] Evaluation after $d_{Emi,Clove}$ = 16 km.

In general, the methodology for virtually testing the real-world driving scenarios was successfully demonstrated in several scenarios for the combined hybrid and the described technical measures. The results indicate that the identified limitations can be mitigated by technical measures and controlled strategy adjustments. In general, the results show that a second generation GPF should always be used in the future to comply with emission limits, as the filtration efficiency increases significantly at the same cost. Concerning the other customer requirements, such as acoustic comfort, CO_2, and pollutant emissions or drivability, a choice can be made between a cost-neutral adaptation of the operating strategy and the use of additional, cost-intensive technologies, depending on the target customer and sales market.

The flexibility of the hybrid powertrain has been optimized and adapted to the needs of the target customer. However, the optimization potential of technologies such as the hot water tank or the electric heated catalyst also depends on the preheating time. Without preheating, for example, the higher thermal inertia results in an emission disadvantage compared to the baseline exhaust aftertreatment system. Intelligent control strategies that, for example, initiate the heating process as soon as the vehicle is unlocked are therefore particularly important for these technologies. In addition, the use of predictive strategies for the thermal conditioning of the catalyst or the combustion engine makes sense when the ICE is switched on after a long electric drive.

6. Summary and Conclusions

To cope with the diversity of requirements in the automotive development landscape, a generic and holistic methodical systems engineering approach was developed in the FVV project *HyFlex-ICE* which can be used to derive a comprehensive catalog of requirements for vehicles. This methodology was demonstrated using the example of a combined plug-in hybrid powertrain. The detailed outcome of the requirement management and the test cases for the vehicle baseline design can be found in the final report of the research project [34]. Using these test cases and evaluation criteria, the baseline powertrain design and operating strategy were derived from the vehicle model developed using statistical test planning in terms of performance, efficiency, cost, and NVH.

The next step was to identify the limitations of the baseline design in real-world driving scenarios. Four real-world driving scenarios with different driving profiles and environmental conditions were derived. In addition, variation parameters such as load, air conditioning, traffic, battery charging state, and driver characteristics were considered. For the evaluation, detailed thermal management and emissions models were coupled with the vehicle model. From a thermal management perspective, driving scenarios with a sporty driving style, a high traffic volume, and extreme temperatures proved to be particularly challenging. In cold climates, a low battery and low engine start temperatures, as well as exhaust aftertreatment, resulted in exceeding the EU7 emission budget and high CO_2 and noise emissions. At high ambient temperatures, the thermal control of electrical components resulted in undesired engine starts, which also led to a deterioration in target values. By identifying the thermal interactions and challenges, the first research question was answered.

Model predictive control (MPC) strategies coupled with advanced hardware adaptations were investigated to optimize and mitigate previously identified limitations. This approach successfully addressed research question two.

The MPC-based energy management system (EMS) offers an adaptable methodology for extending operational targets with minimal computational overhead, enabling the flexible integration of additional considerations. While data-driven prediction is currently envisioned as a prerequisite, the system has been demonstrably effective in multi-criteria optimization for at least two objectives. Beyond reductions in CO_2 emissions and NVH, the MPC strategy achieved significant reductions in total energy consumption and pollutant emissions through operating mode adjustments.

These findings are generally extensible to non-design cycles. However, optimal performance was observed when the MPC operated in an unconstrained ("free oscillating") state. When transitioning to a physical vehicle, the driver influence becomes a significant and inherently unpredictable disturbance factor. We posit that the combination of predictive operating strategies with automated driving holds promise for achieving synergistic effects.

For further optimization, adjustments to the rule-based operating strategy and additional technologies were first examined individually. The results showed significant improvements, but it was not possible to optimize all target variables simultaneously. As a result, four technology packages were defined for different target customers and markets. The COST-NEUTRAL PACKAGE makes it possible to meet all legal requirements simply by adjusting the operating strategy and using a latest-generation gasoline particulate filter. The COLD COUNTRY PACKAGE accelerates the warm-up of key powertrain components such as the HV battery, combustion engine, and catalyst. The HOT COUNTRY PACKAGE includes direct cooling for the electric motor and the HV battery to prevent thermal derating of these components through optimized cooling. The Premium package combines the technologies of the COLD *and* HOT COUNTRY PACKAGES. Instead of the coolant heat storage, the more expensive engine encapsulation is suitable, as CO_2, pollutant, and noise emissions can be optimized equally at a slightly higher cost.

This study presents a holistic methodology for optimal powertrain system design during the earliest development stages. This framework serves as a blueprint for future powertrain development, enabling the early identification and mitigation of limitations encountered in real-world driving scenarios.

The methodology incorporates a novel approach for elaborating limitations arising from real-world driving scenarios, as successfully demonstrated for the combined hybrid powertrain. This approach revealed discrepancies between standardized test procedures (e.g., WLTC) and real-world driving behavior, particularly under extreme operating conditions. Importantly, the methodology facilitates the development of universally applicable solutions through the implementation of targeted technical measures and control system adjustments. These solutions effectively address and mitigate the identified limitations, ultimately enhancing the flexibility of the hybrid powertrain and tailoring its performance to meet specific customer demands.

Author Contributions: Conceptualization, J.K. and J.M.; methodology, J.K. and J.M.; software, J.K., J.M., F.H. and P.H.; validation, J.K., J.M., F.H. and P.H.; formal analysis, J.K.; investigation, J.K., J.M., F.H. and P.H.; resources, M.G. and S.P.; data curation, J.M.; writing—original draft preparation, J.K., J.M. and F.H.; writing—review and editing, M.G. and S.P.; visualization, J.K., J.M., F.H. and P.H.;

supervision, S.P.; project administration, M.G.; funding acquisition, M.G. and S.P. All authors have read and agreed to the published version of the manuscript.

Funding: Parts of this research were funded by FVV eV, project number 1433 "Highly-flexible internal combustion engines for hybrid vehicles" (HyFlex-ICE).

Data Availability Statement: Most relevant data generated or analyzed during this study are included in this published article. Additional data are partly available on request from the corresponding author with the permission of the FVV eV and their members.

Acknowledgments: We thank Martin Nitsche (FVV eV) and Marc Sens (IAV GmbH) as well as the FVV working group for the useful discussions, ideas, and support while conducting the corresponding research project.

Conflicts of Interest: The authors declare no conflicts of interest.

References

1. Gottorf, S.; Fryjan, J.; Leyens, L.; Picerno, M.; Habermann, K.; Pischinger, S. Lean Approach for Virtual Calibration Using Hardware-in-the-Loop and Electronic Control Unit (ECU)-Capable Engine Simulation. *SAE Int. J. Engines* **2021**, *14*, 531–542. [CrossRef]
2. Andert, J.; Xia, F.; Klein, S.; Savelsberg, R.; Guse, D.; Tharmakulasingam, R.; Thewes, M.; Scharf, J. Road-to-Rig-to-Desktop—Virtual Development Using Real-Time Engine Modeling and Powertrain-Co-Simulation. *COMODIA* **2017**, *2017*, A108. [CrossRef]
3. Pischinger, S.; Seiffert, U. *Vieweg Handbuch Kraftfahrzeugtechnik*; Springer Fachmedien Wiesbaden: Wiesbaden, Germany, 2021; ISBN 978-3-658-25556-5.
4. Schäfer, S. *Modellbasierte Steuerung des Kühlkreislaufes einer Brennstoffzelle mit Automatisiertem Test der Software*; VDI-Verlag: Düsseldorf, Germany, 2012.
5. Platner, S.; Kordon, M.; Fakiolas, E.; Atzler, H. Modellbasierte Serien-kalibrierung—Der effiziente Weg für Variantenentwicklung. *MTZ Mot. Z.* **2013**, *74*, 754–761. [CrossRef]
6. Basshuysen, R.; Schäfer, F. *Handbuch Verbrennungsmotor: Grundlagen, Komponenten, Systeme, Perspektiven*, 8th ed.; Springer Vieweg: Wiesbaden, Germany, 2017; ISBN 978-3-658-10902-8.
7. Seibel, J. Optimierte Auslegung von Ottomotoren in Hybrid-Antriebsträngen. Ph.D. Thesis, RWTH Aachen University, Aachen, Germany, 2008.
8. Balazs, A. Optimierte Auslegung von Ottomotorischen Hybridantriebssträngen unter Realen Fahrbedingungen. Ph.D. Thesis, RWTH Aachen University, Aachen, Germany, 2015.
9. Böhmer, M. Simulation der Abgasemissionen von Hybridfahrzeugen für Reale Fahrbedingungen. Ph.D. Thesis, RWTH Aachen University, Aachen, Germany, 2017.
10. Ao, G.-Q.; Qiang, J.-X.; Zhong, H.; Mao, X.-J.; Yang, L.; Zhuo, B. Fuel economy and NO x emission potential investigation and trade-off of a hybrid electric vehicle based on dynamic programming. *Proc. Inst. Mech. Eng. Part D J. Automob. Eng.* **2008**, *222*, 1851–1864. [CrossRef]
11. Chen, Z.; Mi, C.C. An adaptive online energy management controller for power-split HEV based on Dynamic Programming and fuzzy logic. In Proceedings of the VPPC '09, 2009 IEEE Vehicle Power and Propulsion Conference (VPPC), Dearborn, MI, USA, 7–10 September 2009; IEEE: Piscataway, NJ, USA, 2009; pp. 335–339, ISBN 978-1-4244-2600-3.
12. Dextreit, C.; Assadian, F.; Kolmanovsky, I.V.; Mahtani, J.; Burnham, K. Hybrid Electric Vehicle Energy Management Using Game Theory. In Proceedings of the SAE Technical Paper Series, SAE World Congress & Exhibition, SAE International400 Commonwealth Drive, Warrendale, PA, USA, 14 April 2008.
13. Jeon, S.; Jo, S.; Park, Y.; Lee, J. Multi-Mode Driving Control of a Parallel Hybrid Electric Vehicle Using Driving Pattern Recognition. *J. Dyn. Syst. Meas. Control* **2002**, *124*, 141–149. [CrossRef]
14. Danzer, C.; Voigt, T.; Forell, A.; Schreiterer, E.; Kockisch, R.; Sens, M.; Schneider, E. System simulation and visualization of powertrain and mobility concepts. In *Experten-Forum Powertrain: Simulation und Test 2020*; Springer Vieweg: Berlin/Heidelberg, Germany, 2021; pp. 1–15.
15. Teuschl, G.; Ebner, P.; Huss, A.; de Kerpel, N. Simulationsgestütze systemische Entwicklung hybrider Antriebssysteme. *MTZ Mot. Z.* **2021**, *82*, 78–83. [CrossRef]
16. Häußler, L. Generisches Fahrzeug Energiemanagement. Ph.D. Thesis, RWTH Aachen University, Aachen, Germany, 2021.
17. Weiß, F. Optimale Konzeptauslegung elektrifizierter Fahrzeugantriebsstränge. Ph.D. Thesis, Technische Universität Chemnitz, Chemnitz, Germany, 2017. [CrossRef]
18. Maurer, R.; Kossioris, T.; Sterlepper, S.; Günther, M.; Pischinger, S. *Zero-Impact Tailpipe Emission Powertrains: Identify Technical Solutions to Achieve Powertrains with Zero-Impact Tailpipe Emissions under Consideration of a 2030– Legislative Vehicle and Air Pollutant Scenario*; Final Report 1338; FVV e.V.: Frankfurt am Main, Germany, 2023.
19. Pischinger, S.; Genender, P.; Klopstein, S.; Hemkemeyer, D. Aufgaben beim Thermomanagement von Hybrid- und Elektrofahrzeugen. *ATZ Automob. Z.* **2014**, *116*, 54–59. [CrossRef]

20. Uhlmann, T.; Balazs, A.; Lückmann, D.; Müller, A.; Thewes, M.; Sahr, C.; Pischinger, J.; Hellenbroich, G.; Herold, K.L.; Lüdiger, T. High Efficient Gasoline HEV Meeting 2030 CO_2 Targets—The Road towards 59 g/km Fleet CO_2. In Proceedings of the 29th Aachen Colloquium Sustainable Mobility, Aachen, Germany, 5–7 October 2020; Aachener Kolloquium Fahrzeug- und Motorentechnik GbR: Aachen, Germany, 2020. ISBN 978-3-00-064871-7.
21. Eichlseder, W.; Hager, J.; Raup, M.; Dietz, S. Auslegung von Kuehlsystemen Mittels Simulationsrechnung. *Automob. Z.* **1997**, *99*, 638–647.
22. Knauf, B.; Pantow, E. Auslegung eines Kühlsystems mit elektrischer Kühlmittelpumpe. *MTZ Mot. Z.* **2005**, *66*, 878–884. [CrossRef]
23. Eilemann, A. Thermomanagement—Eine Systemaufgabe. *ATZ Automob. Z.* **2015**, *117*, 86. [CrossRef]
24. Banjac, T.; Wurzenberger, J.C.; Katrašnik, T. Assessment of engine thermal management through advanced system engineering modeling. *Adv. Eng. Softw.* **2014**, *71*, 19–33. [CrossRef]
25. Setlur, P.; Wagner, J.R.; Dawson, D.M.; Marotta, E. An Advanced Engine Thermal Management System: Nonlinear Control and Test. *IEEE/ASME Trans. Mechatron.* **2005**, *10*, 210–220. [CrossRef]
26. Lu, L.; Chen, H.; Hu, Y.; Gong, X.; Zhao, Z. Modeling and Optimization Control for an Engine Electrified Cooling System to Minimize Fuel Consumption. *IEEE Access* **2019**, *7*, 72914–72927. [CrossRef]
27. Weiss, M.; Tuncay, V.; Richter, S.; Broz, J. Umfassende Simulation und vernetzte Intelligenz im Thermomanagement. *MTZ Mot. Z.* **2017**, *78*, 42–49. [CrossRef]
28. Kang, H.; Ahn, H.; Min, K. Smart cooling system of the double loop coolant structure with engine thermal management modeling. *Appl. Therm. Eng.* **2015**, *79*, 124–131. [CrossRef]
29. Appel, N.; Brinker, M.; Soetje, H.; Müller, T. Aktives Antriebsstrang-Thermomanagement beim Opel Insignia. *MTZ Mot. Z.* **2020**, *81*, 60–65. [CrossRef]
30. Beste, F. *Thermisch-Elektrische Analyse und Ableitung von Verbesserungsmaßnahmen für den Elektrischen Antrieb eines Plug-In-Hybridfahrzeugs*, 1st ed.; Cuvillier Verlag: Göttingen, Germany, 2016; ISBN 9783736983021.
31. Shutty, J.; Bongards, A.; Kondipati, N.; Ristevski, S. Thermomanagement bei elektrifizierten Antriebssystemen. *ATZ Automob. Z.* **2022**, *124*, 38–43. [CrossRef]
32. Müller, J.; Maurer, R.; Achenbach, J.; Balazs, A.; Knauf, J. Antriebsstrangoptimierung von Hybridsystemen unter Berücksichtigung thermischer Einzelkomponentenwirkungsgrade. In *Experten-Forum Powertrain: Reibung in Antrieb und Fahrzeug 2020*; Liebl, J., Ed.; Springer Vieweg: Berlin/Heidelberg, Germany, 2020; pp. 179–201.
33. Ernstberger, U.; Weissinger, J.; Frank, J. (Eds.) *Mercedes-Benz SL: Entwicklung und Technik*; Springer Fachmedien Wiesbaden: Wiesbaden, Germany, 2013; ISBN 978-3-658-00799-7.
34. Kexel, J.; Müller, J. *HyFlex-ICE: Highly Flexible Internal Combustion Engines for Hybrid Vehicles*; Final Report 1338; FVV: Frankfurt am Main, Germany, 2023.
35. Hick, H.; Küpper, K.; Sorger, H. *Systems Engineering for Automotive Powertrain Development*, 1st ed.; Springer International Publishing: Cham, Switzerland, 2021; ISBN 978-3-319-99628-8.
36. Kexel, J.; Müller, J.; Günther, M.; Pischinger, S. HyFlex-ICE: Highly Flexible Internal Combustion Engines for Hybrid Vehicles. In Proceedings of the International Stuttgart Symposium, Stuttgart, Germany, 4–5 July 2023; Kulzer, A.-C., Reuss, H.-C., Wagner, A., Eds.; Springer Vieweg: Wiesbaden, Germany, 2023; pp. 243–272.
37. Maiterth, J.M. Gesamtkosten- und Emissionsoptimierte Systemauslegung von Nutzfahrzeug-Hybridantriebssträngen. Ph.D. Thesis, RWTH Aachen University, Aachen, Germany, 2022.
38. Granrath, C.; Kugler, C.; Silberg, S.; Meyer, M.-A.; Orth, P.; Richenhagen, J.; Andert, J. Feature-driven systems engineering procedure for standardized product-line development. *Syst. Eng.* **2021**, *24*, 456–479. [CrossRef]
39. Gräßler, I.; Oleff, C. *Systems Engineering: Verstehen und Industriell Umsetzen*; Springer Vieweg: Berlin, Germany, 2022.
40. Werra, M.; Sturm, A.; Küçükay, F. Optimal and prototype dimensioning of 48V P0+P4 hybrid drivetrains. *Automot. Engine Technol.* **2020**, *5*, 173–186. [CrossRef]
41. Maurer, R.; Kossioris, T.; Sterlepper, S.; Günther, M.; Pischinger, S. Achieving Zero-Impact Emissions with a Gasoline Passenger Car. *Atmosphere* **2023**, *14*, 313. [CrossRef]
42. Balazs, A. *Untersuchung zur Optimierten Auslegung von Abschlussbericht Hybridantriebsträngen unter Realen Fahrbedingungen (F1011)*; FVV-Informationstagung: Frankfurt am Main, Germany, 2013.
43. Seibel, J.; Pischinger, S. *Abschlussbericht zum Vorhaben Untersuchung zur Optimierten Auslegung von Ottomotoren in Hybrid-Antriebsträngen (F863)*; Informationstagung Motoren, Heft R 537; FVV: Frankfurt am Main, Germany, 2007.
44. Vaillant, M. *Design Space Exploration zur Multikriteriellen Optimierung Elektrischer Sportwagenantriebsstränge*; KIT Scientific Publishing: Karlsruhe, Germany, 2015; ISBN 978-3-7315-0452-8.
45. Siebertz, K.; van Bebber, D.; Hochkirchen, T. *Statistische Versuchsplanung*; Springer: Berlin/Heidelberg, Germany, 2017; ISBN 978-3-662-55742-6.
46. FEV Software and Testing Solutions. Doe Software with Global Map Optimization Gaussian Process Model. 2022. Available online: https://www.fev-sts.com/fileadmin/user_upload/STS/Brochure-Catalog_2022/FEV-STS_Brochure_xCAL_2022.pdf (accessed on 9 November 2022).
47. Fröhlingsdorf, K.; Doleschal, F. *Final Report: Characteristics for Quantifying the Annoyance of Interference Noise in the Compartment of Vehicles with an Electric Drive System (Interference Noise in the Vehicle Compartment with Electrified Drives), FVV 1369*; FVV Transfer + Networking Event R603; FVV: Würzburg, Germany, 2022.

48. Fröhlingsdorf, K.; Doleschal, F. *Interference Noise in the Vehicle Compartment with Electrified Drives: Characteristics for Quantifying the Annoyance of Interference Noise in the Compartment of Vehicles with an Electric Drive System*; Final Report 1327; FVV: Frankfurt am Main, Germany, 2023.
49. Müller, J.; Besser, N.; Hermsen, P.; Pischinger, S.; Knauf, J.; Bagherzade, P.; Fryjan, J.; Balazs, A.; Gottorf, S. Virtual Development of Advanced Thermal Management Functions Using Model-in-the-Loop Applications. *Energies* **2023**, *16*, 3238. [CrossRef]
50. Müller, J.; Balazs, A.; Knauf, J.; Gottorf, S.; Besser, N.; Fryjan, J. *Advanced Thermal Management Models for X-in-the-Loop Applications—A Gamechanger for Control Unit Function Development*; SAE Thermal Management Systems; SAE: Warrendale, PA, USA, 2021.
51. Fanger, P.O. *Thermal Comfort: Analysis and Applications in Environmental Engineering*; Reprint; Krieger: Malabar, FL, USA, 1982; ISBN 0898744466.
52. ASHRAE. *ANSI/ASHRAE Addendum d to ANSI/ASHRAE Standard 55-2017: Thermal Environmental Conditions for Human Occupancy*; ASHRAE: Peachtree Corners, GA, USA, 2017.
53. Karrar, E. Untersuchungen zum Reibungsverhalten und Ölverbrauch der Kolbengruppe von Verbrennungsmotoren: Lehrstuhl für Verbrennungskraftmaschinen und Institut für Thermodynamik. Ph.D. Thesis, RWTH Aachen, Aachen, Germany, 2009.
54. Baumgarten, H.; Scharf, J.; Thewes, M.; Uhlmann, T. Balazs, A.; Böhmer, M. Simulation-Based Development Methodology for Future Emission Legislation. In Proceedings of the 37th International Vienna Motor Symposium, Vienna, Austria, 28–29 April 2016; Lenz, H.P., Ed.; Fortschritt-Berichte VDI: Düsseldorf, Germany, 2016. ISBN 978-3-18-379912-1.
55. Maurer, R.; Yadla, S.K.; Balazs, A.; Thewes, M.; Walter, V.; Uhlmann, T. Designing Zero Impact Emission Vehicle Concepts. In *Experten-Forum Powertrain: Reibung in Antrieb und Fahrzeug*; Liebl, J., Ed.; Springer Vieweg: Berlin/Heidelberg, Germany, 2020; pp. 75–116.
56. Thewes, M.; Balazs, A.; Yadla, S.K.; Walter, V.; Görgen, M.; Scharf, J.; Sterlepper, S.; Voßhall, T. Zero-Impact combustion engine. In Proceedings of the 28th Aachen Colloquium Sustainable Mobility, Aachen, Germany, 7–9 October 2019; Aachener Kolloquium Fahrzeug- und Motorentechnik GbR: Aachen, Germany, 2019. ISBN 978-3-00-060311-2.
57. Kexel, J.; Müller, J.; Günther, M.; Pischinger, S. Real-World Driving Scenarios for Optimal System Design. *Mendeley Data*, 2024. [CrossRef]
58. Schyr, C.; Spreitzer, H. Digitaler Streckenatlas für die alpine Antriebsstrangerprobung. *Automot. Eng. Partn.* **2004**, *1*, 44–47.
59. Gesundheitsberichterstattung des Bundes. Körpermaße der Bevölkerung Available online: https://www.gbe-bund.de/gbe/pkg_isgbe5.prc_menu_olap?p_uid=gast&p_aid=47463195&p_sprache=D&p_help=0&p_indnr=223&p_indsp=&p_ityp=H&p_fid= (accessed on 18 October 2023).
60. Hopp. *Thermomanagement von Hochleistungsfahrzeug-Traktionsbatterien anhand Gekoppelter Simulationsmodelle*; Springer Fachmedien Wiesbaden: Wiesbaden, Germany, 2016; ISBN 978-3-658-14246-9.
61. Korthauer, R. (Ed.) *Handbuch Lithium-Ionen-Batterien*; Springer: Berlin/Heidelberg, Germany, 2013; ISBN 978-3-642-30653-2.
62. Küpfmüller, K.; Kohn, G. *Theoretische Elektrotechnik und Elektronik: Eine Einführung*, 15th ed.; Springer: Berlin/Heidelberg, Germany, 2000; ISBN 978-3-540-67794-9.
63. Reif, K. *Ottomotor-Management: Steuerung, Regelung und Überwachung*, 4th ed.; Springer Vieweg: Wiesbaden, Germany, 2014; ISBN 978-3-8348-1416-6.
64. Etikyala, S.; Dahlander, P. Soot Sources in Warm-Up Conditions in a GDI Engine. SAE Technical Paper 2021-01-0622. 2021. Available online: https://www.sae.org/publications/technical-papers/content/2021-01-0622/ (accessed on 25 October 2023).
65. Sun, C.; Moura, S.J.; Hu, X.; Hedrick, J.K.; Sun, F. Dynamic Traffic Feedback Data Enabled Energy Management in Plug-in Hybrid Electric Vehicles. *IEEE Trans. Contr. Syst. Technol.* **2015**, *23*, 1075–1086. [CrossRef]
66. Fleming, P.J.; Pursehouse, R.C. Genetic Algorithms in Control Systems Engineering. *IFAC Proc. Vol.* **1993**, *26*, 605–612. [CrossRef]
67. Du, X.; Htet, K.K.K.; Tan, K.K. Development of a Genetic-Algorithm-Based Nonlinear Model Predictive Control Scheme on Velocity and Steering of Autonomous Vehicles. *IEEE Trans. Ind. Electron.* **2016**, *63*, 6970–6977. [CrossRef]
68. Nobis, C.; Kuhnimhof, T. *Mobilität in Deutschland—MiD. Ergebnisbericht: Eine Studie von Infas, DLR, IVT und Infas 360 im Auftrag des Bundesministers für Verkehr und Digitale Infrastruktur (FE-Nr. 70.904/15)*; Bundesministerium für Verkehr und digitale Infrastruktur: Bonn, Berlin, 2018.
69. Kraft, W.; Altstedde, M.K. Use of metallic Phase Change Materials (mPCM) for heat storage in Electric- and Hybrid Vehicles. In Proceedings of the 6th Hybrid and Electric Vehicles Conference (HEVC 2016), London, UK, 2–3 November 2016; Institution of Engineering and Technology: London, UK, 2016. ISBN 978-1-78561-294-7.
70. Lichius, T. Latentwärmespeicher in PlugIn-Hybridfahrzeugen. Ph.D. Thesis, RWTH Aachen University, Aachen, Germany, 2017.
71. Spengler, R.; Essers, U. Einsatzmöglichkeiten von Kunststoff-Teilkapseln an Verbrennungsmotoren. *MTZ Mot. Z.* **1999**, *60*, 164–170. [CrossRef]
72. Röchling Automotive SE & Co. KG. Leichte Kapselung hält Motor länger auf Betriebstemperatur. *Lightweight Des.* **2013**, *6*, 8. [CrossRef]
73. Moos, E. Kostenabschätzung für eine Motorkapselung der Firma Röchling. E-Mail. 2022.
74. Warner, B. High Voltage Coolant Heater.
75. ATT Advanced Thermal Technologies. Battery Heating Solutions. Available online: https://www.thermaltech.at/battery-heating/ (accessed on 25 October 2023).
76. Frenzelit GmbH. Revolutionäre Heizfolien und Heizsysteme aus Innovativem Verbundmaterial. Available online: https://www.frenzelit.com/produkte/hicotec/hicotec-tp-thermal-power (accessed on 25 October 2023).

77. Flextem GmbH. PTC—Heiztechnologie. Available online: https://www.flextem.de/de/folienheizungen/ptc-heizfolien.html (accessed on 25 October 2023).
78. Rose, D.; Boger, T.; Wu, H.; Ingram-Ogunwumi, R. Neue Generation Otto-Partikelfilter. *MTZ Mot. Z.* **2021**, *82*, 60–64. [CrossRef]
79. Rolf Brück. Niedrigstemissionen im Realbetrieb: Lösungen für elektrifizierte Ottomotoren. In Proceedings of the 28th Aachen Colloquium Sustainable Mobility, Aachen, Germany, 7–9 October 2019; Aachener Kolloquium Fahrzeug- und Motorentechnik GbR: Aachen, Germany, 2019.
80. ZVEI-Task Force Spannungsklassen. Spannungsklassen in der Elektromobilität. Available online: https://www.zvei.org/fileadmin/user_upload/Presse_und_Medien/Publikationen/2014/januar/Spannungsklassen_in_der_Elektromobilitaet/Spannungsklassen-Elektromobilitaet.pdf (accessed on 19 October 2023).
81. Kevin Bennion and Gilberto Moreno: NREL. Convective Heat Transfer Coefficients of Automatic Transmission Fluid Jets with Implications for Electric Machine Thermal Management: Preprint. Available online: https://www.nrel.gov/docs/fy15osti/63969.pdf (accessed on 26 October 2023).
82. Shams Ghahfarokhi, P.; Podgornovs, A.; Kallaste, A.; Marques Cardoso, A.J.; Belahcen, A.; Vaimann, T. The Oil Spray Cooling System of Automotive Traction Motors: The State of the Art. *IEEE Trans. Transp. Electrific.* **2023**, *9*, 428–451. [CrossRef]
83. Peters, J.; Peña Cruz, A.; Weil, M. Exploring the Economic Potential of Sodium-Ion Batteries. *Batteries* **2019**, *5*, 10. [CrossRef]
84. Dhaussy, F.; Truong Canh, J.; Raynaud, Y.; El Mejdoubi, A. High Power Battery Systems for Mild-Hybrid with Sodium-ion cobalt free technology. In Proceedings of the 31st Aachen Colloquium Sustainable Mobility, Aachen, Germany, 10–12 October 2022; Aachener Kolloquium Fahrzeug- und Motorentechnik GbR: Aachen, Germany, 2022. Paper No. 05. ISBN 978-3-00-072524-1.
85. Stalp, A.; Heinz, A. Partikelbildung bei DI-Ottomotoren: Systemische Analyse der Partikelbildung an Ottomotoren, FF 1223. 2020.
86. Günter Fischer, P. Betriebspunktorientierte Analyse und Optimierung der Partikelrohemissionen an einem Turboaufgeladenen Ottomotor. Ph.D. Thesis, Otto-von-Guericke-Universität Magdeburg, Magdeburg, Germany, 2020.
87. Ascheberg, C. *Die Sigma Milieus®, das Globale Zielgruppen und Trend System*; Unveröffentlichtes Textmanuskript; Sigma: Mannheim, Germany, 2005.
88. Brokate, J.; Özdemir, E.D.; Kugler, U. Der Pkw-Markt bis 2040: Was das Auto von Morgen Antreibt: Szenario-Analyse im Auftrag des Mineralölwirtschaftsverbandes. Available online: https://www.dlr.de/dlr/Portaldata/1/Resources/documents/2013/DLR-Studie_Pkw-Markt_2040_MQPBDJRL7FdcF45_(1).pdf (accessed on 12 October 2023).

vehicles

Article

Lifting Actuator Concept and Design Method for Modular Vehicles with Autonomous Capsule Changing Capabilities

Fabian Weitz *, Niklas Leonard Ostendorff, Michael Frey and Frank Gauterin

Institute of Vehicle System Technology, Karlsruhe Institute of Technology, 76133 Karlsruhe, Germany; niklas.ostendorff@student.kit.edu (N.L.O.); michael.frey@kit.edu (M.F.); frank.gauterin@kit.edu (F.G.)
* Correspondence: fabian.weitz@kit.edu; Tel.: +721-60845362

Abstract: Novel vehicle concepts are needed to meet the requirements of resource-conserving and efficient mobility in the future, especially in urban areas. In the automated, driverless electric vehicle concept U-Shift, a new form of mobility is created by separating a vehicle into a drive module and a transport capsule. The autonomous driving module, the so-called Driveboard, is able to change the transport capsules independently and is therefore used to transport both people and goods. The wide range of possible capsules poses major challenges for the development of the Driveboard and the chassis in particular. A lifting actuator integrated into the chassis concept enables levelling and, thus, the raising and lowering of the Driveboard and the capsules to ground level. This means that no additional lifting devices are required for changing the capsules or for lowering them to the ground, e.g., for loading and unloading the capsules. To realise this mechanism simply and efficiently, a fully electromechanical actuator is designed and constructed. The actuator consists primarily of a profile rail guide, a steel cable winch, an electric motor, a housing that connects the subsystems and a locking mechanism. The electric motor is used to lift the vehicle and regulate the weight force-driven lowering of the vehicle. This paper describes the design of the actuator and shows the dimensioning of all main components according to the boundary conditions. Finally, the prototype model of the realised concept is presented.

Keywords: lifting actuator; autonomous changing; modular vehicle

Citation: Weitz, F.; Ostendorff, N.L.; Frey, M.; Gauterin, F. Lifting Actuator Concept and Design Method for Modular Vehicles with Autonomous Capsule Changing Capabilities. *Vehicles* **2024**, *6*, 1070–1088. https://doi.org/10.3390/vehicles6030051

Academic Editors: Ralf Stetter, Udo Pulm and Markus Till

Received: 29 April 2024
Revised: 20 June 2024
Accepted: 21 June 2024
Published: 28 June 2024

1. Introduction

In urban areas and on industrial sites, the requirements in terms of manoeuvrability and space requirements for manoeuvring are increasing due to the ever-tighter traffic volume. The increasing lack of parking space also suggests the development of universally usable vehicles. These challenges can be met by a modular vehicle concept. Such a vehicle concept is being developed as part of the U-Shift project. It consists of a driving unit, the so-called Driveboard and application-specific capsules. All drive and energy systems and functions for autonomous driving are integrated into the Driveboard. The appropriate capsule is added to the application-specific configuration of the Driveboard and capsule. The capsules, which are designed as goods transport capsules or passenger cabins, for example, can be connected to the Driveboard or exchanged via a standardised interface. To enable the vehicle to operate autonomously, not only in road traffic but in all areas of operation, a capsule replacement concept is required that allows capsules to be replaced without additional infrastructure or manual operations. Figure 1 shows the Driveboard picking up a passenger capsule.

The capsule change is made possible by the geometry of the vehicle and a special lifting system. The Driveboard is U-shaped, open at the rear and can be lowered by 200 mm. This allows the Driveboard to move under the capsules when lowered, connect with them and then be raised back to the driving level. This function requires lifting actuators that enable the vehicle to be lowered and raised. In order to avoid track changes during the lifting process, a special subframe is also used [2].

Figure 1. Driveboard (on the left) drives up to passenger capsule (on the right) [1].

The selected concept for the lifting actuator is fully electromechanical and works with a cable winch that moves a roller block vertically along a profile rail guide. The lifting actuator connects the frame of the Driveboard, which is lifted, with the subframe, which does not change its level during the lifting process and enables the two components to move relatively to each other. A locking system is used to lock the actuators in the upper and lower end positions for safety and to relieve the stress on the cables. The locking system consists of locking bolts that are actuated by springs to lock and lifting magnets to unlock and connect form-fittingly to the housing. The lifting actuator is driven by a servomotor with an integrated gearbox, which applies the lifting force and controls the weight-driven lowering.

This paper explains the mechanical design in the context of the application, shows the design of the individual components and describes the structure of the linear actuator in detail.

2. Materials and Methods

2.1. Geometric and Physical Boundary Conditions

The lift actuator is the physical and functional interface between the subframe and the frame of the Driveboard. It must, therefore, be positioned between the two structural components and transmit all the forces that occur. Due to the selected actuator concept, the actuators must be positioned between two vertical structures. Figure 2 shows a schematic sketch of the cross-section view of the Driveboard at the front axle and indicates the position of the lifting actuators.

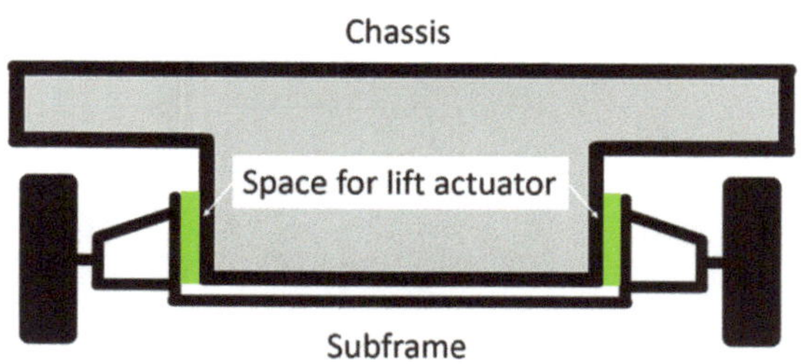

Figure 2. Cross-section view at the front axle of the Driveboard, green: space for lift actuator.

Figure 3 shows the surface provided on the subframe for attaching the lifting actuator. A total of four lifting actuators are installed, two on each of the symmetrically opposite side parts of the subframe. This also has the advantage that the lifting actuators can be built small, and the space between the body and the subframe is kept to a minimum. Figure 4 shows a side view of the driving stool with mounted lifting actuators; the figure also shows the size ratios.

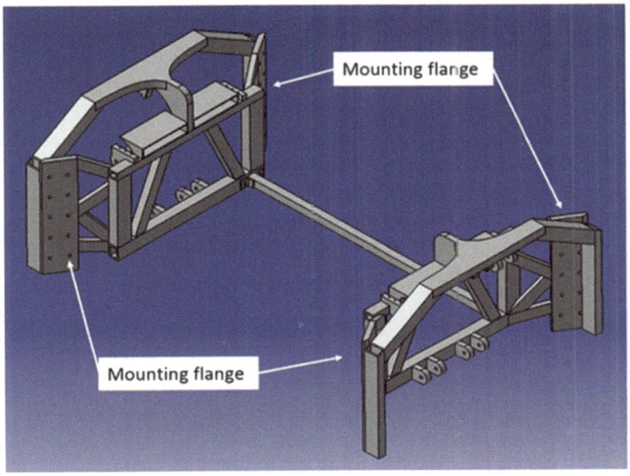

Figure 3. Isometric view of the subframe CAD model.

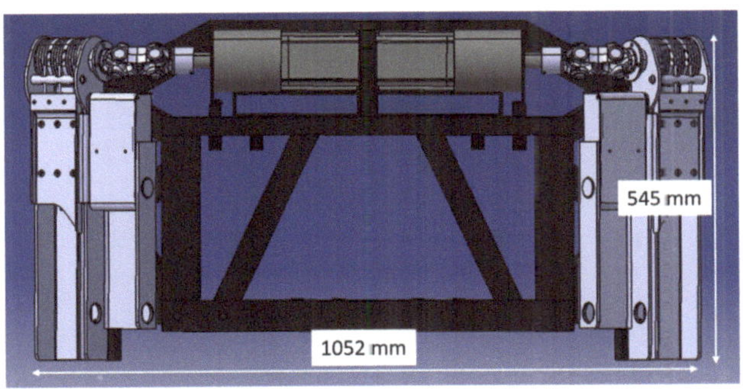

Figure 4. Sideview of the subframe with lifting actuator.

To calculate the forces acting on the lifting actuators, the masses and the mass distribution of the Driveboard (see Table 1) and the maximum accelerations of the standard load cases according to Table 2 are used. The load cases vertical impact (2), longitudinal impact (3) and lateral impact (4) represent high quasi-static loads on the chassis components in all spatial directions. The U-Shift prototype is intended for a low driving speed, but the selected load cases can also potentially occur at low speeds (attenuated). As the wheel accelerations represent limit cases that only rarely occur, the yield strength is used below as the material parameter for the mechanical design of the components. A check of the fatigue strength of the components is recommended when the components are pursued further. By multiplying the wheel accelerations by the wheel load, the forces acting on the vehicle at the wheel contact point are obtained. While 40% of the vehicle mass is on the front axle when the Driveboard is empty, the proportion of the total mass on the front axle is reduced to 30% when the vehicle is fully loaded with a passenger capsule.

Table 1. Driveboard mass and mass distribution.

Configuration/Mass	Total Mass	Front Axle Load	Rear Axle Load
Empty Driverboard	1800 kg	1080 kg	720 kg
Maximum vehicle mass	5000 kg	1500 kg	3500 kg

Table 2. Standard load cases for structural strength [3].

	Standard Load Cases	Acceleration [g]		
		x	y	z
1	Stationary car	0.00	0.00	−1.00
2	Vertical impact 3.0 g	0.00	0.00	−3.00
3	Longitudinal impact 2.50 g	2.50	0.00	−1.00
4	Lateral impact 2.50 g	0.00	2.50	−1.00
5	Cornering right 1.25 g	0.00	1.25	−1.00
6	Braking while cornering	0.75	0.75	−1.00
7	Braking backwards 1.0 g	1.00	0.00	−1.00
8	Accelerate −0.5 g	−0.5	0.00	−1.00
9	Accelerating while cornering 0.7 g	−0.5	0.5	−1.00
10	Diagonal load	0.00	0.00	−1.75
11	Vertical spring in 2.25 g	0.00	0.00	−2.25
12	Vertical spring out 0.75 g	0.00	0.00	−0.75
13	Cornering right 0.75 g	0.00	0.75	−1.00
14	Cornering left 0.75 g	0.00	−0.75	−1.00
15	Braking 0.75 g	0.75	0.00	−1.00
16	Accelerate −0.5 g	−0.5	0.00	−1.00

2.2. Lifting Actuator Concept

Several concepts were created and evaluated according to installation space, weight, efficiency, maintenance and costs with descending weighting [4]. The concept described in this paper was selected and further developed on the basis of the geometric and physical boundary conditions. The basic structure is shown in Figure 5, and the components are labelled in Table 3. A servomotor drives the shaft (1), which is mounted in the housing (2) and connected to the cable winch (3). The three steel cables (4) are rolled up on the cable winch and connected to the adapter (5), which is screwed onto the roller block (6). The roller block is guided by the profile rail (7), which is also bolted to the housing. The locking mechanism (8) is also connected to the adapter, which moves parallel to the roller block in the housing and engages with locking bolts in the neighbouring walls of the housing. The locking bolts are constantly pressed outwards by springs and engage in the holes provided in the housing wall as soon as they are reached. The bolts are retracted by lifting magnets before a lifting operation.

Table 3. Lifting actuator components.

Number	Component
1	Shaft
2	Housing
3	Cable Winch
4	Steel Cables
5	Adapter
6	Roller Block
7	Profile Rail
8	Locking Mechanism Assembly
9	Locking Bolt

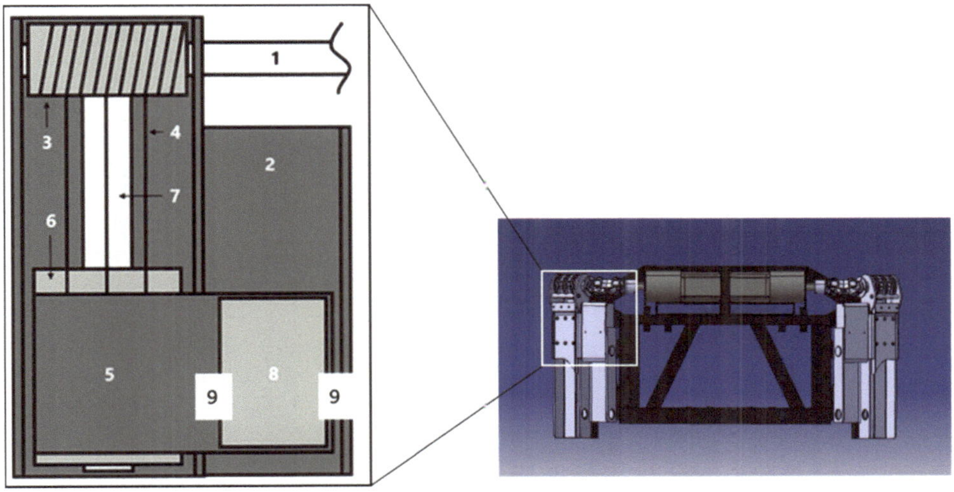

Figure 5. Schematic sketch of the lifting actuator.

2.3. Dimensioning

This chapter describes the methodology for designing the individual components. For simple geometries, strength theory equations can be used, while the underlying standards are used for standardised components. The strength verification of complicated geometries is carried out using simulation. The maximum front axle load and the maximum accelerations in each direction can be used to calculate the forces that act on the front axle and must, therefore, be transmitted via the subframe and the lifting actuators. The maximum forces occurring on a lifting actuator are summarised in Table 4. For the following dimensioning, it is assumed that the forces acting on the front axle are distributed evenly over all four actuators.

Table 4. Maximum forces acting on one lifting actuator.

Axis	x	y	z
Force [N]	9196.88	9196.88	11,036.25

2.3.1. Profile Rail Guide

The profile rail guide must support all forces in the x and y directions. In order to achieve the most compact design possible, a roller block with the following properties is selected:

- Roller recirculation;
- 45° arrangement of the rolling elements;
- No rolling element cage;
- SLS design (slim, long, standard height).

The long design of the roller block is chosen because the vertical construction space is less critical than the horizontal. This also allows higher torques to be absorbed. The static dimensioning is carried out on the basis of DIN ISO 14728-2 [5] and refers to time-independent maximum loads The static load rating C_0 is calculated using the static safety factor S_0 and the statically equivalent load P_C, which correspond to the forces in the x and y directions from Table 4. Table 5, which takes into account the operating conditions of the linear guideway, is used to select a suitable safety factor. Depending on the application, the unknown torque loads and possible contamination of the guideway, the conditions in the second row apply. Conservatively, the highest safety factor of 12 is used for this category.

Table 5. Use case-dependent security factors for linear guides [6].

Conditions of Use	Static Load Safety Factor S_0
Overhead arrangements and applications representing a high hazard potential.	≥12
High dynamic load when at a standstill, contamination.	8...12
Normal dimensioning of machinery and plant without full knowledge of the load parameters or connection details.	5...8
Full knowledge of all the load data. Vibration-free operation is ensured.	3...5
If there are health and safety hazards, paragraph 5.1.3 of DIN 637 is to be observed.	*

The note (*) in the last line of Table 5 indicates that the roller block must not detach from the rail and that this must be prevented by design in applications where there is a potential risk of injury [7]. Due to the cable winch, the roller block cannot leave the profile rail at the upper end. The connecting strut between the side parts of the subframe (see Figure 3) prevents it from leaving the bottom end of the profile rail, as it would collide with the frame of the Driveboard first. As four actuators are installed, which are located at each corner of the U-shaped subframe and are arranged at an angle, the two structural components, the body of the Driveboard and the subframe, cannot be separated from each other even in the case of a complete failure of a profile rail guide. This ensures a critical aspect of operational safety. According to Formula (1) [5]

$$S_0 = \frac{C_0}{P_0} \tag{1}$$

the static load rating results in $C_0 = 110362.56$ N.

The dynamic load rating is calculated in an analogue way to the static load rating in accordance with DIN ISO 14728-1 [8]. In the application of this vehicle, the dynamic equivalent load represents the capsule change in the inclined position. For this scenario, the incline on country roads is considered. The slope downforce that occurs is calculated using the maximum longitudinal incline on country roads. According to the Road and Transport Research Association, this is 8.0% [9] or $\alpha = 4.57°$. The slope acceleration is calculated using Formula (2).

$$a_H = g \times \sin\alpha \tag{2}$$

It results in $a_H = 0.78\frac{\text{m}}{\text{s}^2}$.

As the bearing configuration is 45°, the orientation of the force relative to the profile rail can be ignored [6]. Furthermore, torques are neglected in the calculation. The equivalent dynamic load is therefore calculated according to Formula (3).

$$P = \frac{1}{4} \times m \times a_H \tag{3}$$

to $P = 292.50$ N.

According to the manufacturer's recommendation, a safety factor of $S = 6$ is selected for use in lifting mechanisms [6]. After converting Formula (4), a dynamic load rating can be calculated.

$$S = \frac{C}{P} \tag{4}$$

It results in $C = 1755$ N.

The profile rail guide must meet or exceed the static and dynamic load rating. As the static load is significantly higher, this is considered a selection criterion. The required loads are summarised in Table 6.

Table 6. Requirements for dimensioning of the profile rail guide.

Property	Value	Unit
S_0	12	-
P_0	9196.88	N
C_0	110,362.50	N
P	292.50	N
m	1500	kg
a_H	0.78	$\frac{m}{s^2}$
S	5	-
C	1755	N

2.3.2. Winch and Steel Cables

The cable winches must lift the maximum front axle load of 1500 kg. Due to the configuration of four actuators, each cable winch is subjected to a cable pulling force of $S = 3678.75$ N. The system is designed for one cable per winch. In order to ensure operational safety, even if one cable fails, more cables are installed per cable winch. The width of the cable winch allows the use of three cables per actuator. The design is carried out in accordance with DIN 15020 [10].

In the first step, the drive unit group is determined according to DIN 15020 [10]. This requires the average running time per day to be calculated in hours. Due to the requirements made in the project for the lifting cycles, an average running time of less than 0.125 h/d is achieved. Assuming a heavy load profile results in the engine group "1 D_m". For drive units of this classification, the safety index, which indicates the ratio between breaking force and tensile force, must be at least 3. This condition is checked after determining the minimum cable diameter.

The coefficient c is then determined. Non-elongation-free cables are assumed for this, as the load is guided by the profile rail guide, resulting in a small coefficient. The nominal strength δ_z of the individual wires is assumed to be the lowest permissible value for the drive unit group of 1770 $\frac{N}{mm^2}$. The coefficient can be taken from DIN 15020 [10] and is $c = 0.0710$. The minimum cable diameter can be calculated with the coefficient c and the cable tensile force S according to Formula (5) [10].

$$d_{min} = c \cdot \sqrt{S} \tag{5}$$

The minimum cable diameter is therefore $d_{min} = 4.31$ mm. The value is rounded up to the nearest integer with $d_{min} = 5$ mm.

In addition to the values already listed, two further factors are required to calculate the safety factor. The influence of the load collective is taken into account by k and the filling factor of the cables by f. The calculation of the safety index ϑ based on Formula (6) [10]

$$\vartheta = \frac{k \times f \times \pi \times \frac{1}{4} \times d^2 \times \delta_z}{S} \tag{6}$$

to $\vartheta = 3.69$, which fulfils the requirements for the drive unit.

The minimum cable winch diameter can be calculated using Formula (7) [10].

$$D_{min} = h_1 \times h_2 \times d_{min} \tag{7}$$

This results in $D_{min} = 56$ mm. The unitless coefficients h_1 and h_2 are taken from DIN 15020 [10].

The groove radius in the cable winch for the steel cables is taken from DIN 15020 [10] and is $r = 2.7$ mm.

To prevent the cables from unwinding, the cables are fixed in the drum with a grub screw and wrapped around the drum with more than two wraps in each position of the roller block. A plain bearing is provided for the drive shaft bearing, as this saves space and can absorb high surface pressures. According to DIN 15020, an efficiency of $\eta = 0.96$ is specified for this type of bearing [10]. Table 7 summarises all the requirements for dimensioning the winch and cables.

Table 7. Requirements for dimensioning of winch and steel cables.

Property	Value	Unit
c	0.071	-
S	3678.75	N
δ_z	1770	$\frac{N}{mm^2}$
k	0.85	-
f	0.46	-
ϑ	3.69	
h_1	11.2	-
h_2	1	-
d_{min}	5	mm
D_{min}	56	m

2.3.3. Servomotor

All the information required to calculate the power data for the servomotor is known after the cable winch has been designed. The power can be calculated according to Formula (8).

$$P_E = \frac{F \times s}{\Delta t \times \eta} \tag{8}$$

The required power is $P_E = 255.47$ W.

The required torque can be calculated using Formula (9) and is made up of the weight load per actuator, the effective radius of the cable winch and the efficiency factor.

$$M = \frac{F \times (D_{min} + d_{min})}{2 \times \eta} \tag{9}$$

After inserting all values, the required torque is $M = 116.88$ Nm.

Table 8 summarises all the requirements for dimensioning the servomotors.

Table 8. Requirements for dimensioning of the servomotor.

Property	Value	Unit
F	3678.75	N
s	0.2	m
Δt	3	s
η	0.96	-
D_{min}	0.056	m
d_{min}	0.005	m
P_E	255.47	W
M	116.88	Nm

A motor from Heidrive is selected. The model HMD10-039-048-30 fulfils all requirements with a rated speed of 3000 min^{-1} and a standstill torque of 3.9 Nm. The motor is controlled via the mcDSA-E25XC motor controller from miControl. The position to be approached is specified as the control variable in revolutions, and the integrated sensor system confirms that the specified position has been reached.

2.3.4. Shaft

The shaft transmits the torque from the servomotor to the cable winch, is born on both sides, and carries the cable winch between the bearings. The shaft must be designed for bending, torsion and surface pressure. The bending and torsional torques are included in the combined load. The surface pressure is calculated last. The bending torque is induced by a line load and can be calculated for the present configuration using Formula 10 [11]. The force F represents the cable tension force. The length l is the distance between the bearings, s is the bearing width.

$$M_b = \frac{F \times (s/2 + l/4)}{2} \tag{10}$$

The combined load can be calculated for general cases using Formula (11) [11]. The torsional moment M_t represents the required torque of the servomotor.

$$M_v = \sqrt{M_b{}^2 + 0.75 \times M_t{}^2} \tag{11}$$

The minimum shaft diameter can be calculated using Formula (12) and the permissible bending moment for steel S235 (EN 1.0038) [11].

$$d \geq \sqrt[3]{\frac{M_v}{0.1 \times \sigma_{b\ zul}}} \tag{12}$$

The calculated minimum diameter of $d \geq 20.43$ mm must be increased in order to accommodate a feather key. The resulting geometry is validated with a simulation. All requirements for the dimensioning of the shafts are summarised in Table 9.

Table 9. Requirements for design of the shafts.

Property	Value	Unit
F	3678.75	N
l	38	mm
s	5	mm
M_b	45,064.69	$\frac{N}{mm}$
M_t	116,876.95	$\frac{N}{mm}$
M_v	110,797.08	$\frac{N}{mm}$
$\sigma_{b\ zul}$	130	$\frac{N}{mm^2}$
v	3	-
d	20.43	mm

2.3.5. Locking Bolts

Each lifting actuator has a locking mechanism with two locking bolts that act simultaneously at both sides of the housing. As the locking bolts hold the position of the lifting actuator during the drive, they are each dimensioned with half the vertical force from Table 4. The locking bolts must be designed for bending, shear and surface pressure. As normal stresses and shear stresses cannot simply be added according to the superposition principle, only the shear is considered for dimensioning, as shear stresses represent a higher load. The influence of the bending load is taken into account by a reduced permissible shear stress [11]. The main shear equation is shown in Formula (13) [11]. In this equation, the symbol A refers to the cross-section area. The equation can be rearranged to solve for either the required diameter or the mean shear stress τ_a.

$$\tau_a = \frac{F}{A} \tag{13}$$

This results in $\tau_a = 7.81\ \frac{N}{mm^2}$. The maximum shear stress occurring for circular cross sections is calculated using Formula (14) [11].

$$\tau_{max} = (4/3) \times \tau_a \tag{14}$$

It results in $\tau_{max} = 10.41 \frac{\text{N}}{\text{mm}^2}$. The surface pressure that occurs is calculated using Formula (15), which contains the force acting on a bolt and the projected area.

$$p_a = \frac{F}{d \times s} \tag{15}$$

This results in $p_a = 36.79 \frac{\text{N}}{\text{mm}^2}$. All requirements for dimensioning the locking bolts are summarised in Table 10.

Table 10. Requirements for dimensioning of the locking bolts.

Property	Value	Unit
F	5518.13	N
d	30	mm
τ_a	7.81	$\frac{\text{N}}{\text{mm}^2}$
τ_{max}	10.41	$\frac{\text{N}}{\text{mm}^2}$
s	5	mm
p_a	36.79	$\frac{\text{N}}{\text{mm}^2}$

Red Magnetics ITS-LZ 3869-Z cylinder solenoids are used to actuate the bolts. The lifting magnets pull the bolts into the housing before the lifting process. Springs are used to push the bolts into the catches to disengage them for locking.

2.4. *Topology Optimisation of the Housing*

Topology optimisation is used to calculate the optimum material distribution within a given installation space, taking into account previously set requirements. The iterative optimisation process removes underutilised material until all specifications, such as a mass percentage to be retained, maximum stresses or rigidity, are achieved. The result is a topology that is adapted again in CAD, e.g., to the manufacturing process and then validated by means of an FE analysis [3].

The concept defines the cross-section of the housing as well as the flange surfaces for the profile rail and the bearing mounts for the shaft and locking bolts. The material in the housing is to be reduced by using topology optimisation. The initial geometry for the housing is shown in Figure 6. The maximum available installation space, the design space, is shown in grey, and the non-design space in blue.

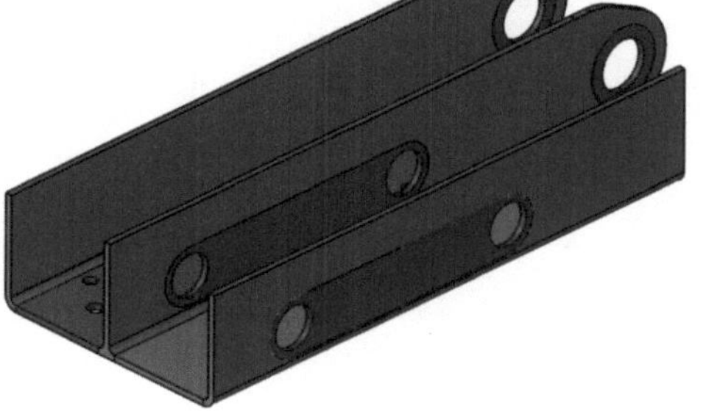

Figure 6. Initial geometry for topology optimisation, grey: designspace; blue: non-designspace.

The result of the topology optimisation with a target mass fraction of 60% and the final CAD model of the component can be seen in Figure 7. Since only limited computing capacities are available, the components are divided into tetrahedral elements with a linear approach function (TETRA4). In order to achieve usable results despite the small number of nodes per element, an average element size of 1 mm is selected, which is small compared to the component dimensions. The outer wall without the locking bearing is almost completely removed. The corner of the housing on the bottom right-hand side has also been removed. The height of the existing walls was reduced. The material around the shaft bearings was completely removed on the side facing off from the stresses. This geometry is not completely carried over but is used as an indicator of where material can be removed.

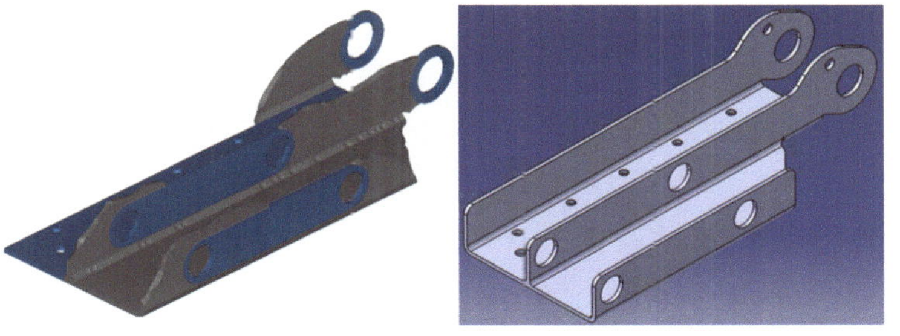

Figure 7. Left: Output geometry after topology optimisation; **right**: CAD model of final component.

2.5. FEA Simulation

For the simulation, settings must be made for the material, the mesh, the boundary conditions and the loads. Structural steel with values defined in Table 11 is used as the starting material for the simulation. All models are calculated using a tetrahedron mesh with a cell size of 1 mm and curve-based fitting.

Table 11. Material properties of structural steel for FEA simulation.

Property	Value	Unit
Young's Modulus	210	GPa
Poisson's Ratio	0.3	-
Density	7.85	$\frac{kg}{m^3}$

2.5.1. Housing

The geometry used for the simulation is derived from topology optimisation and features simplified geometries that are better suited for manufacturing. The housing is made of sheet steel with a wall thickness of 5 mm. The three load cases are calculated independently of each other, as they do not occur simultaneously. All boundary conditions and forces are shown in Table 12.

Table 12. Load cases for FEM Simulation of the housing.

Load Case	Description	Force [N]	Direction (Model)
Constraint	Flange surface restraint via "Clamp" feature and sliding bearing on the backside.	N/A	N/A
Load Case 1	Load on shaft bearing surfaces via "bearing load" feature.	3678.75	Neg. x-axis
Load Case 2	Load on upper locking bolt bearing surfaces via "bearing load" feature.	11,036.25	Neg. x-axis
Load Case 3	Load on lower locking bolt bearing surfaces via "bearing load" feature.	11,036.25	Neg. x-axis

Figures 8–10 show the Von Mises stress distributions of the simulation. It is recognisable that the maximum stress occurs on the bearing surfaces, with secondary points of

stress along the exposed 90° bends, as can be seen in the first two load cases. All maximum stresses and displacements are listed in Table 13. The highest stresses of 41.2 MPa and the highest displacements are generated in Load Case 3.

Figure 8. FEA simulation of housing with Load Case 1.

Figure 9. FEA simulation of housing with Load Case 2.

Figure 10. FEA simulation of housing with Load Case 3.

Table 13. Results of housing simulation.

Load Case	Max. Tension [MPa]	Max Deformation [mm]		
		x	y	z
Load Case 1	11.9	−0.00110	−0.00292	0.00133
Load Case 2	16.3	−0.00247	0.00346	0.00080
Load Case 3	41.2	−0.01020	−0.00233	−0.00134

The maximum deformation is 0.01 mm, and therefore, does not warrant the need for higher strength steel than necessary for the induced stresses. The high stresses in Load Case 3 are due to the lack of support material downward from the bearing surface compared to the other two load cases, especially on the outside wall, which has had material removed according to the topology optimisation and ease of manufacturing in mind/constraints.

Due to these low loads, simple structural steel such as S235 (EN 1.0038) can be used for housing, which has good availability within the EU.

2.5.2. Adapter

The adapter consists of a large flange surface that is screwed onto the roller block and a welded-on U-profile that holds the locking mechanism. In Load Case 1, forces are induced in the z-direction by the bearing surfaces for the locking bolts. Load Case 2 assumes the worst case for the cable attachment and applies a surface load to the model at the outer cable attachment with the lowest surrounding material volume. The exact boundary conditions and loads can be seen in Table 14.

Table 14. Load cases for FEM simulation of the adapter.

Load Case	Description	Force [N]	Direction (Model)
Constraint	Flange surface restraint via "Clamp" feature	unlimited	none
Load Case 1	Load on locking bolt bearing surfaces via "bearing load" feature	11,036.25	Pos. x-axis
Load Case 2	Load on cable end retaining collar via "distributed force"	3678.75	Pos. x-axis

Figure 11 shows the Von Mises stress distribution for Load Case 1. It can be seen that the highest stresses occur in the transition from the U-profile to the flange surface at the upper end. The maximum stress of 436 MPa for material thicknesses of less than 16 mm is still within the load range of S460 with a yield strength of 460 MPa. The largest displacement of approx 0.71 mm also occurs in this load case. The displacement occurs with the free-standing wall and is directed inwards and, therefore, unproblematic, as there is no collision with the housing, which moves relative to the adapter.

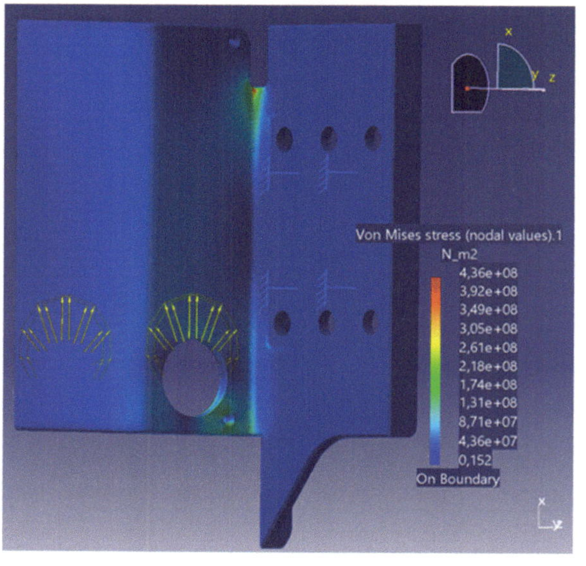

Figure 11. FEA simulation of adapter with Load Case 1.

Load Case 2 is shown in Figure 12. The maximum stresses occur on the open side of the cable attachment. The stresses of 233 MPa are significantly lower than those of Load Case 1 and, therefore, have no further influence on the choice of material.

Table 15 shows the maximum stresses and displacements for both load cases.

Table 15. Results of adapter simulation.

Load Case	Max. Tension [MPa]	Max Deformation [mm]		
		x	y	z
Load Case 1	436	0.71300	0.35700	−0.50100
Load Case 2	234	0.02840	−0.00827	−0.03130

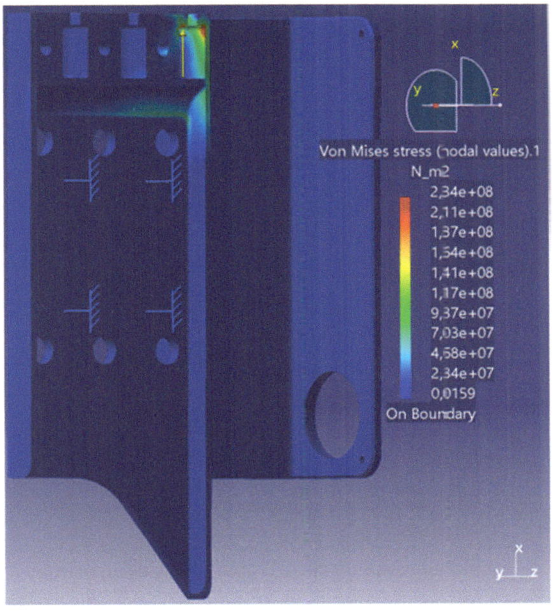

Figure 12. FEA simulation of adapter with Load Case 2.

2.5.3. Shaft

The shaft is simulated with the feather keys. For this purpose, all contact surfaces between the feather keys and the shaft are defined as a sliding connection. The feather keys are defined as firmly clamped. The sliding bearing surfaces are defined as such, and the load of the cable winch is projected onto the top of the shaft. The torque is induced at the shaft end. All constraints and loads are listed in Table 16.

Table 16. Load cases for FEM simulation of the shaft.

Load Case	Body	Description	Force	Direction (Model)
Constraint	Shaft	Bearing surfaces are restraint via sliding contact	unlimited	none
Constraint	Feather Key	Restraint via flange on two perpendicular sides	116.88 [Nm]	Pos. y-axis
Load Case 1	Shaft	Torque applied on one end, distributed force on upper shaft surface	3678.75 [N]	Neg. z-axis

The maximum stresses occur at the end of the contact surface between the shaft and the feather key close to the torque input, as can be seen in Figure 13. It can be assumed that these stress concentrations occur due to the rigidly clamped feather keys and the sharp end of the contact surfaces, which do not optimally represent reality. Nevertheless, the maximum stresses and displacements, which are also summarised in Table 17, are low enough to allow the use of structural steel S235 (EN 1.0038).

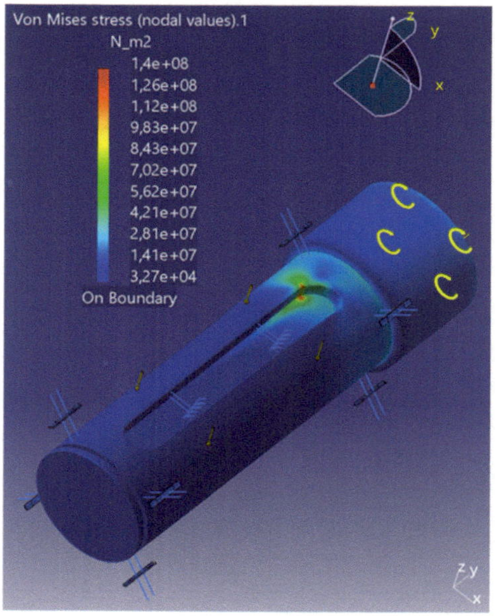

Figure 13. FEA simulation of shaft with Load Case 1.

Table 17. Results of shaft simulation.

Load Case	Max. Tension [MPa]	Max Deformation [mm]		
		x	y	z
Load Case 1	140	−0.00155	0.00020	0.00915

3. Results

Figure 14 shows the final CAD model of the assembly of the lifting actuator. The housing, the adapter and the shaft correspond to the figures already shown in the simulation chapter. The locking bolts are locked in the upper position, which represents the travelling position.

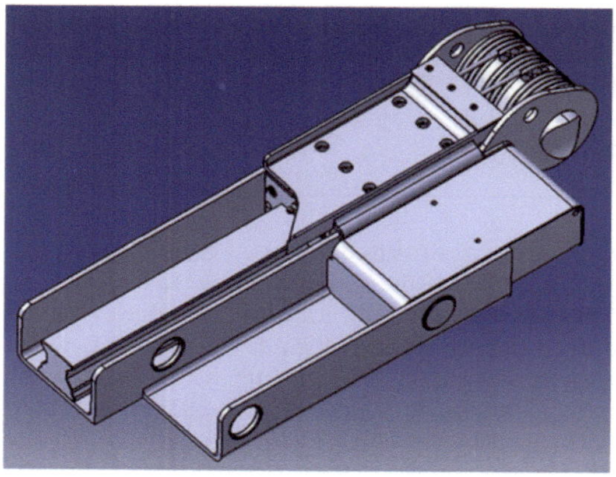

Figure 14. CAD model of final lifting actuator.

Topology optimisation was used to identify areas of the housing with excess material. The optimised housing should nevertheless retain a simple geometry and protect the moving components in the housing. Therefore, the walls were reduced to the height of the adapter, and the entire bottom, which rests against the subframe, was retained. The material around the shaft bearings is also retained to protect the cable winch. The channel for the locking mechanism was shortened at the lower end, and a triangular geometry was removed at the top. The topology optimisation and the FEA simulation suggest that the material thickness of 5 mm can also be reduced. However, this is retained in order to prevent the walls from buckling or twisting and to provide sufficient bearing space.

Two lifting actuators can be seen in Figure 15 and are already mounted on a side part of the subframe. The locking mechanism is locked in the lower position here, which represents the capsule change position. The servomotors are installed above the frame and drive the cable winches by using a cardan shaft due to the angular offset.

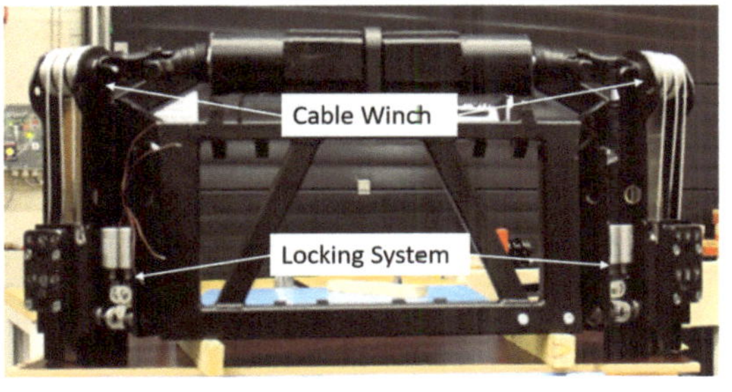

Figure 15. Left side of subframe with attached lifting actuators.

Figure 16 shows a close-up view of a lifting actuator. The actuator is also locked in the lower position in this picture. In this view, the cable winch is not yet mounted, and the plain pressed-in bearings of the shaft and the cable attachment in the adapter are clearly visible.

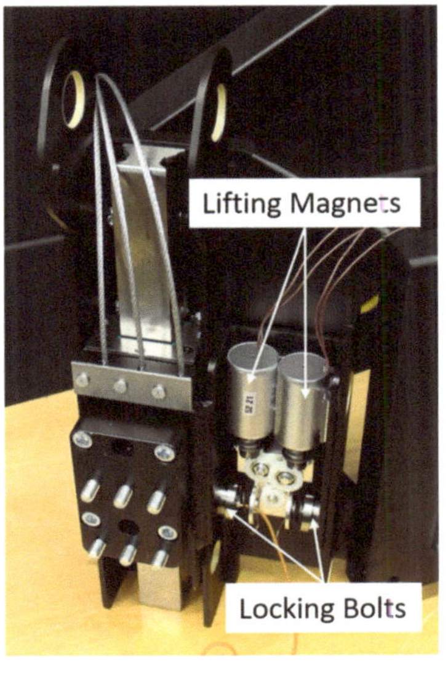

Figure 16. Close up side view of lifting actuator attached to the subframe.

4. Discussion

The design is mostly carried out with the help of standards, manufacturer catalogues and scientific manuals. These materials provide calculations in descending order, with which components can be fully or pre-dimensioned. FEA simulations and topology optimisations were used as further tools for dimensioning. Topology optimisation was used to reduce the mass of the housing, but the focus was also placed on simple geometry. The simulation results show that the geometries can withstand the loads, but by using the housing as an example, even more material could have been saved, and material utilisation could have been increased.

The lifting actuator is designed as a complete unit to allow for separate development. This is particularly useful if other systems have not yet been finalised and, therefore, only the interfaces need to be defined or adapted. However, an integrated development of the lifting actuator is also conceivable and offers greater potential for material, weight and installation space savings. However, this approach to concept and product development is more complex. Further advantages of the separate development are the adjustability of the actuators as well as the possibility of pre-assembly and easier interchangeability due to the flange connection.

The lifting actuator developed enables the lifting process under all the required boundary conditions of the U-Shift project, which could not be realised on the basis of existing systems. The lifting actuator connects the frame of the Driveboard, which is lifted, with the subframe and does not change its level during the lifting process. It enables the two components to move relative to each other and thus prevents changes in track width during the lifting process. For safety and to take the strain off the cables, a locking system is used to lock the drives in the upper- and lower-end positions.

The tests carried out so far have shown that the overall system even exceeds the requirements. It is possible to lift the vehicle in a shorter time and move a higher total mass than originally assumed. In the course of the project, the prototypes will be tested even more extensively in the Driveboard and further developed on the basis of the knowledge gained.

Author Contributions: Conceptualisation, F.W.; methodology, F.W.; validation, N.L.O.; formal analysis, F.W.; investigation, N.L.O.; resources, F.W.; data curation, F.W. and N.L.O.; writing—original draft preparation, F.W. and N.L.O.; writing—review and editing, F.W. and M.F.; visualisation, F.W. and N.L.O.; supervision, M.F. and F.G. All authors have read and agreed to the published version of the manuscript.

Funding: This research was funded by the Ministry of Economic Affairs, Labour and Tourism Baden-Württemberg (Ministerium für Wirtschaft, Arbeit und Tourismus Baden-Württemberg.

Data Availability Statement: Data are contained within the article.

Acknowledgments: We would like to thank the Ministry of Economic Affairs, Labour and Tourism Baden-Württemberg (Ministerium für Wirtschaft, Arbeit und Tourismus Baden-Württemberg) for funding the "U-Shift II" project. The responsibility for the content of this work lies with the authors.

Conflicts of Interest: The authors declare no conflict of interest.

References

1. Deutsches Zentrum für Luft- und Raumfahrt e.V. Projekt U-Shift II (Demonstrator). Picture: DLR, CC-BY-ND. Available online: https://verkehrsforschung.dlr.de/de/projekte/u-shift/u-shift-ii-demonstrator (accessed on 23 June 2023).
2. Weitz, F.; Frey, M.; Gauterin, F. Methodical Design of a Subframe for a Novel Modular Chassis Concept without Knowledge of Final Vehicle Parameters. *SAE Int. J. Commer. Veh.* **2024**, *17*, 91–101. [CrossRef]
3. Ersoy, M. Fahrwerkentwicklung. In *Fahrwerkhandbuch*, 5th ed.; Gies, S., Ed.; Springer Fachmedien Wiesbaden GmbH: Wiesbaden, Germany, 2017; Volume 5, pp. 264–268.
4. Ostendorff, N. Aktorkonzept für die Niveauregulierung in einemneuartigen Fahrzeugkonzept. Bachelor's Thesis, Karlsruher Institut für Technologie, Karlsruhe, Germany, 2021.
5. *DIN ISO 14728-2:2018-10*; Rolling Bearings—Linear Motion Rolling Bearings—Part 2: Static Load Ratings. Beuth Verlag GmbH: Berlin, Germany, 2018.

6. Roller Rail Systems (R999000354). pp. 26–29. Available online: https://www.boschrexroth.com/en/de/media-details/838ed56e-6e53-45c1-a550-ca83e811ca1b (accessed on 19 July 2023).
7. *DIN 637:2016-12*; Rolling Bearings—Safety Regulations for Dimensioning and Operation of Profiled Rail Guides with Recirculating Rolling Elements. Beuth Verlag GmbH: Berlin, Germany, 2016.
8. *DIN ISO 14728-1:2018-10*; Rolling Bearings—Linear Motion Rolling Bearings—Part 1: Dynamic Load Ratings and Rating Life. Beuth Verlag GmbH: Berlin, Germany, 2018.
9. Strassenbau: Grenz- und Richtwerte: Landstraßen. Available online: https://www.bauformeln.de/strassenbau/grenz-und-richtwerte/landstrassen/ (accessed on 19 July 2023).
10. *DIN 15020-1:1974-02*; Lifting Appliances; Principles Relating to Rope Drives; Calculation and Construction. Beuth Verlag GmbH: Berlin, Germany, 1974.
11. Böge, A.; Böge, W. *Handbuch Maschinenbau Grundlagen und Anwendungen der Maschinenbau-Technik*; 24. Auflage; Springer Fachmedien Wiesbaden GmbH: Wiesbaden, Germany; pp. 307–308, 763–790.

vehicles

MDPI

Article

Optimization, Design, and Manufacturing of New Steel-FRP Automotive Fuel Cell Medium Pressure Plate Using Compression Molding

Sharath Christy Anand, Florian Mielke, Daniel Heidrich and Xiangfan Fang *

Institute of Automotive Lightweight Design, University of Siegen, 57076 Siegen, Germany; sharath.anand@uni-siegen.de (S.C.A.); florian.mielke@uni-siegen.de (F.M.)
* Correspondence: xiangfan.fang@uni-siegen.de

Abstract: In this work, a new plastic-intensive medium-pressure plate (MPP), which is part of a fuel-cell system, has been developed together with a steel plate meeting all mechanical and chemical requirements. This newly developed MPP had to achieve the objective of saving weight and package space. The use of compression molding as a manufacturing technique facilitated the use of glass mat thermoplastics (GMT) which has higher E-modules and strength compared to most of the injection molded materials. A steel plate was placed as an insert to help achieve the stiffness requirements. For the development, the existing MPP was benchmarked for its structural capabilities and its underlying functional features. Four different FRP materials were investigated in terms of their chemical and mechanical properties. PP-GMT material, which has both high mechanical performance and resistance against chemicals in the fuel cell fluid, had been chosen. Using the properties of the chosen PP-GMT material, topology optimization was carried out based on the quasi-static load case and manufacturing constraints, which gave a load-conforming rib structure. The obtained rib structure was utilized to develop the final MPP with adherence to the functional requirements of MPP. The developed plastic-intensive MPP exhibits a 3-in-1 component feature with a 55% reduction in package space and an 8% weight reduction. The MPP was virtually analyzed for its mechanical strength and compared with the existing benchmark values. Finally, a press tool was conceptualized and manufactured to fabricate the new plastic-intensive MPP, which was tested in a rig and validated in the FE model.

Keywords: fuel cell; medium-pressure plate; structural optimization; compression molding; GMT; package saving; light weight

Citation: Anand, S.C.; Mielke, F.; Heidrich, D.; Fang, X. Optimization, Design, and Manufacturing of New Steel-FRP Automotive Fuel Cell Medium Pressure Plate Using Compression Molding. *Vehicles* **2024**, *6*, 850–873. https://doi.org/10.3390/vehicles6020041

Academic Editors: Ralf Stetter, Udo Pulm and Markus Till

Received: 11 April 2024
Revised: 18 May 2024
Accepted: 23 May 2024
Published: 25 May 2024

1. Introduction and State of the Art

A globally accepted fact in the current decade is the effect of global warming on climatic changes, which is mainly caused by CO_2 emissions. It was published by Lamb et al. [1] that in 2018, the automotive sector contributed 14% to the overall global CO_2 emissions. This, in turn, creates an emphasis on the necessity to reduce the emissions of vehicles, which translates to a steep rise in the need for vehicles powered with green energy where the carbon emissions are zero. Therefore, the sales of battery electric vehicles (BEVs) have increased in the past decade [2], and projected figures show that the market share of electric vehicles (EV) will be more than 80% in a few decades. Although extensive research has been invested in increasing the energy density of lithium-ion batteries, currently, the energy storage ability still does not meet the demands of many customers. Many consumers consider the charging time of 20 to 30 min for a 200 to 300 km driving range to be still too long [3].

This is where a polymer electrolytic fuel cell (PEMFC) can help plug the gap since the charging time is within 5 min for a driving range of more than 500 km, which is comparable to gasoline-powered vehicles [4–6]. The advantage of a continuous power supply with the availability of PEMFC hydrogen fuel has made it a strong candidate for powering automobiles. Over the years, smaller power-demand vehicles (50–250 KW) powered by fuel cells have been on the automotive market [7]. Fuel cell vehicles can exhibit a high range with smaller increases in the weight of the vehicles in comparison to BEVs [8].

In Figure 1, a simplified schematic sketch of PEMFC, which consists of multiple cells, is shown. Each cell is an assembly of two bipolar plates, bipolar anode and bipolar cathode, which sandwich the membrane electrode assembly. A current collector is placed at the end of these cells in order to collect the electricity. Eventually, the medium-pressure plate (MPP) holds all the single cells inside the stack house. The membrane electrode assembly consists of a proton exchange membrane, catalyst layers and gas diffusion layers. The membrane transports hydrogen protons from the anode to the cathode. The catalyst layer is where the electrochemical reaction takes place, which controls the rate of the reaction based on the necessity of power. The gas diffusion layer is effectively the electrical conductor that transports the electrons to and from the catalyst layer [9]. The reactant gases, hydrogen and oxygen, are supplied to the anode and cathode via the flow ducts of bipolar plates. Electric current is generated by the electrochemical reactions, accompanied by heat development. Therefore, the bipolar plates must be cooled by coolant, for which inlet and outlet ducts must be introduced.

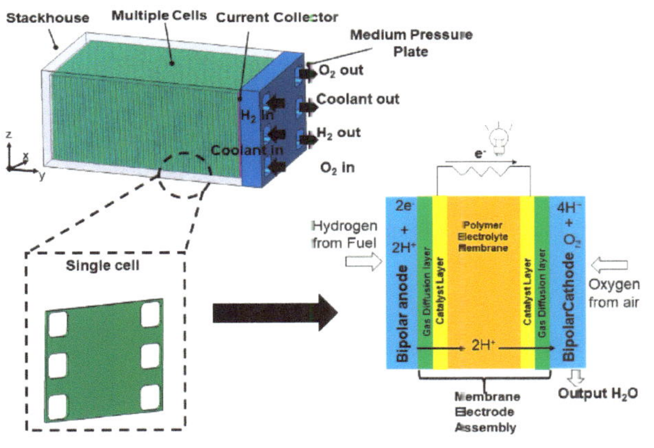

Figure 1. Schematic representation of PEM fuel cell and its working principle.

Lightweight design is an important factor in reducing the fuel consumption of vehicles using combustion engines and/or increasing the driving range of electric vehicles. Various forms of research have been carried out to reduce the weight and redesign the bipolar plates [10–13] to improve the efficiency of PEM fuel cells. Liu et al. performed a 2D shape optimization on the endplate of fuel cells, converted this optimized shape into a 3D structure, and performed FE analysis to check its structural rigidity [14]. A structural study to evaluate the contact pressure of end plates based on bolting strategies was researched by Dey et al. [15].

The body structures of vehicles must often be strongly modified if PEMFCs are to be used as power sources due to their spatial requirements, which results in higher development and manufacturing costs. Package saving in the fuel cell may thus contribute to a faster transition from combustion engines to fuel-cell electric vehicles. Package space saving should ideally be realized in components meant for structural rigidity or support, like Fuel cell housing or the medium-pressure plate, which is the topic of this work. The saved package can be used to increase the number of fuel cells, which helps increase the overall power, and the power output of the stack can be increased [16,17].

1.1. Medium-Pressure Plate

The MPP, commonly manufactured by aluminium casting or steel plate, functions as a manifold through which hydrogen, air and coolant are supplied and regulated to the PEMFC Stack [6]. As the name suggests, it also ensures a large contact force is applied to the bipolar plates, on the stack side, within the stackhouse. Bolted at different locations to the stackhouse, it also needs to guarantee an impervious working condition of the fuel cell stack [15]. The last bipolar plate is in contact with a titanium plate, which acts as the current collector.

The PEMFC stack can be subjected to various loading and stress conditions based on utility in real-time scenarios, where the PEMFC should have a >40,000 h static work lifetime and >5000 h automotive application lifetime [18]. Yan et al. have also reported that the performance of PEMFC is also influenced by temperature and humidity, which affect the charge transfer resistance of the stack [19]. These stress conditions increase the emphasis on the material characteristics of the fuel cell system.

The reference MPP of this work is made of aluminium casting with a layer of glass fibre-reinforced polyphenylsulfide (PPS) material with 30% glass fibre reinforcement per weight, injection molded on the cast aluminium. The FRP layer here acts as an insulation between the aluminium and current collector [20].

A PEM fuel cell system is typically operated between 65 °C up to 125 °C [21] using perfluorosulfonic acid HCOO-CF(CF_3)-O-CF_2CF_2-SO_3H (PFSA) polymers such as Nafion as membranes for their superior conductive and chemical properties [22]. PFSA can also be shortly represented as R_f-CF_2COOH, where R_f represents the non-reactive part of the Nafion molecule. The chemical requirement of MPP is that it should exhibit high resistance to moisture, hydrogen, and acid mediums as there is direct contact with these mediums. Chemical degradation of PFSA is a major issue in PEMFCs. This degradation is due to the dissociation of hydrogen peroxide molecular bonds, which is due to the presence of metal ions like Fe Cu through the corrosion of cell or stack materials [8].

$$\text{Metal ions}\left(M^{(Z+1)}\right) + H_2O_2 \rightarrow M^z + +HO^{\cdot} + H^+ \quad (1)$$

Radical Hydrogen ions get washed out due to the dissociation of hydrogen peroxide (H_2O_2). Oxidative hydroxyl ($HO^{\cdot}$) and hydroperoxyl ($HOO^{\cdot}$) are also produced due to the decomposition of hydrogen peroxide. The oxidative hydroxyl groups react with the carboxylic acid group (–COOH), as shown in Equations (2) and (3), giving rise to a degrading effect on the PEMFC membrane. This phenomenon is termed an "unzipping mechanism" [8,23].

$$R_f - CF_2COOH + HO^{\cdot} \rightarrow R_f - CF_2 + CO_2 + H_2O \quad (2)$$

$$R_f - CF_2 + HO^{\cdot} \rightarrow R_f - CF_2OH \rightarrow R_f - COF + HF \quad (3)$$

$$R_f - COF + H_2O \rightarrow R_f - COOH + HF \quad (4)$$

The hydrogen ion attacks the washed-off fluorine atoms from PFSA and forms a hydrogen fluoride (HF) acidic medium, which is termed free radical attack. Hydrogen fluoride is particularly aggressive towards the FRP materials used to create the insulation layer on MPP, including the glass fibres. The reaction of hydrogen fluoride with glass fibres can subsequently increase the pH values within the fuel cell system, which can lead to detrimental effects. Due to the acidic medium, it was imperative that the FRP should be free of any cracks as the acidic medium will seep into the microstructure of FRP and rapidly degrade it.

Currently, many polymers are used in the manufacturing of PEMFCs. Gaskets within the fuel cell systems may comprise various polymers such as polyester, polyimide, polyethylene naphthalate, and polyethylene terephthalate. Gas diffusion layers are made using porous substrate comprising of fluorinated polymers such as polytetrafluoroethylene (PTFE) and fluorinated ethylene propylene (FEP) [24]. Due to its ability to transport gases or components of dissolved liquids, a widely used polymer in PEMFC is PPS-based FRP material [25].

On the functional side, the MPP has six through holes for various fluid mediums to pass in and out of the stack. For the development of the new plastic-intensive MPP, it was important to capture the entry and exit cross sections of these through holes as it enables a smooth transition of the newly developed MPP into the existing fuel cell systems. However, there were few adaptions made in the geometrical boundary conditions of these through holes, which will be discussed in the following chapter.

On the other side of the medium-pressure plate is the medium interface unit (MIU), which is a combination of MIU housing (see Figure 2) onto which different pumps, sensors and piping systems are bolted that regulate the flow of different fluid medium in and out of the fuel cell stack [26]. Although not prescribed with stiffness requirements, the MIU housing needs to be stiff enough for the various neighbouring components to be bolted upon.

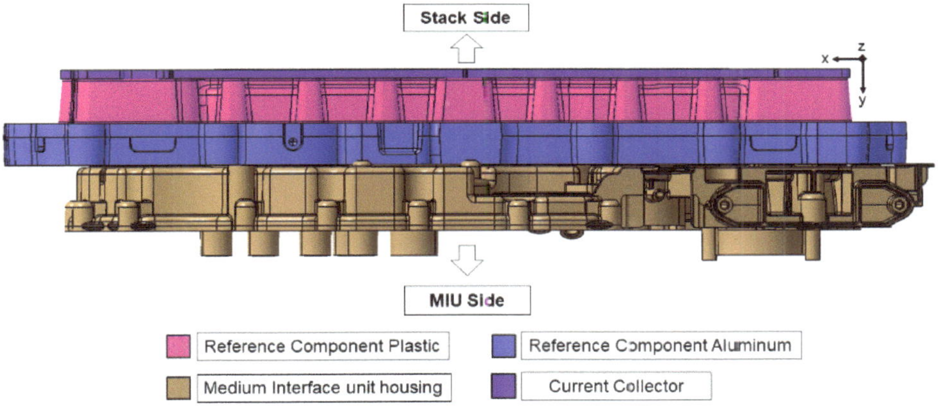

Figure 2. Reference medium pressure plate with medium interface unit.

In this work, the MIU housing will be integrated into the MPP to achieve component integration. Hence, the space occupied by the current MIU housing will be considered for the FRP design space for topology optimization. Due to the larger overall thickness of the MIU housing, it should also be able to carry loads to achieve an improved mechanical property of the entire MDP. Therefore, both material selection and the geometric design of MIU housing are important. Since FRP components are predominantly manufactured with ribs for reinforcements and functionality, the design and placement of these ribs need to be conducted to withstand loads. A topology optimization is one way to predict the rib structure for a given load case.

1.2. Topology Optimization

Topology optimization is basically optimization where a pre-defined design space is created, which is then applied with geometrical and manufacturing constraints based on the functionality of the resulting component. During a topology optimization, an FE simulation is carried out over different iterations along with a design objective. The optimization variables are continuously adjusted over the different iterations until iterative tolerances are achieved [27]. On reaching convergence, the final result is normally a rib geometry with the best possible scatter of material in terms of load-bearing capacity.

The methodology behind topology optimization is a density-based approach incorporating the solid isotropic material with the penalization method (SIMP). In this method, the material density of each element (ρ) of the design space is set as a design variable and varied continuously between 0 and 1 [28]. Here, the densities 0 and 1 correspond to the condition of void or presence of an element, respectively. As the optimization tends to reach convergence, the penalty factor (p) eliminates elements with intermediate densities. In Equation (5) $\underset{\sim}{\boldsymbol{K}}$ is the penalized stiffness matrix, and **K** is the real stiffness matrix [15,28].

$$\underset{\sim}{K}(\rho) = \rho^{P}\mathbf{K} \tag{5}$$

An optimization with an overall goal of attaining a rib structure with a fraction of mass removal from the design space allowed is normally set with a compliance minimization criterion. Compliance, as given in Equation (6), is defined as the flexibility of the component and the inverse of stiffness. This means a structure's global compliance (C) is the addition of the strain energies of the elements of the design space. Reduction of the global compliance maximizes the global stiffness as given in Equation (7), where e are elements in the design space [29].

$$C = \frac{1}{2}u^{T}\mathbf{K}u \tag{6}$$

$$\min C(\{\rho\}) = \sum_{e=1}^{N}(\rho)\left[u^{T}\right][K][u] \tag{7}$$

The culmination of methodology with optimization constraints and objectives is set into an iterative phase, wherein different rib structures are built based on the goal, and eventually, a raw rib structure is formed which is in compliance with the given load case and at the same time adhering to the applied manufacturing constraints.

1.3. Compression Molding

A well-known manufacturing technique used in the automobile industry to produce Fibre-reinforced plastic parts is injection molding. During the injection phase, fibre lengths reduce through shearing of the rotating screw and typically have lengths between 400–700 microns. This fibre length is further reduced through the injection nozzles. Increasing the fibre length can significantly increase mechanical properties, especially strength and energy absorption [30,31]. The property that increases fibre length gives better mechanical properties and paves the way for compression molding.

Compression molding employs similar machines to produce the raw FRP compound with a given percentage of fibre content. As the name suggests, in compression molding, a vertical press machine is used to press the parts into shape rather than a horizontal machine used in injection molding [32]. A schematic representation of compression molding is shown in Figure 3. Due to the delayed introduction of fibres in the extruder, fibre damages are mitigated. Here, the molten FRP taken from the extruder is placed in the open mold cavity and pressed into the cavity contour through press force, where a fibre length of up to 25 mm could be achieved in the final part [33].

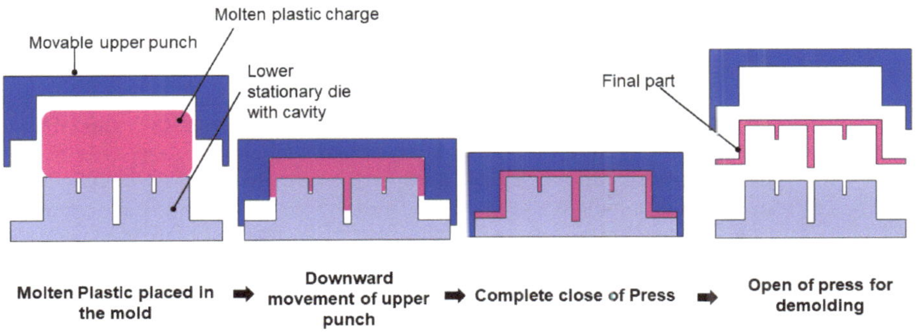

Figure 3. Compression molding schematic.

The advantage of achieving high fibre lengths in parts makes compression molding a useful manufacturing method that will be used for the manufacturing technique for this work.

1.4. Target of the Work

Liu. B et al. [14] have presented research on the virtual shape optimization of pressure plates by studying the shape-related parameters that influence the final design of pressure plates. Similarly, in the works of Liu [14], Lin P et al. [13] and Dey. T et al. [15] have also presented a virtual design solution for pressure plate designs through various optimization methodologies. These studies end with only the virtual design and development of different pressure plates. This opens the avenue for the authors of this paper to not only virtually develop a new design concept for the medium-pressure plate but also extend this development chain into the suitable material selection, manufacturing the new concept and performing mechanical tests.

Furthermore, the benchmarked fuel cell stack and the corresponding MPP shown in Figure 2 are foreseen for Fuel Cell Units of bigger passenger vehicles like sport utility vehicles (SUV). Therefore, the target of the current work is to develop a new steel plus plastic-intensive medium-pressure plate that can save package space and reduce weight. In order to achieve this goal, another kind of FRP material must be used instead of standard PPS with higher strength and Young's modulus. The idea is to use the FRP not only as an insulator but also as a structural component of the pressure plate. Since compression molding enables a higher strength of FRP, this technique should be used, and both material properties and the manufacturing process for the pressure plate should be investigated using this process. In addition, the chemical resistance of the FRP materials must be measured for the final material selection.

The new plastic-intensive MPP should also enable a high component integration besides high strength and rigidity, which should be obtained by using topology optimization. The new space-saving MPP should enable the usage of fuel cells in smaller vehicles instead of larger SUVs.

2. Conceptualization and Process Framework

In order to achieve the above-mentioned goals, certain adaptations to the installation space had to be taken into consideration. These adaptations were decisions made in close cooperation with the OEMs with a futuristic view of the fuel cell system requirements.

The basic idea of the new plastic-intensive MPP is to integrate the MIU housing into the MPP and use FRP material with enhanced E-modules and strength compared to the state-of-the-art. The functionality of MIUs should be combined with their stiffness and strength so that the MPP can be designed with reduced thickness. In addition, since the specific bending stiffness of aluminium is not higher than that of steel, a steel plate should be used to reduce the thickness of MPP by more than 50%.

As shown in Figure 4, the overhang of the MIU housing in the Z-direction and a few sensor hole features were neglected to reduce the complexity of manufacturing the reference medium-pressure plate (Figure 4a) after discussion with the OEM. Opening for the coolant out has been moved to the new position, as shown in (Figure 4b). The cavity for the cooling channel, depicted in orange in Figure 4b, behind the current collector, was deemed unnecessary and thus removed. In addition, it was decided that the bolting locations for various module units had to be recreated as in the reference MIU housing. Gasket lines, where necessary, were also recreated exactly as in the reference MIU housing to facilitate the ease of assembly integration and the use of existing parts.

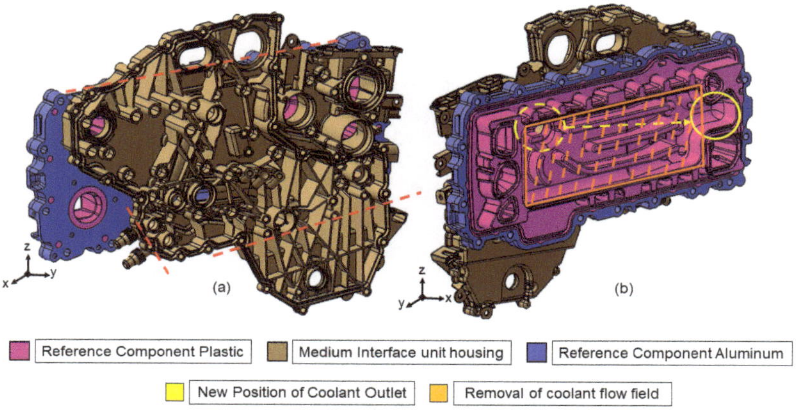

Figure 4. (**a**) Adaptations in the existing medium interface unit housing and (**b**) adaptations in the reference pressure plate.

Once the conceptual adaptations were fixed, a methodology was developed to achieve the goal of this work. The methodology takes into account all the key constraints and boundary conditions like stiffness and function requirements. As shown in Figure 5, the process starts with the reference phase, where stiffness performances of the existing MPP are benchmarked through FE Simulation of the critical load case. Additionally, the package space required for the existing MPP and MIU was taken into account.

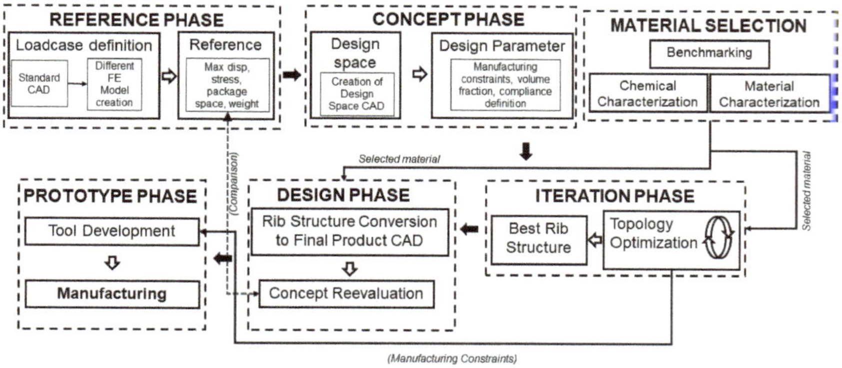

Figure 5. Process framework.

This was followed by the concept phase, where the design space for the topology optimization was first created, and further conditions, such as manufacturing and geometrical conditions, such as rib height, rib thickness, and draw directions, were defined.

These conditions eventually determine the outcome of the rib structure over an iterative FE analysis coupled with optimization. In the material selection and characterization phase, suitable materials were chosen and characterized for their strength and chemical behaviour. The obtained rib structure is taken into the design phase, where it is built around and converted into the final plastic-intensive MPP in a CAD environment, capturing important functional aspects like sealing lines or neighbouring component bolt locations. Here, within the sub-phase of concept revaluation, the final MPP is subjected to the critical load case to see if it meets the expected structural requirements.

For the topology optimization and the final FE calculation of the MPP performance, the properties of the FRP must be determined as FEM material input data. Several FRP materials were thus analysed in comparison to the reference materials.

Once the final CAD was approved, the concepts for tool design for the manufacturing were developed, and the press tool was developed. Here, manufacturing constraints used during the topology optimizations were fed as input for the press tool development. The new plastic-intensive medium-pressure plate was manufactured with the press tool manufactured and taken into commission.

3. Materials

3.1. Material Characterization

As mentioned in Section 1.3, the newly developed MPP in this work will incorporate a higher percentage of FRP material in its construction and use steel plates for stiffness and strength purposes. As mentioned in Section 1, the fuel cell stack works at extreme conditions of temperature and working mediums, which raises the requirements of the materials being used for the manufacturing of MPP. The existing reference MPP has an injection molded short glass fibre-reinforced PPS layer with a 30% fibre content. Within the scope of this work, four different materials described in Table 1 were investigated and characterized. Compared to the reference PPS material, polypropylene (PP) and polyamide (PA) are standard thermoplastic materials and very much cheaper. It could be used to reduce the cost and increase the performance of MPP if their properties are acceptable or even better than the PPS reference.

Table 1. Materials selected for characterization.

Material Name	Sample Norm and Manufacturing Technique	Material Description
PP-X121 F42	DIN EN ISO 527-4 Type 2, Rectangular sample—water-jet-machined	Polypropylene-based glass mat thermoplastic with 42% glass fibre content, including 2 layers of woven fibres
PP-X103 F61	DIN EN ISO 527-4 Type 2, Rectangular sample—water-jet-machined	Polypropylene-based glass mat thermoplastic with 61% glass fibre content, including 2 layers of woven fibres
PA66-GF50	DIN EN ISO 527-2 Type 5A, shoulder sample—water-jet-machined	Polyamide-based thermoplastic with 50% short glass fibre
Reference PPS GF30	DIN EN ISO 527-2 Type 5A, shoulder sample—direct-injection molded	30% short glass fibre content

The materials were analysed in terms of their chemical and mechanical behaviours. The chemical behaviour includes mainly the resistance of the material against the acid medium, and the mechanical properties were measured by quasi-static tensile tests. The same samples were used for both analyses.

Depending on the fibre length of the material, the sample norm was selected. The material architecture of PP-X121 F42 and PP-X103 F61, delivered by Mitsubishi Chemical Advanced Materials, consists of layers of woven fabrics and chopped glass fibres of length 50 mm. In order to avoid damage to the long glass fibres of the PP-based FRPs, rectangular samples of standard DIN EN ISO 527-4 Type 2 were cut using water jet machining. The PA66 50% glass fibre material was manufactured as flat plates using injection molding from which DIN EN ISO 527-2 Type 5A samples were taken out using water jet machining. The Reference PPS material with 30% short glass fibre was directly manufactured as DIN EN ISO 527-2 Type 5A shoulder samples using injection molding.

A major difference here was that the rectangular samples of PP-based materials and the shoulder samples of PA66-based materials had open fibre ends on the waterjet machined edges; however, the injected molded shoulder samples had an injection polymer skin, which acts as a protection layer for the fibres.

For chemical characterization, no common standard exists. Therefore, in this work, the authors proposed the following chemical solutes and testing procedures in cooperation with OEM partners. Firstly, a solution with a mixture of distilled water and concentrated sulphuric acid (H_2SO_4) was prepared until an acidity level or pH level of 3 was 3 reached. To this mixture, 30 mg of Sodium Fluoride (NaF) salt was added per litre. This solution was termed product replacement water and recreates the worst-case working environment scenario in a PEMFC. The addition of NaF salt releases sulphate ions (SO^{4-}), sulphite ions (SO^{3-}), hydrogen sulphate ions (HSO^{4-}), and hydrogen fluoride (HF) acid is also produced.

Three samples from each material in Table 1 were placed inside a neutral, inert container filled with the fuel cell product replacement water at 95 °C. After a period of 720 h (1 month), the pH level of the product replacement water from each sample container was measured using a pH level detector. Here, an increase in pH level means an increase in acidity level, which can have detrimental effects on the fuel cell system. In addition to this, the samples were weighed before and after the immersion process to check the level of water absorption by the samples. High levels of water absorption are bad for electrical isolation, which is one of the major responsibilities of the layer of plastic on the MPP [6].

It can be seen from Table 2 that the pH value of the water rose less for the reference PPS-based FRP material than for the other materials. The reason can be found in the sample preparation and material composition. As can be seen in Table 1, the reference PPS material has less glass fibre content and no open edges since it was directly injection moulded, whereas the other three materials have higher glass fibre content with open edges due to the water jet cutting of the sample. It can be seen that the product water taken from the PP-based materials container has a lower pH value than that of the PA-based materials.

Table 2. pH values, conductivity, and water absorption of the selected materials.

Material	Fibre Orientation	pH Value	Conductivity (μS/cm)	Sample Weight before Immersion (g)	Sample Weight after Immersion Period (g)	Difference (g)	Difference (%)
Product replacement water	-	3	414	-	-	-	-
PP-X121 F42	0°	6.72	266	19.497	19.636	0.139	0.71
	90°	6.76	271	19.412	19.511	0.099	0.51
PP-X103 F61	0°	6.73	257	21.077	21.305	0.229	1.09
	90°	6.52	261	22.098	22.240	0.142	0.64
PA66-GF50		7.25	328	6.122	6.272	0.150	2.45
Reference FRP PPS GF 30		5.49	210	9.877	9.940	0.063	0.64

Considering the water absorption of the different FRP materials, the change in the sample weight shows that the PP-based materials have absorbed less water than the PA-based FRP materials. The values of PP-based materials are quite comparable to those of the reference PPS material.

Furthermore, a second step of chemical characterization was carried out. The chemical characterization was conducted by an Inductively Coupled Plasma Optical Emission Spectroscopy method (ICP-OES) using an Arcos-2 ICP-OES analyser apparatus from Spectro Analytical Instruments, with a micro-mist cross-flow nebulizer. Argon gas was used as a plasma gas. As operational parameters, 1.4 kW plasma power with 30 RPM pump speed and a coolant flow rate of 13 L/min was used. ICP-OES is an analysis technique used to measure the element content in a sample following the principle that ions have the ability to absorb energy and reach excited states, which eventually release light at a certain wavelength [34]. ICP-OES analysis was carried out by taking 5 mL of product water from the stored sample containers after a period of 720 h to determine the washed-out elemental composition.

Table 3 shows the concentration of glass fibre components in product replacement water measured using ICP-OES analyses. Considering the reference PPS material as the benchmark values, the PA66- and PP-based FRP materials were evaluated. PP-based materials had comparable or lesser leaching content of Al, B, Li and Mg elements when compared to the reference PPS material; however, Ca, K and Na have leached out more. A comparison between the PA- and PP-based materials clearly shows that except for Si and Na, all the other elements had leached out at higher quantities by the PA-based materials.

Table 3. Concentration of glass fibre components in product water.

Material	Fibre Orientation	Al [mg/L]	B [mg/L]	Ca [mg/L]	K [mg/L]	Li [mg/L]	Mg [mg/L]	Si [mg/L]	Na [mg/L]	Sr [mg/L]
Product water	-			-	0.30	-	-		15.68	
PP-X121 F42	0°	0.63	0.47	24.12	2.26	0.03	-	40.73	20.42	0.04
	90°	1.16	0.28	26.49	2.71	0.05	1.17	41.77	20.33	0.04
PP-X103 F61	0°	1.29	1.37	27.22	3.74	0.04	1.66	40.78	19.70	0.08
	90°	0.58	1.57	24.02	3.95	0.04		37.82	20.62	0.08
PA66-GF50		2.26	2.55	21.08	7.98	0.05	1.54	90.98	17.90	0.25
Reference FRP PPS GF 30		1.37	1.70	21.20	0.58	0.04	1.58	32.53	17.40	0.06

The above two material characterization studies have shown that PA-based materials show large differences in pH value, conductivity, and water absorption rates, which is not desirable for the working conditions of the fuel cell system. Moreover, through the ICP-OES analysis, it was seen that the leaching of glass fibre elements was also large for the PA-based materials. Hence, it was decided at this stage that the PA-based materials will not be used for the present work. The PP-based material showcased a performance that was slightly different from the reference PPS material. It was thus carried forward and further investigated by tensile tests.

The tensile tests were carried out with three samples each at room temperature, 80 °C and 95 °C. In addition, three samples of each material, which were previously stored in the acidified product replacement water, were also tested at 95 °C. The quasi-static tensile tests were carried out using an Ibertest Testcom-50 test machine with a testing speed of 2 mm/min. For the tensile tests at 80 °C and 95 °C, a thermal climatic chamber was used. The displacements were recorded using an extensometer. Before the start of the temperature-dependent test, the chamber is kept at a set temperature for 90 min to ensure homogenous heat distribution.

The tensile testing results shown in Figure 6a clearly show that with increasing testing temperature the strength of the reference PPS material decreases considerably. The low short fibre content in the reference FRP material does not provide enough strength for the material, especially at higher temperatures. The tensile strength of around 150 MPa at room temperature falls to around 60 MPa at higher temperatures.

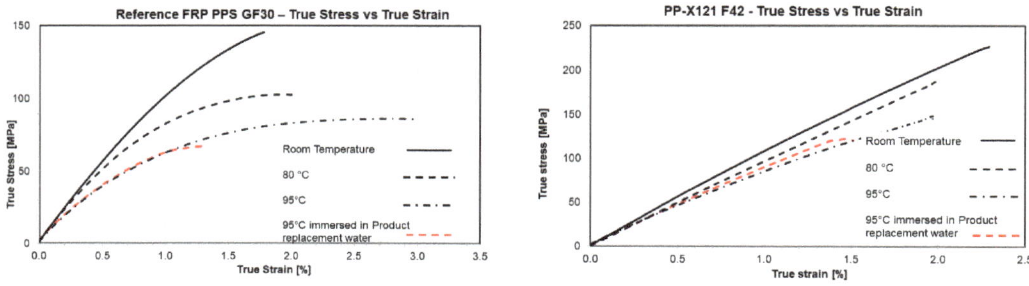

(**a**) (**b**)

Figure 6. Tensile test results at different temperatures (**a**) for reference PPS material, (**b**) for PP-X121 F42 material.

The tensile test for PP-X121 F42 material was conducted on probes made from 2 layers of material pressed together. Due to its higher glass fibre content along with the four layers of woven fabric, the PP-X121 F42 GMT material shows superior performances at all testing conditions in comparison with reference PPS material. At room temperature, the PP-X121 F42 was able to reach a tensile strength of around 226 MPa, and at a higher temperature of 95 °C, the loss in tensile strength was also comparatively small, as can be seen in Figure 6b. The performance of the PP-X121 F42 material in impact load cases is also superior due to the underlying material architecture, as can be seen in [35].

The major mechanical values of both materials are summarized in Tables 4 and 5. In this work, on the basis of the superior performance in the tensile tests, PP-X121 F42 FRP material was selected as the material to create the rib structure of the plastic-intensive MPP.

Table 4. E-modulus and tensile strength of reference PPS material at different temperatures.

Material Properties Reference PPS	Room Temperature	80 °C	95 °C	95 °C Immersed in Product Water
E-Modulus [GPa]	11.5	10.05	7.57	8.31
Tensile Strength [MPa]	145.4	100.7	82.4	67.2

Table 5. E-modulus and tensile strength of PP-X121 F42 material at different temperatures.

Material Properties Reference PP-X121 F42	Room Temperature	80 °C	95 °C	95 °C Immersed in Product Water
E-Modulus [GPa]	10.6	9.88	9.1	9.26
Tensile Strength [MPa]	226.6	187	147.4	122.1

3.2. Architecture of PP-X121 F42

The PPX121-F42 has a complex architecture. As shown in Figure 7, each pre-fabricated extrudate of this fibre-reinforced composite comprises two layers of endless fibres, and the remaining layers have chopped glass fibres of length 50 mm embedded in the polymer matrix, altogether resulting in a 40% weight in fibre content.

The endless glass fibres form a layer of woven fabric with an 80–20% architecture, where there are 80% of the fibres in the longitudinal direction and 20% of the fibres in the transverse direction with a biaxial plain weave. The chopped glass fibres are smeared into the polymer matrix.

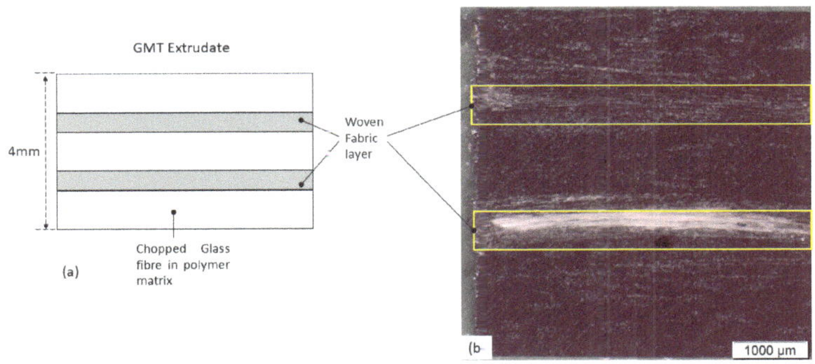

Figure 7. (**a**) PP-X121 F42 GMT schematic layup; (**b**) microscopic scan picture of PP-X121 F42 GMT.

Although the high fibre content of endless and chopped glass fibre has a positive effect on the mechanical properties of the material, it has its drawbacks in product manufacturing and development. Due to the complex architecture of woven fabrics and chopped glass fibres of length 50 mm, when using the extrudate for forming a component with rib structures, it was seen in Figure 8 that the ribs do not get entirely filled with fibres. When the extrudate is pressed into the desired rib structure through compression molding, the fibres flow only up to a certain depth of the ribs. The fibres undergo a phenomenon called the "bridging effect", where, due to the complex fibre architecture, the fibres tend not to flow the complete length of the ribs and stop halfway through the rib [35,36].

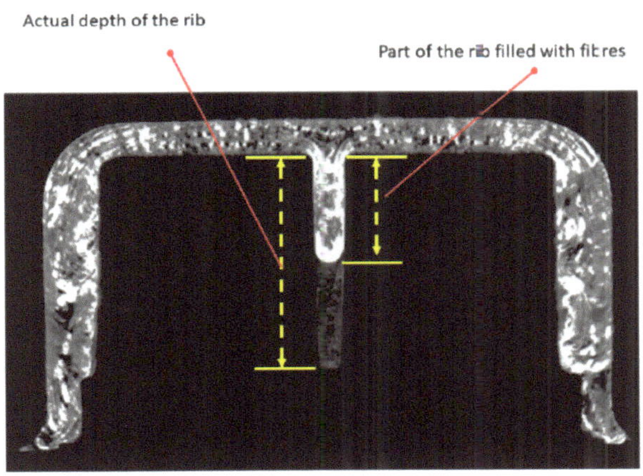

Figure 8. CT ccan of U-profile component with a middle rib [36].

Figure 8 shows a CT scan performed on a compression molded U-profile component with a vertical middle rib. The white areas show the presence of fibres, and the dark areas in the component show the regions only with polymer. This drawback of GMT material was also the reason for not choosing the PP-X103 F61 FRP material, which has an even higher percentage of fibre that hinders the flow. For product development, this fact must be kept in mind. Since the fibre filling of this kind of GMT material cannot be simulated using FEM, many detailed testing and analysis works are required for each specific part to be developed. The material properties for the FE model must be assigned manually for the deeper ribs.

3.3. Steel

For the steel plate, S355MC steel was used to obtain the necessary thickness based on the topology optimization results. Steel plate based on the final CAD design was cut to contour using water jet machining. The through holes for the fluid medium to flow through were cut in accordance with the cross-section of the holes. The steel plate will be placed and inserted into the press tool onto which the plastic GMT is pressed.

4. Fem Methodology and Target Setting

As mentioned in [11], the MPP undergoes high axial loads through the assembly of fuel cell stacks in order to reduce contact resistance and avoid any slippage of stacks under high vehicle accelerations. Prior to designing the new plastic-intensive MPP, the existing MPP needs to be benchmarked for its stiffness and rigidity to obtain the reference requirements for this work.

The geometry considered for the reference FE calculation is shown in Figure 9, which consists of the reference medium-pressure plate of aluminium with a layer of FRP mantled on the aluminium. Here, aluminium is defined as Young's modulus of 70 GPa and yield strength of 200 MPa. The reference FRP PPS material is defined as a Young's modulus of 11 GPa and yield strength of 145 MPa based on the data obtained in Section 3 on reference injection molded PPS FRP material. The MIU is not considered in the FE calculation as it is an additional component which is later screwed onto the MPP and thus does not contribute to the stiffness requirement of the system.

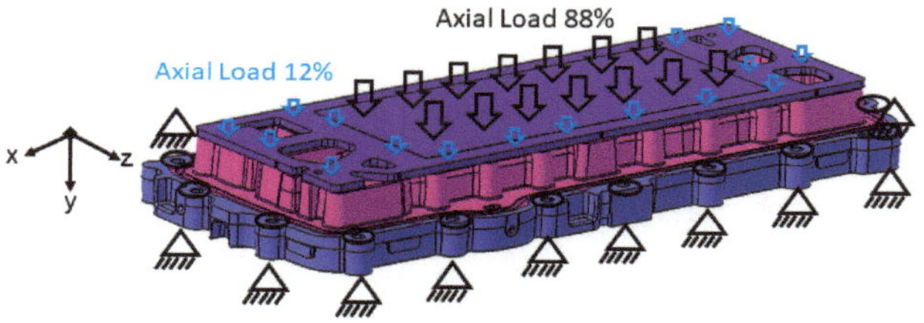

Figure 9. Load case description.

For simplification purposes, the ideal node-to-node connection is applied between the aluminium and FRP. An isotropic elastic–plastic material model was used for both aluminium and FRP layers. On top of this FRP layer is the current collector, which has frictional contact defined by the FRP layer. As depicted, the current collector transfers the axial load onto the medium-pressure plate. The load is split into two parts, whereas a major part of the load (around 88% = 53.7 kN) acts on the middle region of the current collector, where the majority of the force of bipolar cells acts. The remaining part of the load (around 12% = 7.96 kN) acts on the periphery region or on the gasket region. The load implementation was carried out based on the OEM specifications.

The components were meshed with tetrahedral elements with a mesh size of 3 mm. The medium-pressure plate is fixed in the bolt locations using RBE2 elements with the independent node fixed in all degrees of freedom. An implicit conjugate gradient solver with convergence criteria of 1e-6 was set to perform the simulation of this non-linear static load case in Optistruct by Altair. This FE methodology will later be used to evaluate the new plastic-intensive MPP.

It can be seen from Figure 10 that the highest displacement occurs in the middle region of the medium-pressure plate, where the highest axial force of the stack occurs. The stresses are highest around the bolting location in the aluminium part. For the given axial load, the stresses in the aluminium reach a maximum of 120 MPa, and the FRP reaches a maximum of 15 MPa.

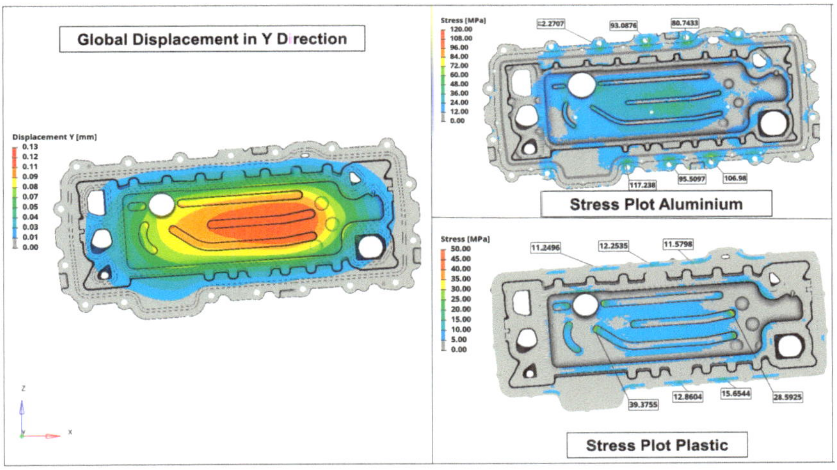

Figure 10. FEM results of reference medium-pressure plate.

5. Development of New Mpp

5.1. Topology Optimization Setup

Once the FE Calculations for the reference MPP were carried out, the design space for the topology optimization was constructed. As mentioned earlier in this work, the new plastic-intensive MPP will need to combine the functionalities of the MIU and perform the operations of MPP by guaranteeing strength requirements and holding the bipolar plates in place. Henceforth, the package space of the existing MIU is also taken into consideration for the topology optimization. Here, a steel plate of thickness of 10 mm was considered because MIU components like pumps need to be directly bolted into the steel plate with a required bolt thread depth of 8 mm in the steel plate. The rest of the design space, as shown in Figure 11, was considered to be FRP material.

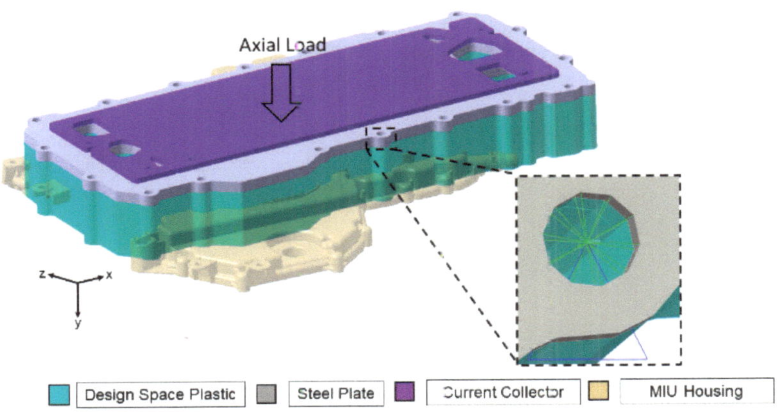

Figure 11. Design space comparison and topology boundary conditions.

For the topology optimization, an isotropic elastic material model for both steel and PP GMT was used. The Young's modulus for steel was considered to be 210 GPa with Poisson's ratio of 0.3 and a density of 7850 Kg/m^3. Since PP-X121 F42 GMT material was chosen during the material investigation in Section 3, its material properties were applied for this optimization. The average Young's modulus is 10.6 GPa, Poisson's ratio is 0.3, and the density is 1500 Kg/m^3 (see also Table 5). Here, a single material topology optimization was set up, and all the design constraints and variables were set up for the FRP design space. As shown in Figure 11, the axial load from the current collector is applied to the steel plate through a tied contact. The FRP design space is attached to the steel plate with node-to-node equivalence. The non-design space elements here define the regions through which fluid medium flows through the fuel cell system and were hence not considered in the domain for topology optimization. RBE2 rigids were used to fix the steel at the 18 bolt locations in all degrees of freedom, similar to the load case as defined in Section 4.

As part of the design constraints, the FRP design space is set to a volume fraction of an upper bound of 0.5. This allows the optimizer to remove only a maximum of 50% of the available design space elements in the iterative process of converging to the load-compliant rib structure design. In addition to this, a compliance variable is applied to the plastic design space. As discussed in Section 1.2, the target is to minimize the compliance or to maximize the stiffness. Minimizing compliance in Equation 6 translates to reducing displacements for a given constant load. Hence, the compliance variable of the plastic design space is called into a designing objective, which sets a target of minimizing compliance with the plastic design space and achieving an even distribution of materials for the given load case. The minimization of compliance can be represented by elaborating Equation (7) as

$$\text{minimze}: \left\{ \omega_1^{\rho} \left[\frac{C(\rho) - \text{Cmin}}{\text{Cmax} - \text{Cmin}} \right]^{q} \right\}^{(1/q)} \tag{8}$$

where ω_1 is the weighting coefficient, and q is the volume fraction, weighted as 1 due to single objective optimization in our case, which is bending. For multiple load cases, the weightage will be divided based on critical criteria, having different values between 0 and 1. $C(\rho)$ is the function of compliance to density, and Cmax is the corresponding compliance response of structure at each iteration. At each iteration, in relation to the given volume fraction and volume distribution, the overall compliance matrix of the structure is solved. Here, each iteration is converged with a particular volume distribution and a corresponding stiffness of the structure [13]. Eventually, total convergence is attained with the most optimal rib distribution and minimal displacements.

Additionally, manufacturing constraints were set to the plastic design phase, which includes a minimum thickness of member, min. dim. = 3.0 mm, a maximum thickness of member, maximum dim. = 6.0 mm, a minimum spacing between the structural members, minimum gap = 15 mm and a draw direction of y = 1. These constraints were selected based on the feasible rib design guidelines prescribed for plastic components.

5.2. Topology Optimization Results

With the boundary conditions as defined in Section 5.1, the topology optimization was carried out. The optimal rib structure needed to withstand the applied load converged after 55 iterations. Figure 12 shows the evolution of material distribution to which the solver converges over these 55 iterations. It is seen that between iterations 0 to 10, the solver first removes the complete material from the plastic design space. From iteration 10 onwards up to 40, the solver distributes material in possible ways to counteract the force by which the system is loaded. Within these iterations, the solver tends to distribute material across the steel plate. The 40th iteration shows longitudinal disjointed rib structures being formed along the Z direction. In this iteration, the rib structures have formed a clear path against the load. However, the optimization continues further as the discontinuous rib structures are not robust enough to withstand the load.

The optimization further continues until iteration 55 to reach convergence. The converged 55th iteration shows continuous stable rib geometries. It can be seen that the final rib structures, with element densities of 1, formed directly beneath the load path, with continuity and where the highest deformations occur. In certain areas, the longitudinal ribs cross paths, depicting the necessity for these ribs to be supported with ribs running across the X-direction of the design space.

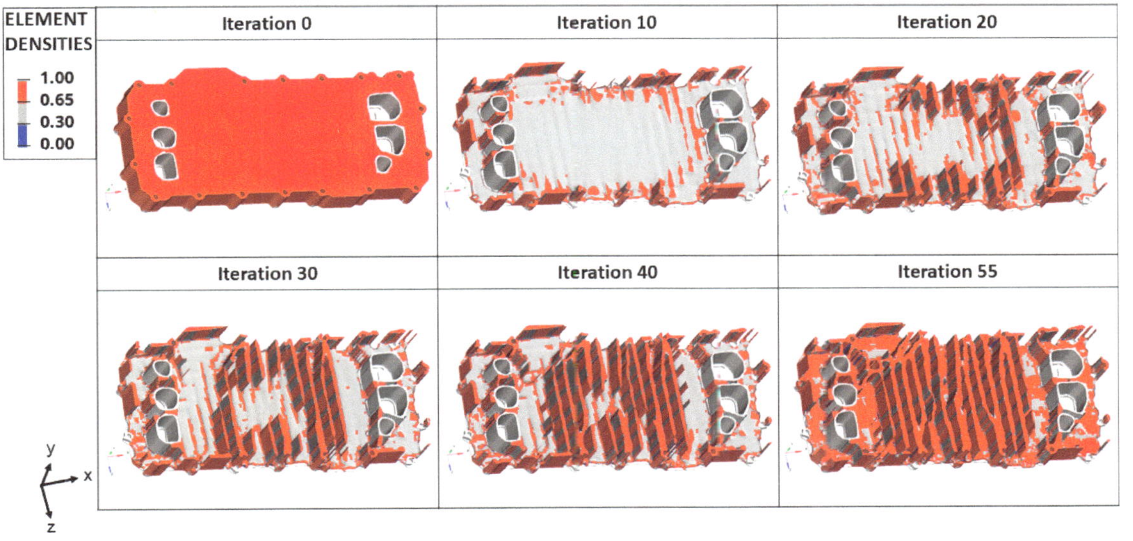

Figure 12. Topology optimization results, view from MIU side.

5.3. Conversion of Rib Geometry to Final Plastic-Intensive MPP

Once the topology optimization results were obtained, the raw rib structure was exported as an STL file and imported into the CAD environment. The obtained raw rib structure had to be utilised in the best possible way in creating the plastic-intensive MPP, whereby the functional aspects of the MPP and the MIU housing were adhered to, and manufacturing constraints were met as well. Additionally, the constraints of the material discussed in Section 3 were also taken into consideration while creating the rib structures. Figure 13 shows the results of topology optimization overlaid with the existing MIU. The obtained rib structure overlaps many existing functional features. This hinders the exact replication of the rib structure in the designing of the plastic-intensive MPP. It also shows that some of the functional features of MIU lie directly on the load path where a high concentration of rib structures was produced from the topology optimization.

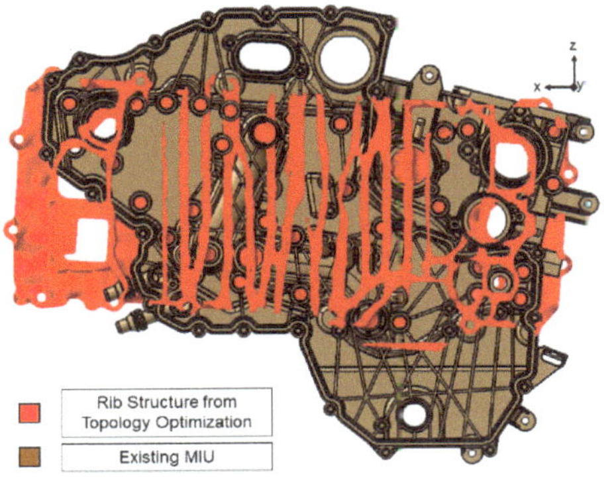

Figure 13. Comparison between topology optimization result and existing MIU housing.

5.4. Plastic-Intensive MPP—Stack Side

Upon discussion with the OEM, we found that the current collector was smaller than the old collector. The new collector, made of aluminium, will contact only the critical area of neighbouring bipolar stacks where the compressive load from the single bipolar stacks is also high. The aluminium current collector needs to be insulated from the Steel plate, and hence, a layer of plastic with rib structures is constructed into which the current collector is embedded. This plastic layer also incorporates the cross sections of the through holes through which the various fluid materials will flow in and out of the fuel cell system. The cross sections for the fluid medium holes on the stack side were directly replicated from the existing medium-pressure plate for ease of integration with the existing fuel cell stack system.

Furthermore, the slots for the gasket around the fluid medium holes were also replicated from the existing medium-pressure plate, which will enable the reuse of existing gaskets. The plastic layer here has a base thickness of 3 mm, and the ribs have a height of 5 mm and draft angle of 1°. As shown in Figure 14, the new current collector has a 3° undercut draft angle, which enables a snug fit with the plastic.

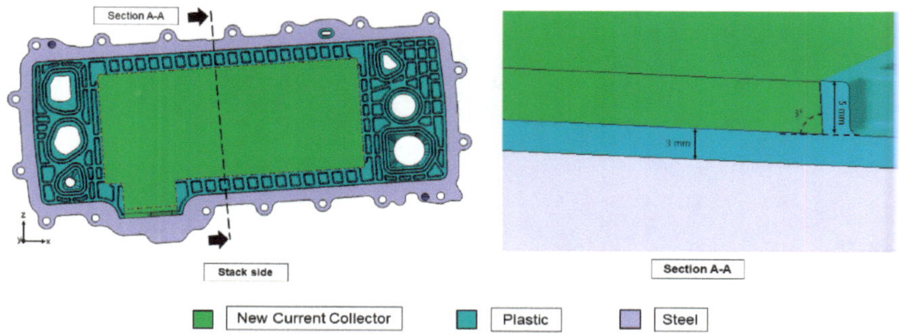

Figure 14. Newly developed MPP—Stack-side design.

5.5. Plastic-Intensive MPP—MIU Side

Figure 15 shows the rib construction of the plastic-intensive MPP on the MIU side. For this, firstly, all the bolting regions for the different pumps and cross sections of different fluid mediums were taken over directly from the existing MIU housing. Around these features the rib structures were designed. Similar to the construction on the stack side, a 3 mm layer of plastic was first constructed onto which the ribs were constructed. The rib structures were restricted to a height of 12 mm, as deeper rib structures would not be filled with fibres and rather filled only with polymer, as shown in Figure 8 [36].

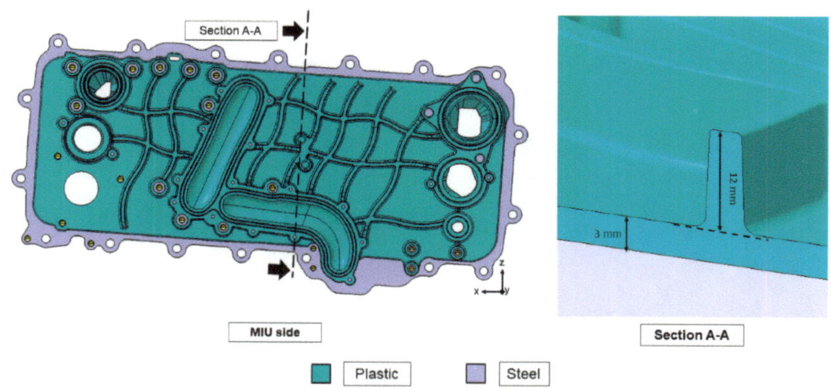

Figure 15. Newly developed MPP—MIU dide.

5.6. Plastic-Intensive MPP Concept Evaluation

Once the final CAD design of the plastic-intensive MPP was done, its stiffness was calculated through FE Simulation. The FE setup and boundary conditions defined in Chapter 16 were used to set up the FE simulation for the plastic-intensive MPP. Only the materials were replaced by the new materials, steel for the reinforcement plate and PP-X121 F42 GMT for FRP rib structures. The steel plate was defined a Young's modulus of 210 GPa and a yield strength of 360 MPa corresponding to that of S355MC grade steel. Although the PP-X121 F42 GMT is an anisotropic material, in this scope of work, the authors have represented it using an elastic–plastic material. The GMT was defined as an isotropic elastic–plastic material model with Young's modulus of 10.6 GPa and the corresponding plastic flow curve of PP-X121 F42. In the plastic-intensive MPP concept, the current collector that applies the load onto the MPP was made smaller than the existing design. Hence, the load was not split as discussed in Chapter 16, Figure 9, but rather completely applied on the middle region where bipolar plates contact the current collector.

Figure 16 shows the displacement and stresses that occur in the plastic-intensive MPP. It can be seen that the displacements in the Y direction have increased to 0.22 mm when compared to the displacements in reference MPP, which is 0.13 mm. The increase in the displacement is attributed to the fact that the complete axial load is applied on the middle region of the plastic-intensive MPP through the new current collector. The stress plot shows that the stresses in the plastic reach a maximum of 50 MPa, which is within the tensile and compressive strength of the PP-X121 F42. The highest stresses are concentrated on the ribs holding the current collector in place and near the bolting locations. The stresses in the steel plate reach a maximum of 300 MPa in the regions near the bolting locations, which is well within the yield stress of the steel material used.

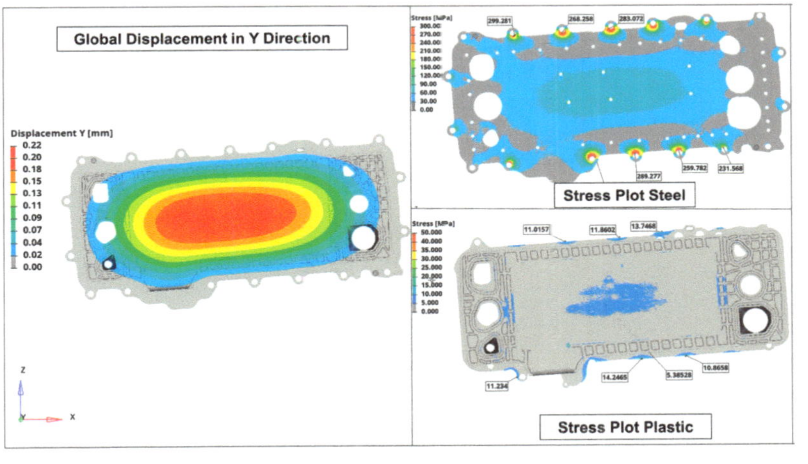

Figure 16. Newly developed MPP—FEM results.

On evaluating the plastic-intensive MPP for its package space, it has an overall height of 45% of the reference plate in the Y-direction, which is 55% less than the existing MPP and MIU housing together, as depicted in Figure 17. Comparing the weights, the plastic-intensive MPP has an 8% weight reduction compared to the existing MPP and MIU housing put together. The saved space in the new MPP can be used to assemble 50 more bipolar plates in the fuel stack, which increases the power of the fuel stack by approximately 15 kW.

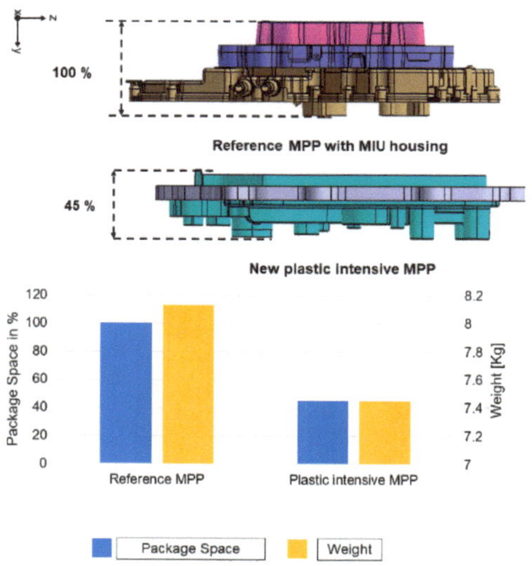

Figure 17. Package space and weight comparison between reference and new MPP.

6. Manufacturing

Once the final design was approved by the OEM, the press tool to manufacture the plastic-intensive MPP through compression molding was designed. The goal of the plastic-intensive MPP design was to achieve a high-component integrated design and manufacturing process. Figure 18 shows a schematic representation of the press tool design for the plastic MPP. It consists of an upper plate and dies for the formation of ribs on the MIU side. The MIU-side die slides into the Down holder cavity, which is connected by gas springs to the upper plate. This forms the upper half of the press tool.

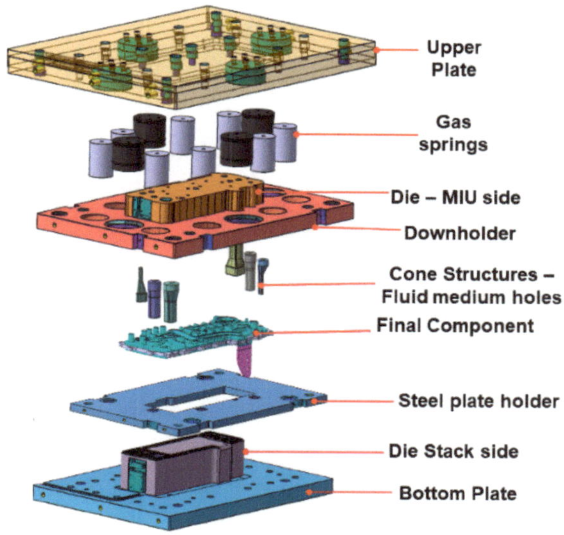

Figure 18. Press tool design for plastic-intensive MPP manufacturing.

The lower half of the press tool has a bottom plate positioned stationary on the press table, and the stack side die, which forms the rib structure on the stack side, is connected to the bottom plate. Cone-like structures, which have the shape and cross-section of the fluid medium holes, were attached to the die correspondingly, which helped form the fluid medium holes.

The press tool, specifically the dies, which come into contact with the plastic GMT, needs to be heated to a certain temperature. For this purpose, both the MIU side and Stack side dies were heated to 80 °C using thermal heating elements placed inside the dies.

Prior to the manufacturing, the required plastic PP-X121 F42 GMT was heated in an infra-red oven to up to 190 °C. The pre-fabricated steel plate was heated to 120 °C in a separate heating plate. The pre-fabricated aluminium current collector was placed into the press tool a few minutes ahead of the actual manufacturing start, which is in contact with the stack side die. As the stack side die was heated up by the thermal heating elements, the aluminium heated up simultaneously.

Once all the components have reached their required temperatures, the manufacturing procedure, as depicted in Figure 19, starts by placing first a layer of heated GMT on top of the aluminium current collector. Then, the heated steel plate was placed and positioned on the steel plate holder. A layer of heated plastic GMT is placed on the steel plate. Finally, the press is closed, where, due to the uncompressed state of the gas springs, the downholder comes in contact with the steel plate first, compressing the gas spring and holding it in position against the steel plate holder. On further pressing, the upper die moves further down due to further compression of the gas springs. This forms the rib structures on the MIU side. Simultaneously here, the stack-side die forms the rib structures on the stack side.

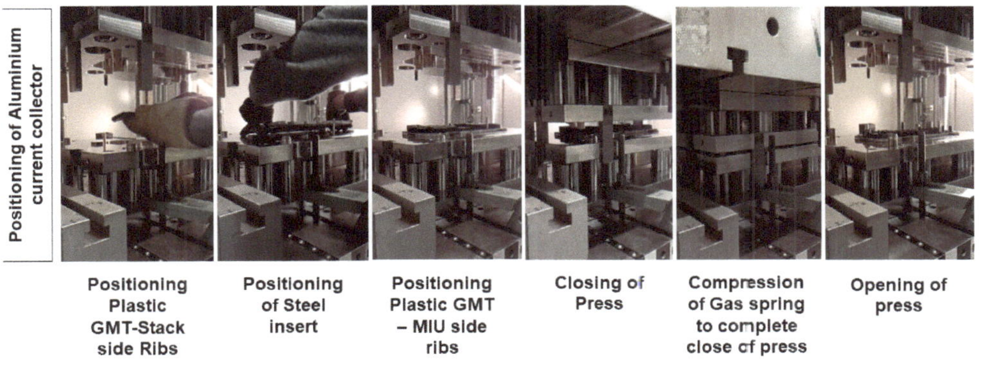

Figure 19. Process steps in compression molding of plastic-intensive MPP.

Once the press is completely closed, it is held in the closed position for up to 50 s to allow the plastic layer to solidify and cool down to reach the mold temperature. Thereafter, the mold is opened, and the final part is taken out and checked for complete filling of ribs surface quality. Based on the quality check, the temperature of the press tool dies, and the press force varies until a good quality part is produced. Figure 20 shows a final part which had a good surface finish on both the stack side and MIU side.

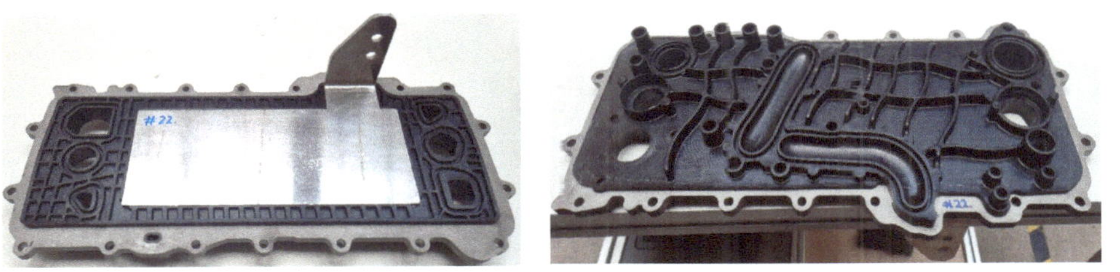

Figure 20. Plastic-intensive MPP after compression molding; stack side (**left**) and MIU side (**right**).

Once the part was taken out of the press tool, it was machined to incorporate the bolt holes into which the different pump systems would be bolted. Once the bolt holes were machined, the bolt supports and plastic inserts from the existing MPP and MIU housing were pressed into the plastic-intensive MPP. The inserts were heated and pressed to a snug fit into their corresponding position, as shown in Figure 21.

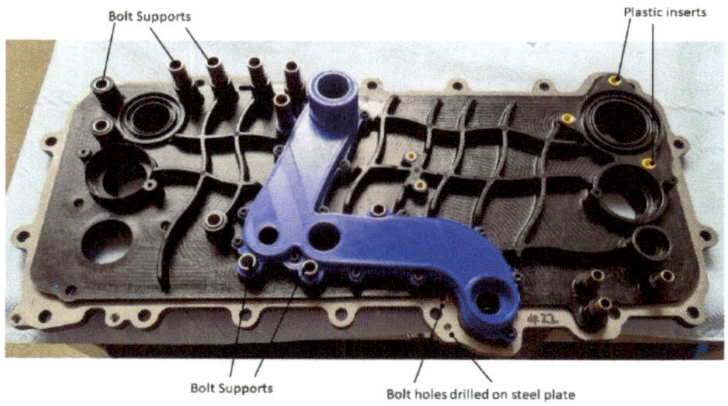

Figure 21. Plastic-intensive MPP after bolt supports and inserts were pressed.

7. Testing and Validation

The manufactured plastic MPP was then tested for its performance to verify the FE model suggested in Section 5.1. The test setup, as shown in Figure 22, consists of the plastic-intensive MPP placed on supports in the length direction at selected regions and held fest on either side in the cross direction. A dial gauge is placed directly beneath the line of the impactor, and its measuring sample is placed in contact with the plastic-intensive MPP on the MIU side.

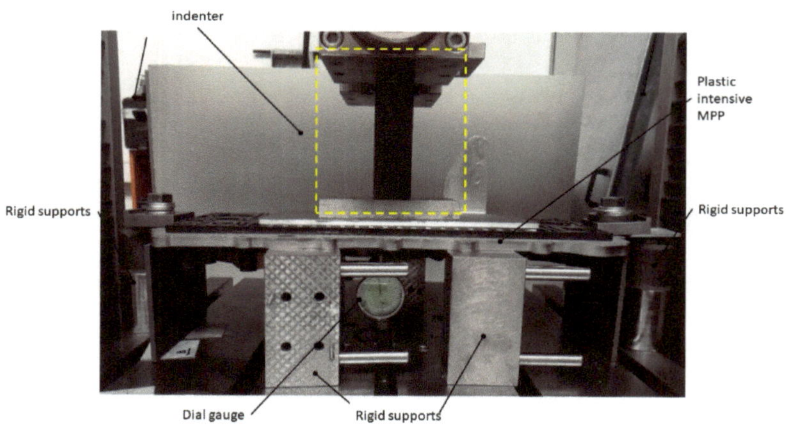

Figure 22. Test Setup for FEM verification.

Instead of a uniformly distributed load on the current collector, a rectangular indenter was used to press the MPP because it can be realized easily using a tensile test machine, as can be seen in Figure 22. The test was done using a Zwick100 tensile and compression testing machine. The displacement was measured using the dial gauge. The force was applied at 1 mm/min, and the movement of the dial gauge was recorded. On testing, the displacement reading noted on the dial gauge was 0.9 mm at the bottom-most point of measurement.

The test setup was also simulated using FEM, where the exact test setup was modelled, as shown in Figure 23a. Here, the FE model was the same as described in Sections 4 and 5.4.

However, instead of the bolting locations shown in Figure 9, the MPP was placed on supports as depicted in Figure 23a, mimicking the testing boundary conditions in Figure 22. The load was not shared amongst the nodes of the current collector as applied in the FE Model in Figure 9 and Section 5.4, but it was applied using an indenter just as in the test setup.

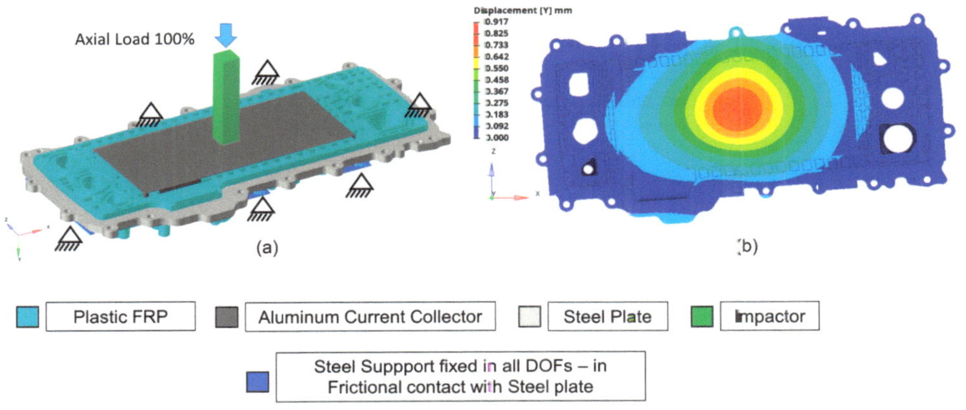

Figure 23. (**a**) Test setup in FEM simulation; (**b**) displacement plot.

The displacement calculated from the FEM simulation, as shown in Figure 23b, is 0.91 mm which produces a minimal error of 2% with the testing as compared in Table 6. The reason for the very high displacement in FEM simulation and testing is the fact that the entire axial load is concentrated through the slender indenter on a very small area of the cross-section on the current collector. Due to the good agreement between test results and FEM, the FE model used in Sections 4 and 5.4 can be considered as verified and thus also the entire development described up to now.

Table 6. Comparison of displacements in testing and simulation.

Displacement Value	In Testing	In Simulation
	0.9	0.91
% Deviation	2%	

8. Conclusion and Outlook

This work was aimed at developing a new plastic-intensive medium-pressure plate with a new manufacturing technique where weight and especially package space reduction were set as end goals. A methodology for this development was developed, which started with benchmarking the existing medium-pressure plate. Quasi-static FE analysis of the existing MPP was carried out and the displacements were determined and set as the target for the new MPP. Different FRP materials were pre-selected and analysed using mechanical and chemical testing. In close discussion with the OEM, design space and functional adaptations for the plastic-intensive MPP were fixed. Eventually, the following goals were achieved through this work.

1. PP-X142 F42 GMT material was chosen for the plastic-intensive MPP. Its mechanical properties are superior, and its chemical properties are comparable to those of the reference PPS material.
2. Using topology optimization, load-optimized rib structures for the MIU side were obtained.
3. An 8% weight reduction and a 55% package space saving were achieved through the new design, which could potentially place nearly 50 more bipolar plates and thus increase the power of the complete PEM-FC stack by 15 KW.

4. The final plastic-intensive MPP design was approved to be feasible for manufacturing: A press tool to manufacture the MPP was so designed that the pre-fabricated steel plate and aluminium current collector were placed as inserts, and the plastic rib structures were formed in accordance with the final design.
5. Manufacturing parameters which affected the final quality of the product were systematically studied within the limited number of manufactured parts.
6. The final plastic MPP design achieved a multi-component 3-in-1 design with the new current collector, medium-pressure plate and medium interface unit housing integrated into one single component. Bolting locations for the various neighbouring components were also taken into consideration.
7. The proposed FE model for MPP design was verified by tests on produced plastic-intensive MPP.

Author Contributions: Conceptualization, S.C.A., F.M. and D.H.; Methodology, S.C.A. and F.M.; Software, S.C.A.; Validation, S.C.A. and F.M.; Investigation, S.C.A., F.M., and X.F.; Writing-Original Draft Preparation, S.C.A.; Writing-Review and Editing, X.F., and S.C.A.; Supervision, X.F.; Project Administration, S.C.A.; Funding Acquisition, X.F. All authors have read and agreed to the published version of the manuscript.

Funding: This research was funded by the German Federal Ministry for Transport and Digital Infrastructure (BMVI), project name: FC-KOMP, grant number: 03B10106D.

Data Availability Statement: The necessary data for this work is contained within this article.

Acknowledgments: The authors would like to thank all partners for their support in the development of this project.

Conflicts of Interest: The authors declare no conflicts of interest.

References

1. Lamb, W.; Wiedmann, T.; Pongratz, P.; Andrew, R.; Wiedenhofer, D.; Minx, J. A review of trends and drivers of greenhouse gas emissions by sector from 1900 to 2018. *Environ. Res. Lett.* **2021**, *16*, 073005. [CrossRef]
2. Cano, Z.P.; Banham, D.; Ye, S.; Hintennach, A.; Lu, J.; Fowler, M.; Chen, Z. Batteries and fuel cells for emerging electric vehicle markets. *Nat. Energy* **2018**, *3*, 279–289. [CrossRef]
3. Gröger, O.; Hubert, A.G.; Suchsland, J.P. Review-Electromobility: Batteries or Fuel Cells? *J. Electrochem. Soc.* **2015**, *162*, 2605–2622. [CrossRef]
4. Path to Hydrogen Competitiveness. Available online: https://hydrogencouncil.com/wp-content/uploads/2020/01/Path-to-Hydrogen-Competitiveness_Full-Study-1.pdf (accessed on 14 March 2023).
5. Pundt, M.; Kirchner, M.; Stremlau, T.; Märker, G. Integrating a fuel cell system into a vehicle. *ATZ Worldw.* **2018**, *120*, 38–41. [CrossRef]
6. Pertschy, F. Das Comeback der Brennstoffzelle. Available online: https://www.automobil-produktion.de/technologie/das-comeback-der-brennstoffzelle-790.html (accessed on 20 February 2023).
7. Nakagaki, N. The Newly Developed Components for the Fuel Cell Vehicle, Mirai. *SAE Tech. Pap.* **2015**. [CrossRef]
8. Thomas, C.E. Fuel Cell and Battery Electric Vehicles Compared. *Int. J. Hydrogen Energy* **2009**, *34*, 6005–6020. [CrossRef]
9. Pollet, B.G.; Franco, A.A.; Su, H. Proton Exchange membrane Fuel cells. In *Compendium of Hydrogen Energy*; Elsevier: Amsterdam, The Netherlands, 2016; pp. 3–56.
10. Giampieri, A.; Ling-Chin, J.; Ma, Z.; Smallbone, A.; Roskilly, A.P. A review of the current automotive manufacturing practice from an energy perspective. *Appl. Energy* **2020**, *261*, 114074. [CrossRef]
11. Li, X.; Sabir, I. Review of bipolar plates in PEM fuel cells: Flow-field designs. *Int. J. Hydrogen Energy* **2005**, *30*, 359–371. [CrossRef]
12. Maheshwari, P.H.; Mathur, R.B.; Dhami, T.L. Fabrication of high strength and low weight composite bipolar plate for fuel cell applications. *J. Power Sources* **2007**, *173*, 394–403. [CrossRef]
13. Lin, P.; Zhou, P.; Wu, C.W. Multiple Objective topology optimization of end plates of proton exchange membrane fuel cell stacks. *J. Power Sources* **2011**, *196*, 1222–1228. [CrossRef]
14. Liu, B.; Wei, M.Y.; Ma, G.J.; Zhang, W. Stepwise optimization of endplate of fuel cell stack assembled by steel belts. *Int. J. Hydrogen Energy* **2016**, *41*, 2911–2918. [CrossRef]
15. Deshpande, J.; Singdeo, D.; Ghosh, P.C. Study of PEM Fuel Cell End Plate Design by Structural Analysis Based on Contact Pressure. *J. Energy* **2019**, *2019*, 3821082.
16. Youssef, M.E.; Amin, R.S.; El-Khatib, K.M. Development and performance analysis of PEMFC stack based on bipolar plates fabricated employing different designs. *Arab. J. Chem.* **2018**, *11*, 609–614. [CrossRef]

17. Syampurwadi, A.; Onggo, H.; Indriyati; Yudianti, R. Performance of PEM fuel cells stack as affected by number of cells and gas flow-rate. *IOP Conf. Ser. Earth Environ. Sci.* **2017**, *60*, 012029. [CrossRef]
18. Bharti, A.; Natarajan, R. Proton exchange membrane testing and diagnostics. In *PEM Fuel Cells. Fundamentals, Advanced Technologies and Practical Applications Book*; Elsevier: Amsterdam, The Netherlands, 2022; pp. 137–171.
19. Yan, X.; Hou, M.; Sun, L.; Liang, D.; Shen, Q.; Xu, H.; Ming, P.; Yi, B. AC Impedance characteristics of a 2 kW PEM fuel cell stack under different operating conditions and load changes. *Int. J. Hydrogen Energy* **2007**, *32*, 4358–4364. [CrossRef]
20. Fu, Y.; Hou, M.; Yan, X.; Hou, J.; Luo, X.; Shao, Z.; Yi, B. Research Progress of aluminium alloy endplates of PEMFCs. *J. Power Sources* **2007**, *166*, 435–440. [CrossRef]
21. Haider, R.; Wen, Y.; Ma, Z.F.; Wilkinson, D.P. High temperature proton exchange membrane fuel cells: Progress in advanced materials and key technologies. *R. Soc. Chem.* **2021**, *50*, 1138–1187. [CrossRef]
22. Tellez Cruz, M.M.; Escorihuela, J.; Solorza-Feria, O.; Conpan, V. Proton Exchange Membrane Fuel Cells (PEMFCs): Advances and Challenges. *Polymers* **2021**, *13*, 3064. [CrossRef]
23. Yang, D.; Tan, Y.; Li, B.; Ming, P.; Xiao, Q.; Zhang, C. A Review of the Transition Region of Membrane Electrode Assembly of Proton Exchange Membrane Fuel Cells: Design, Degradation and Mitigation. *Membranes* **2022**, *12*, 306. [CrossRef]
24. Yandrasits, M.A.; Hicks, T.M.; Pierpont, M.D. Manufacturing of Fuel Cell Membrane Electrode Assemblies Incorporating Photocurable Cationic Crosslinkable Resin Gasket. U.S. Patent No. 11/962,848, 25 June 2009.
25. Mitchell, J.; Fuller, J.T.; Gacek, T. Sulfonated PPS Fuel Cell Electrode. U.S. Patent No. 13/492,310, 12 December 2013.
26. Barbir, F. *PEM Fuel Cell, Theory and Practice*; Academic Press: Cambridge, MA, USA, 2005; pp. xiii–xv.
27. Roper, S.W.K.; Lee, H.; Huh, M.; Kim, I.Y. Simultaneous isotropic and anisotropic multi-material topology optimization for conceptual-level design of aerospace components. *Struct. Multidiscip. Optim.* **2021**, *64*, 441–456. [CrossRef]
28. Choi, W.H.; Kim, J.M.; Park, G.J. Comparison study fo some commercial structural optimization software systems. *Struct. Multidiscip. Optim.* **2016**, *54*, 685–699. [CrossRef]
29. Giele, R.; Groen, J.; Aage, N.; Andreasen, C.S.; Sigmund, O. On approaches for avoiding low-stiffness regions in variable thickness sheet and homogenizaiton-based topology optimization. *Stuructural Multidiscip. Optim.* **2021**, *64*, 39–52. [CrossRef]
30. Moritzer, E.; Heidrich, G.; Hirsch, A. Fibre Length Reduction during Injection Molding. *AIP Conf. Proc.* **2019**, *2055*, 070001.
31. Fu, S.; Lauke, B. Effects of Fiber length and fiber orientation distributions on the tensile strength of short-fiber-reinforced polymers. *Compos. Sci. Technol.* **1996**, *56*, 1179–1190. [CrossRef]
32. Fang, X.F.; Kloska, T. Hybrid forming of sheet metals with long Fiber-reinforced thermoplastics (LFT) by combined deep drawing and compression molding process. *Int. J. Mater. Form.* **2020**, *13*, 561–575. [CrossRef]
33. Tatara, R. Compression Molding. In *Applied Plastics Engineering Handbook*; William Andrew Publishing: Norwich, NY, USA, 2011; pp. 289–311.
34. Khan, R.S.; Sharma, B.; Chawla, P.A.; Bhatia, R. Inductively Coupled Plasma Optical Emission Spectrometry (ICP-OES): A Powerful Analytical Technique for Elemental Analysis. *Food Anal. Methods* **2022**, *15*, 666–688. [CrossRef]
35. Jayakumar, S.; Anand, S.; Hajdarevic, A.; Fang, X.F. Experimental and FE analyses of the crushing and bending behaviors of GMT and hybrid-formed Al-GMT structures. *Thin-Walled Struct.* **2023**, *186*, 110648. [CrossRef]
36. Kuhn, C.; Walter, I.; Taeger, O.; Osswald, T.A. Experimental and Numerical Analysis of Fiber Matrix Separation during Compression Molding of Long Fiber Reinforced Thermoplastics. *J. Compos. Sci.* **2017**, *1*, 2. [CrossRef]

vehicles

MDPI

Article

Evaluation of SiL Testing Potential—Shifting from HiL by Identifying Compatible Requirements with vECUs

Rudolf Keil [1,2,*], Jan Alexander Tschorn [2], Johannes Tümler [3] and Mehmet Ercan Altinsoy [1]

1 Institute of Acoustics and Speech Communication, TUD Dresden University of Technology, Helmholtzstraße 18, 01062 Dresden, Germany; ercan.altinsoy@tu-dresden.de
2 Vehicle Dynamics Department, Volkswagen AG, Berliner Ring 2, 38440 Wolfsburg, Germany; jan.alexander.tschorn@volkswagen.de
3 Anhalt University of Applied Sciences, Bernburger Straße 55, 06366 Köthen, Germany; johannes.tuemler@hs-anhalt.de
* Correspondence: rudolf.keil@tu-dresden.de or rudolf.keil@volkswagen.de

Abstract: Due to the increasing complexity of vehicle software, it is becoming increasingly difficult to comprehensively test all requirements. This inevitably means that alternative test methods, e.g., simulation-based methods, must be used more frequently. However, the challenge involves identifying appropriate requirements that can be technically tested in a simulation environment initially. The present work is aimed at evaluation and optimization of the effectiveness of software-in-the-loop (SiL) simulations in the testing process of vehicle software. The focus is on supporting the testing process by shifting specific test cases from hardware-in-the-loop (HiL) test benches to SiL-based simulations. For this purpose, a systematic approach was developed to analyze and categorize requirements, enabling precise and efficient allocation of test cases. Furthermore, a detailed review and recommendation for improving the ProSTEP iViP standard for virtual electronic control units (vECU) was carried out. The developed matrix associates the defined requirement clusters with different classifications of vECUs, facilitating the identification of suitable test environment types for conducting specific test cases. By assigning test cases to appropriate vECU levels, the testing processes can be targeted and cost-optimized. Finally, the theoretical results were evaluated in an SiL simulation environment. It was observed that a significant part of the requirements could effectively be tested using a vECU. These findings confirmed the potential of SiL simulation environments to not only support, but also enhance, the testing process for vehicle software by providing a cost-effective and flexible complement to traditional HiL test benches.

Keywords: automotive software testing; virtual electronic control units; software-in-the-loop; AUTOSAR standard; validation and verification

Citation: Keil, R.; Tschorn, J.A.; Tümler, J.; Altinsoy, M.E. Evaluation of SiL Testing Potential—Shifting from HiL by Identifying Compatible Requirements with vECUs. *Vehicles* **2024**, *6*, 920–948. https://doi.org/10.3390/vehicles6020044

Academic Editors: Ralf Stetter, Udo Pulm and Markus Till

Received: 18 April 2024
Revised: 22 May 2024
Accepted: 23 May 2024
Published: 29 May 2024

1. Introduction

The ongoing digital transformation in the automotive industry necessitates continuous adaptation and development of testing and validation procedures for automotive software. With the introduction of advanced functionalities such as advanced driver assistance systems (ADAS), autonomous driving (AD), and cyber security, the complexity of automotive software is increasing rapidly. Modern vehicles may comprise up to 100 electronic control units (ECUs) and around 100 million lines of code that must work together seamlessly to ensure safe and reliable operation of the vehicle [1]. By 2030, the number of lines of code per vehicle is expected to increase to up to 300 million [2]. As the scope of software expands, the corresponding testing effort increases [3]. Compounding the challenge, there is a trend towards reducing the number of prototypes. This paradigm shift highlights the growing importance of virtual testing methods, positioning them as essential supplements and alternatives to traditional approaches like test driving [4–6]. These methods offer significant advantages, including frontloading, cost efficiency, scalability of test environments, and faster development cycles [7]. The issue is clearly evident: as the number of functions in cars increases, so does the testing effort. Meanwhile, the number of prototypes continues to be reduced, making virtual testing methods, as mentioned in [6,8], increasingly indispensable. Virtual testing methods are commonly understood to be "X-in-the-Loop" (XiL) systems. These are categorized in the literature as model-in-the-loop (MiL), software-in-the-loop (SiL), hardware-in-the-loop (HiL), and vehicle-in-the-loop (ViL). Figure 1 illustrates that in the early phases of vehicle development, tests are predominantly carried out in a virtual environment. As development progresses, real components are increasingly integrated into the tests. With the test drive phase, the testing takes place entirely in the real environment, leading to an increased test accuracy due to the use of real components [9].

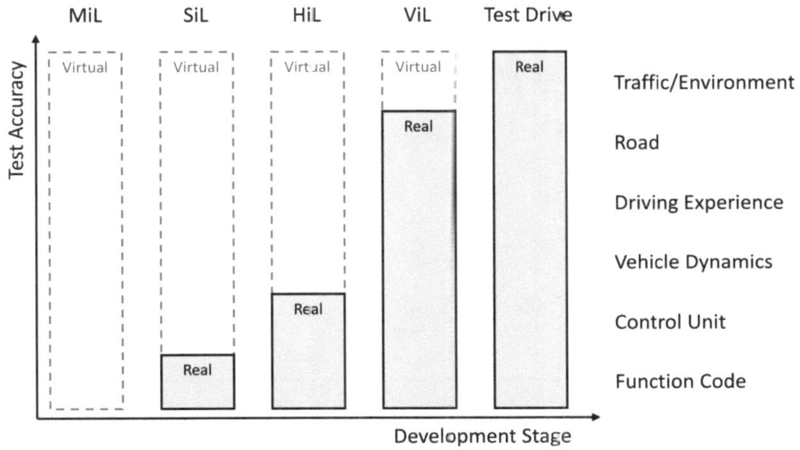

Figure 1. Overview of vehicle testing methodologies. The illustration has been modified and extended according to [9].

In the survey by Altinger et al. [10], it was reported that approximately 17 to 23% of software tests in the areas of research, pre-development, and series development are already being conducted using SiL. Hardware-specific tests with the whole vehicle, HiL, or PiL still constitute the majority of tests. Since hardware resources are often limited for prototypes, PiL or HiL-specific tests, software testing sometimes needs to be conducted in an MiL or SiL environment, to ensure comprehensive test coverage. However, the implementation of SiL environments depends on overcoming technical limitations, an aspect that is often not mentioned in the existing literature. Furthermore, as indicated by [11], the scope of these sources is limited to functional and behavioral aspects, simulation modeling, and optimal control modeling, which represent only a fraction of the relevant areas in the broad field of software development and testing. As a result, an SiL cannot test all the functionalities that a HiL is capable of assessing. This work aims to bridge this gap by devising a methodological framework to determine the feasibility of implementing functional and non-functional requirements within an SiL environment, thereby optimizing the testing process efficiency.

Virtual ECUs, as a crucial element within the SiL environment, are software-based simulations of physical ECUs used in vehicles. Instead of physical hardware, the functions of the ECUs are simulated in a virtual environment, so that developers and testers can examine the software and its interactions with other systems without having to rely on the target hardware. The ECU software is configured to be cross-compiled on the simulation host (typically an Intel x86 server or desktop) [7]. The ProStep iViP standard classifies vECUs into five different abstraction levels [12]. Figure 2 illustrates the development of vECU abstraction levels in the context of realism and accuracy. It becomes evident that, as the abstraction level increases (from vECU Level 0 to vECU Level 4), the degree of realism in the simulation also increases. This is visualized by a shift from MiL through SiL to HiL. While vECU Level 0 is primarily located in the MiL domain, Levels 1 to 4 reflect the SiL domain, with a progressive approximation to real hardware conditions in the HiL domain. The end of the spectrum represents the physical ECU—a physical ECU containing the compiled source code original to the target hardware. The deltas (Δ0 to Δ3) depicted in Figure 2 represent the discrepancies for individual vECUs, indicating how many requirements and associated test cases cannot be tested compared to a physical ECU due to technical limitations. The higher the level of a vECU, the smaller the delta and, consequently, the more requirements or test cases that can be tested. This difference is based, among other factors, on the extent of unchanged source code, allowing for various levels of realism to be represented in the simulation. The amount of unchanged source code also enhances the credibility of the simulation environment. This is because changes and adaptations of the source code for a different target hardware represent an additional source of errors, which in the worst case may have nothing to do with the original software.

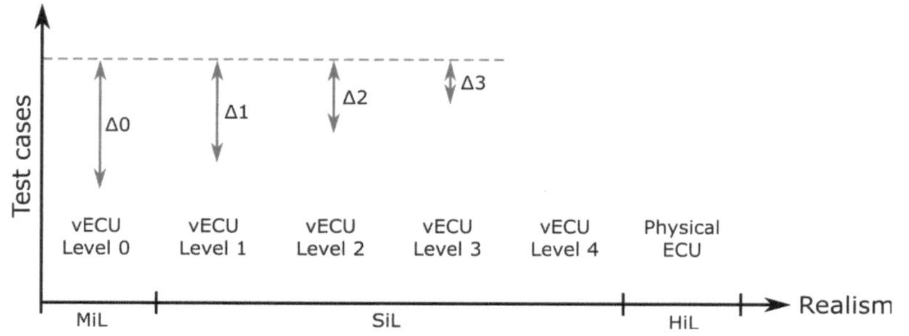

Figure 2. vECU-Levels and their associated realism from MiL to HiL.

vECUs can be in proprietary formats (e.g., Silver by Synopsys as in [13], VEOS by dSPACE as in [14], or MATLAB/Simulink by MathWorks), each offering specific advantages, depending on the application and system requirements. Determining which test cases can be efficiently and realistically represented in an SiL environment remains a challenging task. In particular, for test cases with specific hardware dependencies, the SiL method has its limitations [15]. This is because testing certain requirements may require specific hardware components that cannot be adequately simulated in a purely virtual environment or require significant effort to do so.

The presented paper introduces an approach to address this challenge by presenting a generic classification of requirements for ECU software, with a specific focus on a brake control system as an example. The aim was to develop a clear decision method that identifies which requirements are suitable for verification in an SiL environment and which are better tested with other methods such as HiL. This systematic approach aims to, not only make the testing process more efficient, but also enhance the quality and reliability of test results. Finally, the developed classification and the resulting action strategy are validated using an example, not only to assess their applicability but also to contribute to the further establishment and optimization of SiL-based testing and validation procedures in the automotive sector.

2. Related Work and Research Questions

In the field of SiL, various papers have been published pursuing various research objectives. A comprehensive overview is provided in the work by Clausen et al. [11]. Their paper discusses, among other topics, the application domains in which SiL environments have been utilized. In a systematic literature review, 17 of the 88 research papers were associated with the automotive domain. Noteworthy is the detailed examination of application areas that have benefited from SiL environments. These studies spanned sub-domains such as ADAS, AD, and control systems for steering and braking, focusing primarily on functional and behavioral analysis, simulation modeling, control model optimization, and concurrent testing during development. Another significant aspect of the automotive industry, not mentioned in the paper, is the calibration of controller parameters, as detailed in the sources by [13,16].

However, the limitations of SiL discussed in the papers analyzed by Clausen et al. [11] are not exhaustive. Specific limitations mentioned include the development process, aspects of real operational conditions, validation, interoperability, real-time requirements, scalability, and a cost–benefit analysis. In particular, challenges in replicating real operational conditions were noted, requiring clear differentiation depending on the implementation of the SiL environment. In this context, the white paper by ProSTEP iViP [12] offers deeper insights into various approaches specific to the automotive sector that are implementable in an SiL context. Different abstraction levels of vECUs are defined, providing guidelines on how closely the original software is represented. However, the assignment of which tests can be conducted with the various vECU abstraction levels remains an open question. This leads to the following key research questions that this paper aims to answer.

RQ1: What issues are associated with the abstraction levels for virtual ECUs, and how might these be improved? It must be emphasized again whether the existing definitions are sufficient for the use cases, or whether they need to be refined.

RQ2: What software testing requirements can be identified in the automotive sector? To better assign restrictions, it must first be defined which clusters are present in software testing in the automotive sector.

RQ3: How can the abstraction levels of vECUs be meaningfully assigned to different testing areas in the automotive sector? Given that not every SiL offers the same conditions, a gradation must be made to better explore the limits of an SiL.

3. Requirement-Based Testing and Standardization of vECUs

Our methodical approach is based on the following two concepts: requirement-based testing and the associated testing process, as well as the AUTOSAR standard, which also serves as a basis for classifying vECUs. The following Section 3.1 covers the fundamentals of requirement-based testing and discusses the structured process through which software is systematically evaluated and verified. In Section 3.2, the AUTOSAR standard is introduced, which defines the interoperability and modularity of software components. Building on this, Section 3.3 delves into the ProSTEP iViP standard, which defines various vECU levels.

3.1. Requirement-Based Testing in the Automotive Sector

Requirement-based testing is a type of testing that aims to determine the extent to which a system meets specified functional and non-functional requirements. There are different types of software tests, as introduced in [17]. The goal is to verify whether the finished software product complies with and conforms to the specified requirements. This approach is intended to uncover inconsistencies in the software early on, in areas such as performance, reliability, extensibility, user-friendliness, and security [18]. Each requirement should be associated with a test result to ensure traceability. Requirement-based testing, as a testing method, is embedded in a validation process [19].

The validation process aims to systematically detect potential bugs in the software, forming the basis for creating reports on the release and development status, which are essential for the subsequent software release process. This process can be understood as a cycle with three central components (see Figure 3). It is initiated by creating a test plan and specifying the tests. Subsequently, the tests are executed. The results of these tests represent the discrepancies between the desired target state and the actual state. If errors are identified during test specification or test execution, appropriate corrective actions are taken.

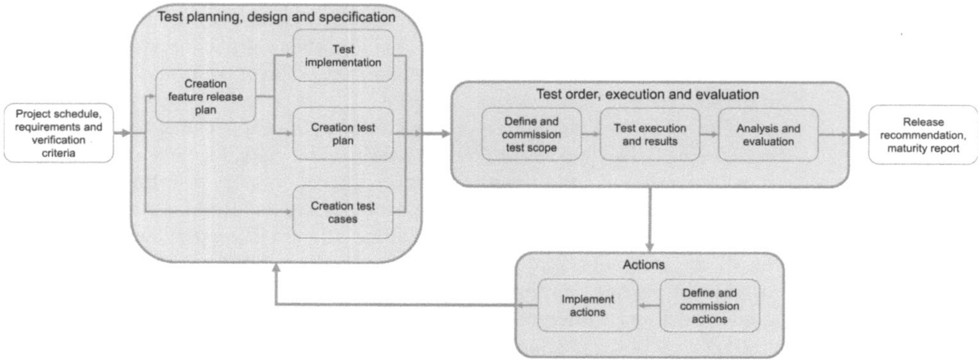

Figure 3. Schematic representation of a test process. The illustration has been modified and extended according to [20].

In the test plan, based on the project and feature release plans, the scheduling of individual test executions for the respective features with their abstraction levels takes place. Test executions should be created according to the requirements specified in the feature release plan, including release timeframes. For conducting tests in specific test environment types, a test case description is required. In the case of requirement-based testing, this test case description is based on the underlying requirements and their associated verification criteria. ISO 26262 serves as the authoritative guideline, offering differentiated recommendations for test methods based on the automotive safety integrity level (ASIL) [21].

The specification of test cases should logically follow from the test design, to ensure traceability back to the foundational requirements. The instructions for creating these specifications and implementing the test cases are defined within the test concepts of individual test instances. Different test instances provide various test environments, such as overall test drive, system HiL, ECU HiL, or SiL. The determination of the type of test environment in the verification criteria of the test design is intended to indicate where specific types of tests should logically occur. The responsibility for assessing the test environment lies with the test manager. Ideally, the required properties of the test environment should be defined before provisioning, based on the verification criteria. In the testing process, decisions are made based on the verification criteria, to allocate specific test cases to the corresponding test environments.

The relevance of this scientific research is demonstrated by the necessity for proficiently allocating requirements to an appropriate testing environment. Due to the finite availability of test benches and prototypes for software testing, it is imperative to validate as many requirements as feasible within a scalable testing environment, such as SiL, to maximize efficiency and coverage. A essential point is the variance in the abstraction levels of vECUs, which complicates the specific assignment of test cases. The appropriate assignment of test environments in the verification criteria and the early definition of their properties are important, in order to efficiently utilize the different types of test environments. This presents the challenge of developing a systematic method for evaluating and classifying test environments (in this paper, SiL) that aims to maximize test coverage and make more efficient use of test resources.

3.2. *AUTOSAR Classic Platform*

The Automotive Open System Architecture (AUTOSAR) is an international standard for software architectures in vehicles. The aim of the standard is to manage the increasing complexity in the development of ECU software. Another key idea is to separate the functional software from the underlying hardware, to facilitate a seamless exchange of software components across different hardware platforms, vehicle models, and even different original equipment manufacturers (OEMs) [22]. In the past, software was specifically written for the hardware it would run on [23]. With the AUTOSAR standard, programmers do not need expertise on how the hardware is structured and can generate the source code independently (see Figure 4). By separating the application layer from the underlying hardware architecture, vECUs can be developed, while keeping the functional software unchanged [24].

Figure 4. Legacy SW and AUTOSAR-based SW implementation [25].

This standard for software architecture, consisting of a set of specifications, application interfaces, and a methodology, already exists in two variants: AUTOSAR Adaptive and Classic. The AUTOSAR Adaptive Platform, based on a dynamic, POSIX-compliant operating system, is designed for applications with higher computational requirements, such as automated driving and infotainment systems [26]. For a comprehensive understanding, it is noted that this paper exclusively focuses on AUTOSAR Classic, because the standards have different structures and various application areas, necessitating a separate approach.

The architecture of the AUTOSAR Classic Platform (see Figure 5) comprises three software layers running on a microcontroller: the application layer, the runtime environment (RTE), and the basic software (BSW). The application layer, which contains the functional software, is mostly hardware-independent. Communication between software components and access to BSW are facilitated through the RTE, which serves as the complete interface for applications. The individual layers will be described in more detail below [22].

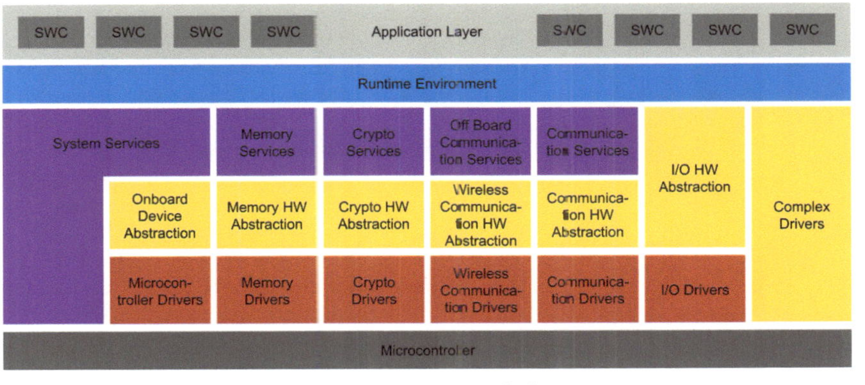

Figure 5. AUTOSAR Classic Platform software structure [22].

1. Application Layer
   The application layer is at the top and includes all parts of the functional software that are mostly hardware-independent. The application layer is divided into individual software components (SWCs), each carrying a functional implementation. According to the AUTOSAR standard, SWCs can be categorized into the following groups [27]:
   - Sensor/Actuator Software

- Composite Software
- NVBlock Software
- Parameter Software
- Application Software
- Service Software

Due to the abstraction principle of AUTOSAR, software programmers do not need to specify internal interfaces for connecting to other components or determine which data need to be sent to other ECUs via bus systems when developing an SWC.

2. Runtime Environment
All interfaces of SWCs communicate through the RTE (see Figure 5, in blue), which serves as an implementation of the virtual functional bus (VFB) within an ECU. The RTE acts as a kind of transport layer and forms a virtual communication platform that enables interaction between individual SWCs and the connection to the underlying basic software layer, including the operating system. This allows for abstract and hardware-independent communication. The RTE standardizes communication mechanisms, thus facilitating the portability and reuse of software components across different hardware platforms [28].
3. Basic Software
In the subsequent layer, the BSW is divided into three further sublayers. Directly adjacent to the RTE is the service layer (see Figure 5, in purple). This layer includes, among other things, the operating system and provides modules for memory management, communication control, as well as a range of other services for administration and monitoring. Apart from the operating system, these components are designed to be largely hardware-independent.
Following the service layer is the ECU abstraction layer (see Figure 5, in yellow), which represents the abstracted layer of the driver interface. This segment includes modules for input/output hardware abstraction and interface modules for various bus systems and memory. Specific hardware dependencies begin to emerge at this level, particularly in terms of the selection and configuration of data transmission channels. Complex device drivers (CDD) are a special case that are not standardized. The primary goal of CDDs is to enable the implementation of complex sensors and the control of actuators by directly accessing the microcontroller via specific interrupts or extended microcontroller peripherals or external devices.
The bottom layer of the BSW is the microcontroller abstraction layer (MCAL), which contains internal hardware drivers for internal microcontroller peripherals such as a serial peripheral interface (SPI), watchdog, controller area network (CAN), or analog–digital converter (ADC), etc. The software is directly dependent on the specific hardware implementation, leading to a close integration of software and hardware [22].
Software libraries (e.g., fixed-point mathematical, bit handling, atomic multicore-safe operations, etc.) provide standardized algorithms for basic software modules (BSWM) and SWCs. These include functions in C code that can be in either source or object code and can be used by the BSW, SWCs, the RTE, or complex drivers [29].
In addition to the three software layers (service layer, ECU abstraction layer, and microcontroller abstraction layer), a vertical division into functional subgroups can be made, which will be briefly described below [30].

- Input/Output (I/O): Standardized access to sensors, actuators, and ECU peripherals.
- Memory: Standardized access to internal and external (non-volatile) memory.
- Crypto: Standardized access to cryptographic primitives, including internal and external hardware accelerators.
- Communication: Standardized access to in-vehicle network systems, ECU onboard communication systems, and internal ECU software.

- Off-board Communication: Standardized access to Vehicle-to-X communication, wireless network systems in the vehicle, and ECU-external communication systems.
- System: Provision of standardizable services (operating system, timers, fault memory) and ECU-specific services (ECU state management, watchdog manager), as well as library functions.

This concept ultimately serves to abstract the microcontroller's hardware from the application software. Developers can thus develop their applications independently of the hardware, making the software more portable and reusable. Building on this section, various standards for vECU levels have emerged, which will be discussed in the following section.

3.3. Standardization of vECUs

In general, two different types of vECUs are distinguished: host-compiled vECUs and target hardware-compiled vECUs. Host-compiled vECUs are characterized by the process of cross-compiling ECU software on a simulation host, which typically uses an Intel x86 architecture [7]. This methodology allows for a flexible development environment, but without direct involvement of an ECU hardware model, which requires abstraction of lower, hardware-dependent layers through the simulation of software APIs at defined interfaces in the software stack. The resulting virtual prototype does not represent the complete production code. Certain software layers must be modified or replaced with simulation equivalents to ensure compatibility and functionality within the virtual environment.

In contrast, target hardware-compiled vECUs use compilers tailored to the specific microcontroller or system on chip (SoC). These types of vECUs include detailed hardware simulation models that provide a comprehensive representation of the ECU hardware and peripherals, allowing the execution of production software without any modifications. The software is compiled exactly as intended for the physical ECU in the actual vehicle, requiring extensive and precise modeling of all ECU hardware components.

These two vECU architectures offer their own advantages and use cases, depending on the specific requirements of the development and testing process in the automotive industry. The choice between host and target hardware-compiled vECUs is significantly influenced by factors such as the required accuracy, development speed, available hardware, and specific goals of the software validation. Furthermore, a granular differentiation of vECUs is made with regard to certain scopes of unchanged source code. Table 1 provides an overview of the standard and it is briefly explained below.

Table 1. Abstractions levels of vECU types [31].

Level vECU	Application SW	Runtime Environment	Basic Software	OS, Drivers, Complex Drivers
Level 0 vECU	Controller Model	-	-	-
Level 1 vECU	Production Code	Simulated Code	-	-
Level 2 vECU	Production Code	Production Code	Simulated Code	-
Level 3 vECU	Production Code	Production Code	Production Code	Simulated Code
Level 4 vECU	Binary Code	Binary Code	Binary Code	Binary Code

- Level 0 vECU
  A Level 0 vECU includes only the behavioral model itself (e.g., as a MATLAB Simulink model) or the C code generated from the behavioral model (e.g., as a functional mock-up unit (FMU) or dynamic link library (DLL)). A Level 0 vECU does not contain production code and is limited to the individual SWCs or functions of the application

layer. The extent of the application layer is irrelevant. Behavioral models focus on testing the functional and logical behavior at an early development stage. For other tests, such as software interface tests, more complex vECU types that are closer to the physical ECU's production code are required. A Level 0 vECU also falls into the MiL category, because it uses a behavioral model (see Figure 2).

- Level 1 vECU
  A Level 1 vECU includes the production code of the application software. The BSW is removed and supplemented with basic runtime and I/O simulation code, to enable the operation of the application software. The simulated code is specifically created or generated for the vECU. In the case of AUTOSAR Classic, this includes the RTE and functionalities that allow the vECU to send and receive data, for example. The vECU can include all parts or only a subset of the application software for a physical ECU.
- Level 2 vECU
  Based on a Level 1 vECU, a Level 2 vECU additionally implements simulated BSW functionalities. The basis for the simulated BSW is the production code, which is abstracted and modified at a high level. Unlike Level 1 vECUs, which primarily communicate at the signal level, Level 2 vECUs can enable advanced communication capabilities that operate at both the bus and network levels, such as simulating common vehicle communication protocols like CAN or Ethernet.
- Level 3 vECU
  A Level 3 vECU enhances the BSW with the original production code. The essential condition is that both the application and BSW layers are implemented in a hardware-independent manner to ensure smooth integration and functionality within the vECU. In terms of the AUTOSAR standard, hardware dependencies should only occur in the components of the MCAL, the operating system, and certain parts of complex device drivers. Although Level 3 vECUs include a larger coverage of the production code, they can still be complemented with simulated software components, to optimize the completeness and functionality of the vECU. This is relevant for functions that cannot be directly adopted from the production code due to hardware dependencies.
- Level 4 vECU
  A Level 4 vECU contains the entire production code compiled directly for the target system, i.e., the physical ECU. It is also possible to include hardware dependencies. This integration allows for the inclusion of all software layers, including the MCAL, the operating system, and complex device drivers, while also considering hardware dependencies. A significant advantage of Level 4 vECUs is that the software uses the same build process for both the simulation and the physical ECU. This eliminates the need for code changes between the vECU and the physical ECU. It also ensures direct transferability of test results and insights between the virtual and real worlds [12].

Synopsys has extended the vECU standard by introducing another sub-level within Level 4 vECU, as depicted in their white paper. This differentiated approach manifests itself in two specific sub-levels (see Table 2). Level 4a and Level 4b, with Level 4b corresponding to ProSTEP iViP's Level 4 classification. Level 4a represents a new approach, allowing certain software functions to be bypassed, aiming to increase flexibility and simulation speed [7].

Table 2. Abstractions levels of vECU types according to [7].

Level vECU	Application SW	Runtime Environment	Basic Software	OS, Drivers, Complex Drivers
Level 4a vECU	Binary Code	Binary Code	Binary Code + Bypass	Binary Code + Bypass
Level 4b vECU	Binary Code	Binary Code	Binary Code	Binary Code

The VDA also addressed the standard in their recommendation VDA 710, but further specified the aspect of signal and message-oriented communication. A signal-oriented implementation facilitates direct communication at the runtime environment (RTE) level. Specifically, data and values are transmitted in a signal-oriented manner through this high-cut interface. In contrast, the low-cut implementation covers the communication layer of the BSW. The signal transmission is message-oriented, characterized by appropriate signal quantization and targeted message packaging within the vECU. For Level 2 vECUs, it is explicitly mentioned that no specific interfaces have been defined as a reference for orientation so far [32].

4. Materials and Methods

The aim of this section is to present a clear and structured process for the analysis and consolidation of requirements and vECU standards in the automotive industry. Figure 6 provides a schematic representation of the applied approach.

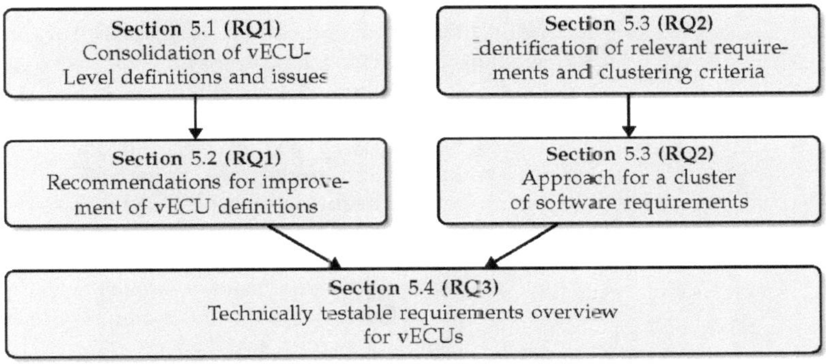

Figure 6. Schematic representation of the applied methodology.

For the identification of requirements, an extensive literature review was conducted to consolidate the existing vECU standards and definitions. This review not only considered common standards but also delved into the differences, challenges, and issues arising from various definitions. From the analysis of existing standards and the identified practical problems currently encountered, recommendations for improvement were developed. These recommendations aim at refinement of vECU definitions, to better align them with the practical requirements of the automotive industry.

To answer the first research question, several SiL environments were compared, to identify differences. Various vECUs were utilized in this process. It was examined whether and to what extent SiL environments are used across different departments. To ensure a meaningful analysis, while keeping the effort manageable, four different SiL environments from three distinct domains were investigated, incorporating five different vECUs. Some of the SiL environments from the chassis domain were partially extended and commissioned in-house. A consistent environment, including a residual bus simulation, had to be created to ensure error-free operation of the vECUs. The four SiL environments are summarized in Table 3.

Table 3. Overview of analyzed SiL environments.

Environment	vECUs	Tools	Additional Models
Chassis 1	Electronic Stability Control	MATLAB/Simulink + CarMaker	Brake Hydraulic + Brake Booster
Chassis 2	Electronic Stability Control	MATLAB/Simulink + CarMaker	Brake Hydraulic + Brake Booster
Chassis 3	Electronic Stability Control + Zone Controller	Silver + CarMaker	Brake Hydraulic
Powertrain	Engine Control Module	Silver	Battery Management + Coolant Circuit – Electric Machine

The first two SiL environments have a similar setup and were implemented in a MATLAB/Simulink environment. Since these are two brake control systems, a vehicle dynamics model was also implemented and supplemented with a brake hydraulic model as an s-function. For better integration of the vehicle dynamics model with MATLAB/Simulink, CarMaker for Simulink was used. An electronic stability control (ESC) with a behavioral model of a brake booster was integrated as the vECU.

The third environment also encompassed a brake control system, but it was implemented in Silver. This environment includes an ESC and a zone controller capable of implementing more extensive driving functions. Additionally, a residual bus simulation and brake hydraulic model was included. While the ESC was implemented as an FMU in Silver, a zone controller was specifically created for Silver. This environment was again operated in co-simulation with CarMaker.

The fourth SiL environment was from the powertrain domain and was also implemented in Silver. The engine control module (ECM) was supplemented with an additional behavioral model, such as an electric machine, battery management, or coolant circuit.

For further analysis of the SiL environments, the differences in commissioning and actual handling were examined. Standard functional tests were conducted, and it was also investigated to what extent advanced tests, such as diagnostics, could be performed. Communication with suppliers was established, to understand how the vECUs were created and to what extent they could also be classified by suppliers. The focus of the analysis was always on the properties and limitations of the vECUs.

To obtain a comprehensive and representative set of requirements for physical ECUs, a data analysis of project-specific specification documents and test case catalogs was conducted. By systematically reviewing these documents using specific clustering criteria, relevant requirements could be identified. The specific criteria for clustering were partially based on the number of requirements related to similar properties. Additionally, relevant test cases for specific software releases were considered. This approach aimed to answer the second research question.

Based on the requirements, an assignment to the vECU levels was carried out. Ultimately, an analysis was conducted to determine the technical prerequisites a vECU must meet in order to test a specific requirement cluster. Finally, this methodical approach provides a solid foundation for evaluating and categorizing testing processes using vECU standards in the automotive industry. The results of this method are detailed in Section 5.

5. Results

In this section, the results are summarized, divided into four subsections, each addressing specific aspects of vECUs and the virtual software testing process, along with the associated challenges and solutions.

The first subsection focuses on practical problems with the established ProSTEP iViP standard as they arise in real-world application scenarios. The analysis of discrepancies between theory and practice aimed to shed light on a new approach, including recommendations and modifications, which will be discussed in the second subsection. The proposed adjustments and additions not only serve as suggestions but also form the basis for the final subsection.

The third subsection presents systematically developed clusters of requirements for software validation. These clusters are the result of in-depth analysis, expert interviews, and practical experience. The resulting categorization of requirements aims to assess the technical constraints of virtual methods in the validation process.

The concluding subsection provides a comprehensive overview of which of the defined requirement clusters can be technically tested with different levels of vECUs. It delves into the technical prerequisites that vECUs must meet to validate the categorized requirements. Furthermore, it discusses the remaining restrictions resulting from volatile development methods. This holistic perspective offers a methodical approach to assess the suitability of various vECU configurations for specific requirements and provides valuable insights for the further development and optimization of testing processes in the automotive industry.

5.1. Analysis and Identification of Issues with the Standard for vECUs (RQ1)

The analysis draws on practical experience from day-to-day operations and conversations with suppliers involved in the development of vECUs. Various vECUs were used and analyzed, some created with different development methods and executed in various simulation environments, including proprietary formats or using the FMI standard. This diversity spans across five different suppliers and covers vECUs in areas such as the chassis and drivetrain. An overview of the analyzed vECUs is presented in Table 4. Some suppliers provided their vECUs in two different formats (FMU and DLL). For anonymization purposes, the names of the suppliers were changed.

Table 4. Overview of analyzed vECUs.

Tier 1	Level vECU [1]	Domain in Vehicle	File Format
Supplier A	Level 3	Chassis	DLL + FMU
Supplier B	Level 2–3	Chassis	FMU
Supplier C	Level 2–3	Chassis	FMU
Supplier D	Level 3	Powertrain	DLL + FMU
Supplier E	Level 2	Zone controller	DLL

[1] Assessment from the supplier.

The following subsection will take a closer look at this variance and show how the current definitions could be sharpened to meet the challenges and requirements of this dynamic landscape.

5.1.1. Uncertainties in the Classification of vECUs

While the first two abstraction levels of vECUs are relatively straightforward to interpret, the distinction between Level 2 and Level 3 in the vECU classification is often unclear and can lead to confusion. The uncertainties become apparent in practice, because developers lack clear criteria for when the transition from one level to the next begins. Frequently, this results in a blurry classification, as seen with two of the five suppliers who classified their vECUs as hybrids between vECU Level 2 and 3. This challenge is further exacerbated by the introduction of the new FMI standard 3.0, which is opening up new possibilities for vECUs [33]. vECUs provided as FMUs, due to their lack of message-oriented communication according to the standard, are typically assigned a maximum of vECU

Level 2. However, theoretically, most parts of the software, with few exceptions, could even be presented as production code within an FMU, possibly extending up to the MCAL layer.

The restrictions on signal-oriented communication arise from the technical requirements of the FMI standard. It is, however, possible to implement simulation models as C code for hardware components. Furthermore, within the FMU, signal processing can take place to prepare signals for the communication interface, as schematically illustrated in Figure 7. Within the FMU, signals coming from outside are processed to be prepared for the communication stack as protocol data units (PDU).

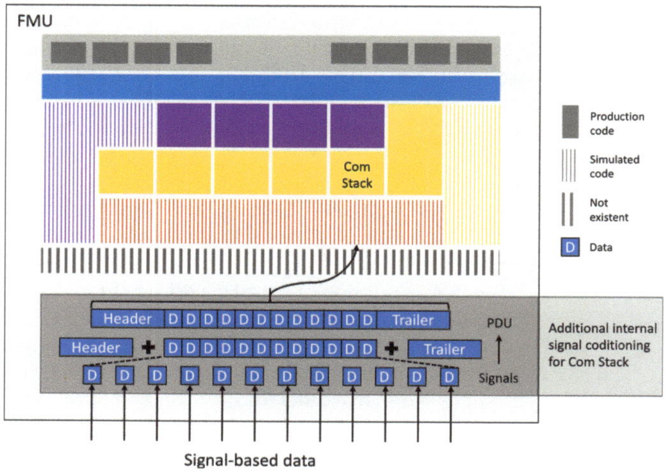

Figure 7. Mechanisms for implementing message-oriented communication within the FMU.

The FMI standard allows for only signal-oriented communication outward, which implies that, in practice, only a high-cut interface can be provided, as mentioned in [32]. This might lead one to believe that implementing the BSW functionalities within this scope is not feasible. However, a solution arises through the implementation of additional software within the FMU, which serves as a supplement to the actual ECU software. This adaptation allows for the augmentation of signals with information and processing them into PDUs. This approach involves packaging data into PDUs, adding information, and scheduling, which is particularly relevant, as bus signals are usually sent cyclically. In a typical scenario, an error management would detect when the prescribed cycle times, for example, of CAN messages, are not adhered to, which should lead to a diagnostic trouble code (DTC). However, this contradicts the continuous reading of inputs within an FMU, which occurs at a specified step size. To bridge this discrepancy, two approaches are possible: either the error management is deactivated or bypassed, or an attempt is made to assign a specific cycle time to each message within the FMU. The second method allows messages to be correctly processed in the lower software layers of the COM stack. This is an important step to ensure the functionality of the FMU in line with existing communication protocols and requirements. Some proprietary software formats offer advanced solutions for simulating the communication interface by integrating the possibility of a virtual CAN bus, enabling message-oriented communication. Two examples of such software solutions are VEOS from dSPACE and Silver from Synopsys. Another interesting approach is that the FMI standard can be used as an interface for vECUs, where the actual vECU is provided in a different proprietary format. This can bypass certain restrictions of the FMI standard, especially in version 2.0.

Interestingly, a supplier classified their vECU as Level 2 due to these communication limitations, even though almost the entire software code was adopted unchanged, which would normally meet the requirements of a Level 3 vECU. Two different factors are being mixed: the amount of unchanged source code in the vECU, and the additional technical functions provided by the supplier. Important aspects such as providing a virtual CAN bus or access to non-volatile memory (NVM) for configuration also need to be considered.

In summary, it can be noted that the classification of vECUs is complicated by the absence of clear, hard boundaries and definitions. This ambiguity has become evident in practice. In particular, the concepts of high-cut and low-cut interfaces introduce additional uncertainties into the determination and classification of vECUs.

5.1.2. The Variance in the Technical Scope of vECUs

Regardless of the specific implementation of the original source code, it is the technical capabilities and features of a vECU that play a crucial role, both in terms of validating the simulation environment and ensuring testability according to specific requirements. The technical capabilities significantly impact how effectively and comprehensively a vECU can be used to perform the necessary tests and simulations.

In practice, it is evident that different suppliers develop their vECUs differently, leading to a variance in functionality, even with similar vECU classifications. This disparity can arise, among other factors, because the implemented source code does not necessarily correlate with the functionality of the vECU. A prominent example of this is error management, which exists as source code, but the developer does not provide a means to read diagnostic trouble codes (DTCs). These differences in development result in varying capabilities and functions of vECUs, which, in turn, have direct implications for their applicability in different test scenarios. Some examples of technical capabilities that can be considered during the development of vECUs include the following:

1. Communication interface
   Common communication interfaces in the automotive sector include CAN, local interconnect network (LIN), Flexray, and Ethernet. When it comes to the communication interface for vECUs, there is a fundamental distinction between signal-oriented and message-oriented communication, each of which has specific application areas and advantages. Signal-oriented communication is often sufficient for component testing, especially when all involved simulation models do not require message-oriented communication within the same environment. In this method, signals are directly injected into the vECU, without more complex communication protocols such as arbitration, checksum calculation, or message counters. They are typically processed immediately at a higher software level. This type of communication is well-suited for virtual test environments, where direct signal transmission is sufficient. The advantage is that, initially, less effort is required to create the vECU. In contrast, message-oriented communication involves the implementation of additional protocol and security aspects, such as identifiers, checksum calculation, and message counters. This type of communication theoretically allows the vECU to communicate with physical ECUs if a compatible hardware interface is available. Often, proprietary formats that provide such virtual communication interfaces are required for implementation. Typically, the hardware components of the communication interface also need to be emulated. A significant advantage of this method is the ability to test specific fault patterns in the bus system, such as incorrect cycle times, timeouts, or errors in checksum calculation. vECUs that support message-oriented communication are particularly well-suited for HiL test benches, where they can easily be used as residual bus simulations, since communication can be processed at the same level as with physical ECUs through hardware interfaces. A prerequisite for this is that they can be simulated in at least real time. This enables realistic simulation and testing of ECUs under various conditions and scenarios.
   For completeness, the last possibility is mentioned again, which is typical for Level 4 vECUs, for example, when a hardware emulation of a CAN transceiver is implemented. In this method, communication takes place at the lowest level of the communication protocol, the physical layer. However, the disadvantage is that the

signals exist as physical bits in a virtual simulation environment and are difficult to process further. A direct transmission medium is required. Analogous to the open systems interconnection (OSI) model, communication occurs in the first layer.

2. Memory access
The implementation of non-volatile memory is essential, especially when integrating the BSW of the memory stack. It enables the management and utilization of non-volatile memory within the vECU. Non-volatile memory, typically in the form of an electrically erasable programmable read-only memory (EEPROM), plays a critical role in physical ECUs. It is used for storing essential parameters, calculated values, coding, and diagnostic trouble codes (DTCs), regardless of the power supply status. This type of memory is particularly required for information that extends beyond a single driving cycle, such as stored DTCs that can be read in a workshop or by a dealer. During system startup, in the initialization phase, and during normal operation, this non-volatile memory undergoes several cyclic redundancy checks to ensure the integrity of the stored data and the memory cells themselves. Since it significantly contributes to the storage of relevant parameters and DTCs, simulating this component is of great importance, especially in the context of on-board diagnostics (OBD). In the implementation of non-volatile memory in vECUs, a simple text file is often used to store the data as hexadecimal values. These pieces of information are organized and stored according to memory addresses. This approach facilitates handling and the gives possibility to manually verify or modify data if necessary.
The implementation of non-volatile memory and the ability to modify it form the basis of the configuration and calibration of the vECU. Furthermore, meaningful validation of the vECU can be carried out based on this information by utilizing a dump of an EEPROM from a physical ECU for validation purposes.
3. Calibration
Calibration allows for the adjustment of an ECU's functions through application parameters, which can include control parameters or characteristic curves, typically stored in non-volatile memory. Especially for virtual tests of the powertrain or chassis, it is essential that the vECU contains the same parameters as those used in the physical vehicle. Various parameters can significantly influence behavior. A typical example of parameter data in the powertrain area, specifically in the ECU, is the engine map. In principle, writing a parameter set is possible by providing non-volatile memory in the form of a text file. However, it would be very cumbersome to save the corresponding parameter datasets as hexadecimal numbers at the correct memory addresses. Therefore, some suppliers offer toolchains that can read in the parameters in the appropriate file formats during the initialization phase. The ability to adjust parameters in a simulation also opens up new possibilities for application in general. In real-world testing, changing and calibrating data is very time-consuming. In contrast, it is conceivable to perform optimization procedures in a simulation where parameters are optimized according to certain criteria. The determined parameters could then be used in the actual vehicle. This would not only save time and costs but also provide the opportunity for more precise and efficient calibrations.
Another important aspect is the adjustment and reading of specific internal signals for certain test cases to monitor the internal states of the ECU during runtime. These internal signals are usually stored in the microprocessor's RAM and can be read, for example, using the Universal Measurement and Calibration Protocol (XCP). For this purpose, the software supplier must ensure that, for instance, an XCP over Ethernet port is provided. In general, these capabilities go hand in hand with the implementation of the first two points (communication interface and memory access). Nevertheless, it is not a given that these capabilities are actually implemented.

5.2. Approach and Recommendation for Improving the Standard (RQ1)

Given the various special cases observed in practice, a new approach is proposed here to create a better foundation for the final section and the assignment to the requirement clusters. One main issue is that the ProSTEP iViP standard blends the scope of unchanged source code with the technical functions of a vECU, leading to unclear classifications from some suppliers, especially in connection with the FMI standard. Below, the ProSTEP approach to classification is taken up and expanded with information on the technical scope. However, the definitions of the levels should exclusively refer to the scope of unchanged source code and the target hardware for which the source code is compiled. In particular, with regard to the ambiguities between Level 2 and Level 3 vECUs, the following explanation is intended to provide a better overview. As a result, the following recommendations (see Table 5) mainly relate to the share of production code. The percentage breakdown refers to the source code, but simply counting lines of code is not an effective method of evaluation. A clearer method is the categorization into SWCs and BSWMs. The number of SWCs varies depending on the specific project and the scope of implemented functions. In the context of ECUs developed according to the AUTOSAR Classic Platform specification, implemented, for example, with Vector Microsar software, an approximate guideline for the number of BSW modules in such a project can be assumed to be around 40 modules. This value serves as a rough reference for the number of BSWMs that can be expected in a typical project of this kind.

Table 5. Overview of the vECU classification with details of the unmodified source code.

Level vECU	Application Software	RTE	Basic Software	OS, Drivers, CDDs
Level 0	Controller Model >1%	-	-	-
Level 1	Production Code >1%	Production/ Simulated Code	-	-
Level 2	Production Code 90–100%	Production Code 100%	Production Code < 90%	Simulated Code
Level 3	Production Code 100%	Production Code 100%	Production Code 90–100%	Simulated Code
Level 4	Binary Code 100%	Binary Code 100%	Binary Code 100%	Binary Code 100%

- Level 0 vECU
  For Level 0 vECUs, everything remains unchanged, so any portion of the source code or SWCs can be implemented as a behavioral model.
- Level 1 vECU
  For Level 1 vECUs, there is already production code in the application layer, but this proportion can vary. However, to enable communication between SWCs, the RTE and possibly some portions of the BSW need to be implemented, although these are minimal and not listed in Table 5.
- Level 2 vECU
  A Level 2 vECU should primarily consist of unchanged source code in the application layer, with minimal exceptions. Therefore, a benchmark of at least 90% production code is used as a reference. This value is taken as a guideline because it can be easily achieved, and a tolerance range is allowed in case some SWCs are missing. The RTE must be implemented as a complete communication interface. In contrast to the ProSTEP standard, an adjustment is made in the BSW. The majority of BSW can be implemented as simulated source code, but there is the possibility to supplement the BSW with production code for specific functions. It is conceivable that functional blocks in a vertical orientation (see Figure 5), such as the COM or memory stack, could

be realized up to the MCAL. To implement such software components, portions of the hardware peripherals may need to be simulated.

- Level 3 vECU
  A Level 3 vECU should primarily consist of production code, with the entire application layer present. All BSWMs that do not require a direct hardware interface should also be in their original source code. However, it has been observed in practice that the implementation of certain modules, such as crypto services, can be associated with significant effort. Therefore, a certain degree of variability is allowed here. None of the five suppliers, for example, integrated the cyber security functionality into the vECU. For Level 3 vECUs, simplified hardware simulations may also be required. Specifically, memory access for the memory stack must be implemented. Since the communication stack is implemented up to the lowest level of the basic software, a low-cut interface should normally be present as well. In this case, the vECU expects PDUs as input data, for example.
- Level 4 vECU
  The definition of a Level 4 vECU remains unchanged, as at this level, the entire ECU hardware is emulated. This level of simulation allows for the use of the unmodified binary code of the ECU. While the inclusion of the Synopsys definition is theoretically possible, it represents a proprietary configuration specifically developed for their toolchain and has not seen widespread adoption.

In addition to specifying the unmodified source code, which is represented by the level classification and provides a rough indication of how closely a vECU corresponds to the original software, an extension with technical content will be introduced. This extension encompasses two essential technical areas that include a variety of functions. The problem identified in Section 5.1 is that the quantity of source code does not necessarily correlate with the technical functional scope. Therefore, it is proposed that suppliers should have the option to declare the specific functional scopes using an extended nomenclature. This is particularly relevant for certain test cases that rely on memory access or the communication interface. These functional scopes are mainly relevant for Level 2 and Level 3 vECUs, as they require a certain proportion of basic software functions. The additions will have no impact on Level 0 and Level 1 vECUs, as they do not implement any components of the basic software. Consequently, only Level 2 and Level 3 will be considered further, as they are the most relevant in practice.

To specify the implementation of a communication interface for message-oriented communication, the letter "c" for "Communication" is added to the nomenclature. For example, an implementation that meets the requirements mentioned above would be referred to in the nomenclature as "Level 2c vECU". The addition of "c" signifies that the vECU has the necessary interface for message-oriented communication at the network level. In Figure 8, the communication stack is shown in more detail to illustrate where the PDUs are externally injected. With a comprehensive implementation of the communication stack, for example, in a CAN protocol, the frames are injected either between the CAN interface and CAN driver or below the CAN driver. In the case of signal-oriented communication, this processing chain is missing, and the signals are injected directly into the RTE. As mentioned above in the FMU example, a vECU that has implemented the communication stack may still only include signal-oriented bus communication.

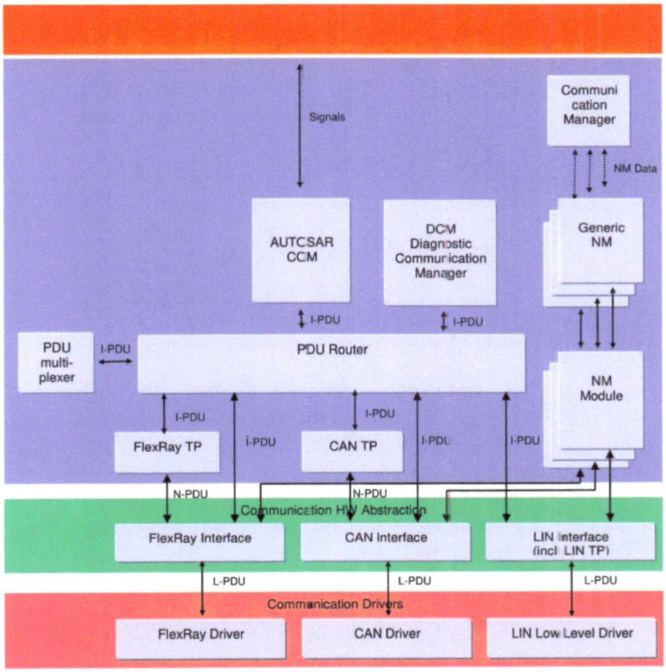

Figure 8. Detailed illustration of the communication stack in AUTOSAR [34].

Due to these requirements, the FMI standard is not suitable for message-oriented communication, because it does not support the corresponding external communication. Instead, a proprietary format is needed, which may also need to simulate parts of the driver or hardware. Such a communication interface is relevant for manipulating the bus system, as it constitutes a significant part of the test cases. Additionally, implementation of the XCP interface should be considered to enable real-time measurement of internal signals. This interface is essential for monitoring the vECU's internal parameters during operation, allowing for in-depth analysis of system behavior. The second significant technical implementation relates to NVM, which is essential for storing parameter data, encoding, and DTCs. This memory is relevant for numerous technical functions. To symbolize that the relevant aspects of non-volatile memory have been implemented, the nomenclature will be extended with the letter "m" for "Memory", analogous to the communication interface. The following functions should be ensured by this designation:

- The coding of the vECU is feasible
- Parameters on the vECU can be rewritten
- DTCs can be read

Although the implementation of NVM is already necessary to meet the requirements of a Level 3 vECU, significant challenges exist in making it accessible and usable for users. Without specific knowledge of the storage locations of individual pieces of information within an NVM, a document filled with hexadecimal numbers is hardly usable for the user. Therefore, it is crucial to provide users with simple tools or scripts for accessing and interacting with the memory. Implementation of non-volatile memory could also be considered for a Level 2 vECU, requiring certain parts of the BSW to be implemented. This results in three additional levels, which are briefly described in the following Table 6.

Table 6. Proposed new abstraction levels of vECU types.

Level vECU	Non-Volatile Memory	Message-Oriented Communication	Adaptable as a FMU	Adaptable as Proprietary Format
Level 2	–	–	–	–
Level 2c	–	✓	–	✓
Level 2m	✓	–	✓	✓
Level 3	✓	–	✓	✓
Level 3c	✓	✓	–	✓

The additional nomenclature allows for a clear and strict differentiation between the extent of unchanged source code and technical functions. In a Level 2c vECU, the scope of the original source code for the BSW may be limited, but the COM stack can still be implemented for specific use cases. In contrast, a Level 2m vECU could also have certain limitations in production code, preventing it from reaching Level 3. However, the implementation of NVM can be highly relevant for some test cases, which is why this designation is included.

Three out of five suppliers delivered a Level 3 vECU according to their definitions, all of which also implemented an NVM with fault memory, encoding capabilities, and parameterization. Therefore, the scope of a Level 3 vECU should adhere to these criteria by default. However, message-oriented communication is not necessarily guaranteed, even if the original source code is fully implemented up to the MCAL. This is because additional signal processing can occur within an FMU, resulting in the vECU exhibiting signal-oriented communication externally. The drawback is that this additional processing is unrelated to the original software. It is a potential source of errors, which should be considered and potentially validated during vECU development. Due to these restrictions, Level 3c and 2c vECUs cannot be implemented with the FMI standard.

If the proposed approach were applied to the vECUs mentioned at the beginning (Section 5.1), it would provide a more precise classification. The vECUs could be better assigned to a level, and it would be evident which vECUs provide a low-cut interface for communication. According to this approach, the levels would look as shown in Table 7.

Table 7. Overview of analyzed vECUs with new classification.

Tier 1	Level vECU [1]	Level vECU [2]	Domain in Vehicle	File Format
Supplier A	Level 3	Level 3c Level 3	Chassis	DLL FMU
Supplier B	Level 2–3	Level 3	Chassis	FMU
Supplier C	Level 2–3	Level 2m	Chassis	FMU
Supplier D	Level 3	Level 3c Level 3	Powertrain	DLL FMU
Supplier E	Level 2	Level 2c	Zone controller	DLL

[1] Assessment from the supplier. [2] Assessment with new approach.

The results obtained here provide an initial approach and recommendations for further refining the definition of vECUs. The identified problems and challenges are limited to the analyzed vECUs and reflect the diversity of approaches for creating such vECUs, including highlighting special cases. However, a more precise definition of vECUs can help establish clear requirements and standards for vECUs in the future. This becomes particularly relevant when an OEM plans to use vECUs for specific test cases as part of software releases. Clear criteria and prerequisites will need to be established for this purpose.

5.3. Requirement Cluster for Software Validation (RQ2)

The release of software for use in a vehicle is based on a software release process that ensures the requirements for the product, especially the software, are met. As explained in Section 5.1, the verification criteria define which requirements are addressed by which types of test environments. This, in turn, depends on which technical functions are implemented in a simulation environment with vECUs and how much original source code is implemented in a vECU. It does not make much sense to use a vECU for test cases that contain a small amount of production code. Therefore, the development of a new method for the testing process using vECUs starts with the creation of requirement clusters. The generation of a cluster takes into account various criteria. A crucial criterion for clustering is the location of the software components to be tested according to the AUTOSAR standard, as shown in the corresponding figure (see Figure 5). This standardization allows for a systematic identification of software components and their functions within the vehicle architecture. A crucial aspect in requirement analysis is the availability of software components. If specific parts of the software are missing, testing of these components is not possible. This insight is fundamental because it directly affects the testability and thus the reliability of the entire testing process. The clustering carried out primarily provides an overview of whether and what fundamental limitations exist regarding the testing of software components. In addition to considering the AUTOSAR architecture, other criteria that are specific to the validation of vECUs are included in the cluster formation. These criteria include, among others, time-critical requirements that must ensure that the vECU reflects valid timing behavior. In relation to control functions, such tasks within the vECU should be properly mapped. Safety requirements, classified according to ASIL, are also relevant. Furthermore, requirements for dynamic functions and logical functions located in the application layer are taken into account in the evaluation. These factors play a significant role in defining the test strategy and help reduce the complexity of the testing processes and increase efficiency.

The resulting requirement clustering is therefore not only a methodological basis for systematic test development but also a strategic tool to ensure the integrity and performance of vehicle software. This systematic approach creates a basis for future test scenarios that can secure both the technical and safety-relevant aspects of ECU software. The basis for the requirement analysis was created through expert interviews, project work experience, and a systematic review of the individual requirement documents from which the requirements arise. Taking the brake control system as an example, this resulted in approximately 18,000 requirements that needed to be tested throughout the project. However, for certain release processes, especially at the beginning of a project, not all of these requirements were used, because the functions were developed in the different steps. The goal of the analysis was to assign these requirements to specific clusters, to better align them with the vECU classifications. The following sections will introduce and describe the identified clusters, which cover nearly all requirements.

1. Flashing
The commissioning of an ECU is an important step that precedes actual testing. This process involves various steps, including flashing the software, encoding, and writing a dataset. During the flashing process, the actual software is transferred to the flash memory of the ECU's microcontroller. This requires the presence of a suitable bootloader on the microcontroller. The main goal of this process is to ensure that the software can be flashed onto the hardware without any issues. The most common method is flashing via the on-board diagnostics (OBD) interface, which is standardized in modern vehicles [35].
2. Coding
Coding an ECU refers to the ability to adapt the software so that it is suitable for a specific vehicle variant and functions according to the specific equipment features.

Typically, the software on an ECU has a wider scope, providing various variants and parameters for different vehicle versions. This, of course, assumes that the corresponding hardware is installed in the vehicle.
A good example of this is the trailer stability function in electronic stability control (ESC), which is available as an optional feature in some vehicle variants when a trailer hitch is present. The function is generally present in the software for all variants but is only activated through appropriate coding. This can also affect the bus communication when additional ECUs are installed in the vehicle. Coding is an essential part of the initial commissioning process, because the software would not function properly without the correct coding. During the testing process, all possible and relevant codings are checked to ensure that they can be coded correctly. At the same time, non-configurable vehicle variants should not be codable. Coding is also done via OBD.

3. Dataset
Another important part of commissioning is the dataset stored in the ECU's EEPROM. This dataset includes various control parameters for functions and is usually not modified during a test campaign. Changing and writing the dataset is more relevant for the calibrators who tune the vehicles and make individual adjustments. However, a well-tuned dataset still plays a crucial role in evaluating the software. Especially in the chassis area, proper parameterization can have a significant impact on vehicle behavior. Parameterization allows adapting the software's behavior and performance to specific requirements and conditions, which is essential for assessing and optimizing the software. Writing the datasets can also be performed via OBD.
4. Cyber Security
Cyber security requirements, according to [36], are a response to the increasing potential for cyberattacks on the vehicle network. To ensure data integrity, authenticity, and timeliness, additional security measures are taken. These measures include data source authentication, message integrity verification, and ensuring confidentiality through encryption of relevant data segments. In this context, specific critical CAN messages containing safety-critical signals are encrypted to prevent manipulation from external sources [37]. Security mechanisms must also be thoroughly tested, to ensure their effectiveness in protecting against cyberattacks.
5. Diagnostics
The diagnostics category includes all tests that can be controlled via OBD. Standardization covers not only emission-related electronic modules in a vehicle but rather the entire electronics. This area accounts for a significant portion of test cases and offers a variety of functions that are particularly important for workshops, developers, and maintenance work. The software implements legally required features, including selective monitoring of emission-related systems and monitoring of all major ECUs whose data are accessible through the software. Error messages can be retrieved later through standardized interfaces [35].
The standardized interface used in this context is unified diagnostic services (UDS). UDS is a diagnostic communication protocol used in the field of automotive ECUs and is specified in ISO 14229 [38]. This protocol enables efficient and standardized communication between diagnostic tools and the various ECUs in the vehicle [39]. To perform this function, some components in the basic software (BSW) must be implemented, including the diagnostic communication manager (DCM), which is located in the service layer (see Figure 9).

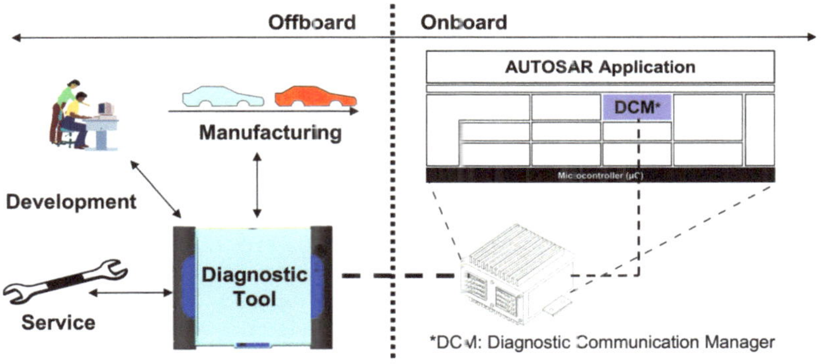

Figure 9. Diagnostic communication between offboard and onboard [40].

There are numerous functions that fall under this cluster, and it is not possible to detail each one individually. However, some of the most important functions include reading DTCs. OBD defines an extensive list of standardized DTCs, which serve to identify and diagnose vehicle issues more easily. Additionally, relevant input and output data can be monitored in real time. OBD provides real-time data on various vehicle parameters, including engine speed, speed, air temperature, fuel consumption, and more. The selection of this cluster was based on its relevance to the technical requirements of a vECU. It constitutes a significant portion of the test cases, which will be explained in more detail later.

6. Electrical tests

   The electrical tests are divided into component tests and software tests. Component tests are carried out, for example, to determine the durability of components under certain voltage conditions or short-circuit currents. In contrast, software tests must ensure that the software responds correctly to specific electrical events such as short circuits, over-voltage, or under-voltage conditions and detects them. It is checked whether the corresponding DTCs appear in the fault memory.

7. Interfaces

   The requirements for interfaces encompass all physical connections that an ECU has outside of typical bus communications. These often include sensors and actuators directly connected to the ECU. During the testing phase, it is ensured that these interfaces function correctly. An example of this would be the ESC, where wheel speed sensors play a relevant role. These sensors capture wheel speeds and can transmit measurement values through their protocol in the case of active wheel speed sensors. Another example would be the electronic parking brake, which is controlled by the ESC and serves as an actuator. Finally, human–machine interfaces like control or display interfaces are also included, often supported by specific protocols. These specific interfaces must be tested for certain requirements. Due to the variety of protocols used, the range of requirements for these interfaces can be quite broad.

8. Bus communication

   The bus communication cluster encompasses all tests related to the correct implementation of requirements in the bus system. Various aspects are verified, such as compliance with prescribed cycle times for message transmission. It ensures that transmitted values fall within the specified minimum and maximum value ranges. Furthermore, it checks whether the ECU is capable of detecting errors like incorrect message counters or errors in the cyclic redundancy check (CRC) calculation. If such errors occur and a predefined monitoring time has been exceeded, the ECU should report a corresponding DTC.

9. Network management

   The network management cluster primarily focuses on verifying requirements related to monitoring and controlling the internal states of the ECU. This includes ensuring that relevant network management messages are transmitted correctly. Furthermore, it checks whether the ECUs enter and wakes up from the sleep mode at the correct times, as specified. A malfunction in this area can lead to the failure to achieve proper bus sleep in the vehicle network. This, in turn, can result in various network participants keeping each other active, leading to unnecessary power consumption and communication issues.

10. Functions
The functions cluster includes all the classic application functions that need to be specifically developed for each ECU and project. While other clusters may contain generic requirements, the requirements in this area are tailored to the specific functionality of the ECU. These functions can be generally categorized into three categories relevant for testing: conditional functions, control functions, and routine functions. Conditional functions can be further divided into if conditions, statement conditions, and logical functions. Conditional functions are expected to implement a specific response based on certain conditions.
Control functions are functions that represent a control loop. They set a reference value and respond to feedback and/or disturbances using a controller. Examples include the lambda control in an internal combustion engine or ESC in dynamic driving situations. Figure 10 illustrates a general control loop.

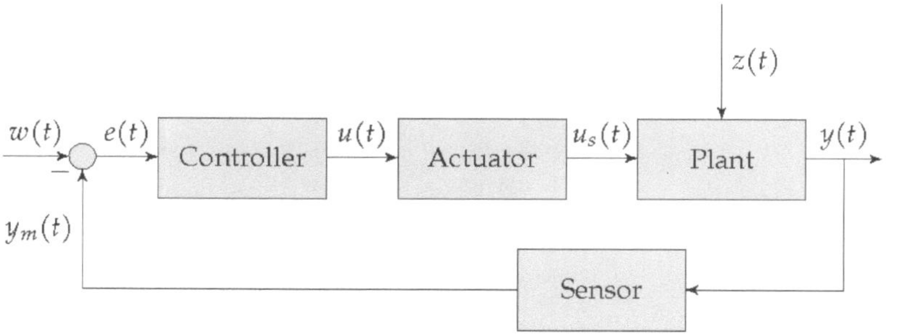

Figure 10. Illustration of a simple control loop, modified and extended according to [41].

A routine function, implemented as a function, influences one or more variables as input variables to change other variables as output variables based on the given control arrangement. In this case, the output variable does not affect the input variable, due to the lack of feedback. As an analogy to regulation, consider the example of the accelerator pedal position as an input variable taken by the driver, which represents an open-loop system. Setting a cruise control to maintain a certain speed would be a closed-loop system with feedback and, therefore, a form of regulation.
Conditional functions are also a focus of the validation process. In particular, with conditional functions, a strict approach is applied following the specified specifications, to verify whether certain output variables exist when certain input variables are provided. A simple example of this is the brake disc wiping function in ESC, which requires certain conditions (e.g., activation of the wipers) before it is activated. Subsequently, a cyclic routine is executed, which builds up slight brake pressure in the brake cylinders for a short period to dry the brake discs during wet weather conditions.
Control functions can be evaluated and tested both objectively based on specific evaluation criteria and subjectively. To continue with the cruise control example, a subjective evaluation may involve assessing whether the speed would be adjusted too abruptly through rapid deceleration or acceleration. An objective criterion could be that the control system is stable and can compensate for control deviations without oscillations.

5.4. Mapping of Requirements That Can Be Tested with vECUs (RQ3)

In this section, a systematic overview is provided based on the findings from the previous sections. It outlines the specific requirement areas that can be technically tested with different levels of vECUs. This evaluation is particularly relevant for refining the requirements that an OEM should impose on its suppliers, especially when software-related tests are to be conducted in SiL simulation environments. The assessments presented in the matrix are based on experiences gained from projects involving vECUs.

The matrix (Table 8) provides insights into the technical testability of various requirements depending on the level of the vECU. Flashing software onto ECUs requires specific hardware, including at least one microcontroller, to implement the software code. This requirement is only met by a vECU at Level 4. For coding, the use of NVM is necessary. Therefore, this function is only suitable for vECU levels that can simulate NVM, which is available from Level 2 m. The same requirement applies to parameterization and datasets. The implementation of cyber security measures is associated with significant effort in practice. This effort also depends on whether the security features were implemented in the original system on the software or hardware side. For this reason, such tests should either be conducted on real hardware or with emulated hardware. Diagnosis functions, on the other hand, require NVM and certain parts of the BSW. Furthermore, the implementation of the UDS protocol is relevant for equating the handling of these requirements with a HiL environment. Due to the number of requirements in diagnosis, they should receive special attention. Electrical tests are only possible if the hardware is either emulated or physically present, which is why these requirements strictly do not fall within the scope of vECUs. Interface tests are highly dependent on the hardware. Depending on the implemented hardware components in a Level 3 vECU, some of these tests can be conducted. Bus communication is an important aspect of many test scenarios. Most requirements in this area can be fulfilled as long as a virtual communication interface is available, which applies to all vECU levels with a "c" in the designation. Network management, due to the need to emulate electrical circuits, is also a challenge. However, the states that occur during active operation can be simulated when a communication interface and the corresponding BSW modules are available. The user layer, which includes the logic of the actual software, can be partially tested at a logical level even with a Level 0 vECU, although the complete original source code is usually not available. This also applies to Level 1 vECUs, which often do not contain the entire source code of the user layer. Starting from Level 2, nearly the entire user layer becomes available, allowing for the testing of such requirements. With each increase in the level within the vECU matrix, as you move further to the right, more production code is present. This ensures that the real software exhibits the same behavior and minimizes potential sources of errors that could arise from simulated code and software cuts in the vECU.

The evaluation of a Level 3 vECU for ESC was validated through random testing. Representative test cases from various requirement clusters, usually tested in HiL tests or directly in the vehicle, were used. The test cases were conducted within a simulation environment using the vECU. It was shown that the vECU was capable of covering the majority of test cases. This included testing different coding variants and datasets, which can be easily implemented on the vECU. With current architectures, cybersecurity requirements cannot be met. The necessary components are not integrated into the vECU. In the diagnostic area, there are also some limitations since the UDS protocol is not implemented, resulting in only those diagnostic aspects that the vECU developers have provided interfaces for being testable. Another obstacle arose in testing the hardware interfaces to wheel speed sensors, which were not feasible due to the lack of physical interfaces. In these test cases, the protocol of the wheel speed sensors was checked. However, an additional simulation of the wheel speed sensors would be required. In terms of network management, the vECU was capable of modifying states through specific CAN messages of the network. When testing the application layer, almost all requirements can be covered, but a comprehensive simulation environment, including some track models, is necessary. In summary, it can be stated that approximately 80% of the requirements could be tested using this vECU. The most extensive test scopes were found in the diagnosis, bus communication, and function test areas. These results underscore the efficiency of the vECU in the testing process but also highlight the limitations imposed by the simulation level and the availability of specific protocol implementations.

Table 8. Overview of which requirements can be technically tested with vECUs.

Requirements	**Level 0**	**Level 1**	**Level 2**	**Level 2c**	**Level 2m**	**Level 3**	**Level 3c**	**Level 4**
Flashing	–	–	–	–	–	–	–	✓
Coding	–	–	–	–	✓	✓	✓	✓
Dataset	–	–	–	–	✓	✓	✓	✓
Cyber Security	–	–	–	–	–	–	–	✓
Diagnostic	–	–	–	–	◐	✓	✓	✓
Electrical tests	–	–	–	–	–	–	–	✓
Interfaces	–	–	–	–	–	◐	◐	✓
Bus communication	–	–	–	✓	–	–	✓	✓
Network management	–	–	–	–	–	✓	✓	✓
Functions	◐	◐	✓	✓	✓	✓	✓	✓
Amount of production code								

✓ Should be completely possible; ◐ Should be partially possible; – Not possible.

6. Discussion of Limitations

In this section, the limitations of this method are highlighted, and suggestions for future approaches are provided.

6.1. *AUTOSAR-Specific Limitations*

The method presented in this paper is specifically designed for the AUTOSAR Classic Platform. This standard provides a robust framework for the development of embedded software in vehicle ECUs. However, a clear limitation arises for ECUs that do not adhere to the AUTOSAR standard. For such ECUs, the applied method is less suitable, because the specific architecture and software structure of non-AUTOSAR systems cannot be easily assessed using the procedures described here. Therefore, the method cannot be directly applied to the AUTOSAR Adaptive Platform without modifications, as it is based on different principles that enable a more dynamic and flexible software architecture. Within this study, only AUTOSAR-compliant ECUs and hybrid solutions that include both AUTOSAR components and proprietary software elements were considered.

6.2. *Variance*

Another limiting factor is the variance in the production of vECUs. The results of this study clearly show that there is no uniform approach to creating vECUs. Different manufacturers use different tools and apply varying degrees of resources and efforts to simulate hardware components. As a result, the quality and scope of vECUs vary significantly, directly affecting the amount of original source code that can be implemented and the number of technically testable requirements. The analysis of five different ECUs confirmed that each ECU was developed using different tools and approaches, partially limiting the comparability and transferability of the results.

6.3. Credibility

The final limitation pertains to the validation of the technical feasibility compared to a real system. Empirical investigation into the extent to which vECUs accurately replicate the logic and timing behavior of physical ECUs was not conducted, as demanded by Hansen et al. [42]. Physical ECUs often work with multiple processor cores that process tasks in parallel, handle interrupts, and operate in real time. On the other hand, PCs, which are commonly used for running vECUs, generally execute only a single task at any given moment. Furthermore, microcontrollers often operate in microsecond time frames and control analog signals with response times of less than 1 ms—a scenario that is difficult to simulate with a vECU. These aspects are central to understanding the performance and limitations of vECUs, and they should be more comprehensively examined in future studies, to assess and understand their impact on test results.

7. Conclusions

In the present paper, the growing importance of virtual test methods for software validation in vehicle ECUs has been emphasized. In this context, vECUs are of central importance, but a clear methodology for assessing their suitability for certain software validation tasks is still lacking. In particular, it was unclear which requirements can be tested using vECUs and what limitations may arise. To address this gap, this paper proposed a revision of the existing vECU standard according to ProSTEP iViP and introduced new approaches and additions to enable a better classification and assessment of vECUs.

The study was based on the examination of five different vECUs provided by various suppliers. In parallel, comprehensive analysis and categorization of various requirements were conducted, serving as the basis for requirement-based testing and considering different types of test environments. By using a matrix that assigns various requirement clusters to corresponding vECU abstraction levels, a better allocation of test cases to specific test environment types could be achieved. The results of the study showed that a significant portion of requirements can be virtually implemented through the targeted use of SiL simulation environments. This ultimately reduces costs and time in the testing process.

Finally, an empirical verification was conducted using an SiL simulation environment that implemented a Level 3c vECU for a brake control system. The results of this sample confirmed the findings discussed earlier, indicating that approximately up to 80% of requirements can be successfully tested using such a vECU. This confirmation underscores the practicality of the proposed methodology and demonstrates its potential to enhance the efficiency and effectiveness of virtual testing processes.

In the future, the accuracy of the results obtained in this study could be further enhanced by implementing automation of the simulation environment. Such automation would allow for systematic execution of test cases and objective evaluation of results. This could enable precise quantification of the requirements that can actually be realized in an SiL environment. Additionally, automation provides the opportunity to directly compare results from the SiL environment with results from HiL tests. This comparison is particularly valuable as HiL test benches incorporate real hardware components and the physical interfaces of the vehicle, serving as a good reference for validation. By comparing results from both environments, the validity and limits of SiL-based simulations can be determined even more accurately.

Author Contributions: Conceptualization, R.K.; methodology, R.K.; formal analysis, R.K.; investigation, R.K.; resources, R.K.; data curation, R.K.; writing—original draft preparation, R.K.; writing—review and editing, R.K., J.A.T., J.T. and M.E.A.; visualization, R.K.; supervision, J.A.T., J.T. and M.E.A. All authors have read and agreed to the published version of the manuscript.

Funding: This research project was supported by Volkswagen AG. The results, opinions, and conclusions expressed in this publication are those of the authors and do not necessarily represent the views of Volkswagen AG. The APC was funded by the Open Access Publication Fund of TU Dresden.

Institutional Review Board Statement: Not applicable.

Informed Consent Statement: Not applicable.

Data Availability Statement: In the context of this study, virtual electronic control units and data from suppliers were analyzed, as well as the internal specifications of the original equipment manufacturer. These data are subject to a confidentiality agreement and cannot be provided. The necessary and relevant results are presented in the paper.

Conflicts of Interest: Authors Rudolf Keil and Jan Alexander Tschorn were employed by the company Volkswagen AG. The remaining authors declare that the research was conducted in the absence of any commercial or financial relationships that could be construed as a potential conflict of interest.

Abbreviations

ADC	Analog Digital Converter
ASIL	Automotive Safety Integrity Levels
AUTOSAR	Automotive Open System Architecture
BSW	Basic Software
BSWM	Basic Software Module
CAN	Controller Area Network
CDD	Complex Device Driver
COM	Communication
CRC	Cyclic Redundancy Check
DCM	Diagnostic Communication Manager
DLL	Dynamic Link Library
DTC	Diagnostic Trouble Code
EEPROM	Electrically Erasable Programmable Read-Only Memory
ESC	Electronic Stability Control
FMI	Functional Mock-up Interface
FMU	Functional Mock-up Unit
HiL	Hardware-in-the-Loop
LIN	Local Interconnect Network
MCAL	Microcontroller Abstraction Layer
Mil	Model-in-the-Loop
NVM	Non-volatile Memory
OBD	On-Board-Diagnose
OEM	Original Equipment Manufacturer
OSI	Open Systems Interconnection
PDU	Protocol Data Unit
RTE	Runtime Environment
SiL	Software-in-the-Loop
SoC	System on Chip
SPI	Serial Peripheral Interface
SWC	Software Component
UDS	Unified Diagnostic Services
VDA	Verband der Automobilindustrie
vECU	virtual Electronic Control Unit
VFB	Virtual Functional Bus
XCP	Universal Measurement and Calibration Protocol

References

1. Deicke, M. Virtuelle Absicherung von Steuergeräte-Software mit hardwareabhängigen Komponenten. In *Wissenschaftliche Schriftenreihe "Eingebettete, Selbstorganisierende Systeme"*; Universitätsverlag Chemnitz: Chemnitz, Germany, 2018; Volume 16.
2. Amringer, N.; Asemann, P. Simulation of Virtual ECUs in the context of ECU Consolidation. In Proceedings of the 9th AutoTest Technical Conference, Modena, Italy, 28–30 June 2023.
3. Muslija, A.; Enoiu, E. *On the Correlation between Testing Effort and Software Complexity Metrics*; PeerJ Preprints: London, UK, 2018. [CrossRef]
4. Abdellatif, H.; Gnandt, C. Use of Simulation for the Homologation of Automated Driving Functions. *ATZelectronics Worldw.* **2019**, *14*, 68–71. [CrossRef]

5. Knauss, A.; Schroder, J.; Berger, C.; Eriksson, H. Software-Related Challenges of Testing Automated Vehicles. In Proceedings of the 2017 IEEE/ACM 39th International Conference on Software Engineering Companion (ICSE-C), Buenos Aires, Argentina, 20–28 May 2017; pp. 328–330. [CrossRef]
6. Donà, R.; Ciuffo, B. Virtual Testing of Automated Driving Systems. A Survey on Validation Methods. *IEEE Access* **2022**, *10*, 24349–24367. [CrossRef]
7. Synopsys. Accelerating Development of Software Defined Vehicles with Virtual ECUs. Available online: https://www.synopsys.com/content/dam/synopsys/verification/white-papers/virtual-ecu-wp.pdf (accessed on 6 October 2023).
8. Sievers, G.; Seiger, C.; Peperhowe, M.; Krumm, H.; Graf, S.; Hanselmann, H. Driving simulation technologies for sensor simulation in sil and hil environments. In Proceedings of the DSC, Antibes, France, 5–7 September 2018.
9. Park, C.; Chung, S.; Lee, H. Vehicle-in-the-Loop in Global Coordinates for Advanced Driver Assistance System. *Appl. Sci.* **2020**, *10*, 2645. [CrossRef]
10. Altinger, H.; Wotawa, F.; Schurius, M. Testing methods used in the automotive industry: Results from a survey. In Proceedings of the Proceedings of the 2014 Workshop on Joining AcadeMiA and Industry Contributions to Test Automation and Model-Based Testing, New York, NY, USA, 21 July 2014; JAMAICA 2014; pp. 1–6. [CrossRef]
11. Clausen, C.S.B.; Jørgensen, B.; Ma, Z. A scoping review of In-the-loop paradigms in the energy sector focusing on software-in-the-loop. *Energy Inform.* **2024**, *7*, 1–40. [CrossRef]
12. Prostep IVIP. Smart Systems Engineering: Requirements for the Standardization of Virtual Electronic Control Units (V-ECUs). Available online: https://www.ps-ent-2023.de/fileadmin/prod-preview/WhitePaper_V-ECU_2020_05_04-EN_shortversion.pdf (accessed on 8 March 2023).
13. Vahldiek, M. Einsatz einer Software-in-the-Loop-Umgebung zur virtuell gestützten Applikation des Motorstarts eines hybriden Ottomotors. In *Experten-Forum Powertrain: Simulation und Test 2020*; Liebl, J., Ed.; Springer: Berlin/Heidelberg, Germany, 2021; pp. 87–103.
14. Urbina, M.; Owda, Z.; Obermaisser, R. Simulation Environment Based on SystemC and VEOS for Multi-core Processors with Virtual AUTOSAR ECUs. In Proceedings of the 2015 IEEE International Conference on Computer and Information Technology; Ubiquitous Computing and Communications; Dependable, Autonomic and Secure Computing; Pervasive Intelligence and Computing, Liverpool, UK, 26–28 October 2015; pp. 1843–1852. [CrossRef]
15. Alhasan, W. Evaluating Challenges, Benefits, and Dependability of Virtual and Physical Testing of Embedded Systems Software. Ph.D. Thesis, Mälardalen University, School of Innovation, Design and Engineering and Malardalen University, Eskilstuna, Swiden, 1 January 2024.
16. Riccio, A.; Monzani, F.; Landi, M. Towards a Powerful Hardware-in-the-Loop System for Virtual Calibration of an Off-Road Diesel Engine. *Energies* **2022**, *15*, 646. [CrossRef]
17. Alaqail, H.; Ahmed, S. Overview of Software Testing Standard ISO/IEC/IEEE 29119. *Ijcsns Int. J. Comput. Sci. Netw. Secur.* **2018**, *18*, 112–116.
18. Sarwar, T.; Habib, W.; Arif, F. Requirements based testing of software. In Proceedings of the 2013 Second International Conference on Informatics & Applications (ICIA), Lodz, Poland, 23–25 September 2013; pp. 347–352.
19. Witte, F. *Testmanagement und Softwaretest*; Springer Fachmedien Wiesbaden: Wiesbaden, Germany, 2016. [CrossRef]
20. Spillner, A.; Rossner, T.; Winter, M.; Linz, T. Test Management: Test Process Fundamentals. Available online: https://www.luigicardarella.it/test-management-test-process-fundamentals/ (accessed on 27 January 2024).
21. *ISO 26262*; Road Vehicles—Functional Safety. International Standard of Organisation: Geneva, Switzerland, 2018.
22. AUTOSAR. Classic Platform. Available online: https://www.autosar.org/standards/classic-platform (accessed on 27 January 2024).
23. Avin Systems. AUTOSAR Migration and Integration. Available online: https://www.avinsystems.com/services/autosar-migration-and-integration/ (accessed on 27 January 2024).
24. Arestova, A.; Martin, M.; Hielscher, K.S.J.; German, R. A Service-Oriented Real-Time Communication Scheme for AUTOSAR Adaptive Using OPC UA and Time-Sensitive Networking. *Sensors* **2021**, *21*, 2337. [CrossRef]
25. Lee, K.; Park, I.; Sunwoo, M.; Lee, W. AUTOSAR-ready Light Software Architecture for Automotive Embedded Control Systems. *Korean Soc. Automot. Eng.* **2013**, *21*, 68–77. [CrossRef]
26. AUTOSAR. Adaptive Platform. Available online: https://www.autosar.org/standards/adaptive-platform (accessed on 27 January 2024).
27. AUTOSAR. Software Component Template: Release R21-11. Available online: https://www.autosar.org/fileadmin/standards/R21-11/CP/AUTOSAR_TPS_SoftwareComponentTemplate.pdf (accessed on 27 January 2024).
28. AUTOSAR. Specification of RTE Software: Release R20-11. Available online: https://www.autosar.org/fileadmin/standards/R20-11/CP/AUTOSAR_SWS_RTE.pdf (accessed on 23 January 2024).
29. AUTOSAR. Requirements on Libraries: Release R20-11. Available online: https://www.autosar.org/fileadmin/standards/R20-11/CP/AUTOSAR_SRS_Libraries.pdf (accessed on 23 January 2024).
30. AUTOSAR. Layered Software Architecture: Release R22-11. Available online: https://www.autosar.org/fileadmin/standards/R22-11/CP/AUTOSAR_EXP_LayeredSoftwareArchitecture.pdf (accessed on 23 January 2024).

31. Keil, R.; Tschorn, J.A.; Tümler, J.; Altinsoy, M.E. Optimization of Automotive Software Tests by Simplification of the Bus Simulation. In Proceedings of the 2023 IEEE International Workshop on Metrology for Automotive (MetroAutomotive), Modena, Italy, 28–30 June 2023; pp. 72–77. [CrossRef]
32. Verband der Automobilindustrie e. V. VDA 710; Software-in-the-Loop (SiL) Standardisierung. Berlin, Germany, 2022.
33. FMI. Functional Mock-up Interface Specification. Available online: https://fmi-standard.org/docs/3.0/ (accessed on 27 January 2024).
34. AUTOSAR. Specification of CAN Transport Layer: R21-11. Available online: https://www.autosar.org/fileadmin/standards/R21-11/CP/AUTOSAR_SWS_CANTransportLayer.pdf (accessed on 27 January 2024).
35. Malekian, R.; Moloisane, N.R.; Nair, L.; Maharaj, B.T.; Chude-Okonkwo, U.A.K. Design and Implementation of a Wireless OBD II Fleet Management System. *IEEE Sens. J.* **2017**, *17*, 1154–1164. [CrossRef]
36. UNECE, WP. Uniform provisions concerning the approval of vehicles with regards to cyber security and cyber security management system. *Regulation* **2021**, *155*, 1.
37. Vector. Lösungen für Automotive Cybersecurity. Available online: https://www.vector.com/de/de/produkte/solutions/safety-security/automotive-cybersecurity/#c59044 (accessed on 15 January 2024).
38. Bidkar, S.; Patil, S.L.; Shinde, P. Virtual ECU Development for Vehicle Diagnostics Software Testing using UDS Protocol. In Proceedings of the 2021 Asian Conference on Innovation in Technology (ASIANCON), Pune, India, 27–29 August 2021; pp. 1–6. [CrossRef]
39. Vector. UDS Diagnose. Available online: https://www.vector.com/de/de/produkte/solutions/diagnose-standards/uds-unified-diagnostic-services-iso14229/# (accessed on 15 January 2024).
40. Specification of CAN Interface. Release R21-11. Available online: https://www.autosar.org/fileadmin/standards/R21-11/CP/AUTOSAR_SWS_CANInterface.pdf (accessed on 15 January 2023).
41. Schäuffele, J.; Zurawka, T. *Automotive Software Engineering: Grundlagen, Prozesse, Methoden und Werkzeuge Effizient Einsetzen*, 6th ed.; ATZ/MTZ-Fachbuch, Springer: Wiesbaden, Germany, 2016. [CrossRef]
42. Hansen, S.T.; Gomes, C.Â.G.; Najafi, M.; Sommer, T.; Blesken, M.; Zacharias, I.; Kotte, O.; Mai, P.R.; Schuch, K.; Wernersson, K.; et al. The FMI 3.0 Standard Interface for Clocked and Scheduled Simulations. *Electronics* **2022**, *11*, 3635. [CrossRef]

vehicles

MDPI

Article

Exploring Factors Influencing Electric Vehicle Purchase Intentions through an Extended Technology Acceptance Model

Zhiyou Sun and Boyoung Lee *

Department of Industrial Design, Hanyang University, ERICA Campus, Ansan 15588, Republic of Korea; sun4888@hanyang.ac.kr

* Correspondence: ondodam@hanyang.ac.kr

Abstract: Recently, with climate deterioration and environmental pollution, consumers are becoming more and more aware of the use of sustainable energy. In particular, the demand for electric vehicles that use sustainable energy is also increasing. In addition, due to the simple driving principle of pure electric vehicles, many electric vehicles developed by electronics companies are continuously being launched. Electric vehicles not only use renewable energy to protect the environment but also save on various usage expenses, so they are expected to become the main products in the mobile travel equipment market in the future. This study aims to explore the impact of product design dimensions on electric vehicle (EV) purchase intentions, provide a theoretical basis for companies' differentiation strategies, and reflect the impact of product design on purchase intention. This study uses Davis's TAM combined with environmental awareness (EA) for analysis; an online survey was conducted on Chinese (n = 468) and Korean (n = 409) consumers, both male and female, aged 20–60 years and above. We found that, for Chinese consumers, the aesthetic and symbolic dimensions do not affect perceived usefulness and perceived ease of use, but they do affect environmental awareness, while the functional dimension affects not only perceived ease of use and usefulness but also environmental awareness. For Korean consumers, the aesthetic, functional, and symbolic dimensions all affect perceived ease of use and environmental awareness, but perceived usefulness is only affected by aesthetics and environmental awareness. Through simulation analysis, the results show that perceived ease of use, usefulness, and environmental awareness all directly affect purchase intentions. Perceived ease of use and environmental awareness are particularly important for Chinese consumers, while Korean consumers pay more attention to the test drive experience and environmental awareness. The results show that electric vehicle manufacturers should develop new technologies for the Chinese market to attract consumers, while in the Korean market, they should improve perceived usefulness through test drives and pay attention to environmental awareness. Specific statistical data show that both Chinese and Korean consumers assign importance to the impact of environmental awareness on purchase intention, proving the importance of environmental awareness. The results of this study will be of great reference value to electric vehicle manufacturers, policymakers, and consumer behavior researchers, helping them to better understand the role of product design in improving the market acceptance of electric vehicles.

Keywords: TAM model; structural equation modeling; product design; electric vehicle purchase intention

Citation: Sun, Z.; Lee, B. Exploring Factors Influencing Electric Vehicle Purchase Intentions through an Extended Technology Acceptance Model. *Vehicles* **2024**, *6*, 1513–1544. https://doi.org/10.3390/vehicles6030072

Academic Editors: Ralf Stetter, Udo Pulm and Markus Till

Received: 22 July 2024
Revised: 12 August 2024
Accepted: 22 August 2024
Published: 30 August 2024

1. Introduction

Electric vehicles are improving the environment and modern people's lives, with various appearances and functions. Modern people's definition of cars is also constantly changing. For example, the electric car released by China's Xiaomi in 2024—a case in which a company that manufactures electronic products successfully transitioned into the automotive industry—has attracted global attention. With people's increasing awareness of environmental protection, electric vehicles, as mobile tools that can reduce pollution and effectively save nonrenewable energy, have gradually become a new development trend in the automobile industry [1].

According to Statista, a global statistical data and market research platform, global sales of new electric vehicles increased significantly by 55.1% in 2022, with sales exceeding 10.25 million units [2]. Expected trends show that EU countries will push for the implementation of CO_2 emission standards and steadily move toward zero-emission vehicle regulations. This indicates that the development trend of electric vehicles will continue. In other words, more consumers will use electric vehicles in the future, and various demands have been generated for them. In order to meet these consumer needs, electric vehicles with various high-tech technologies and unique shapes have emerged.

In other words, in order to satisfy consumers, the demand for electric vehicles with unique styling, environmental protection symbolism, and advanced technology continues to grow [3,4]. In the future, when the electric vehicle market reaches the saturation stage, product design will become the only competitive advantage of each automobile manufacturer [5].

Therefore, it is of great significance to study consumers' demand and consumption cognition for electric vehicles. However, to date, most of the research on electric vehicles has been about battery and charging technology [6], consumer characteristics [7], supervisory norms with respect to electric vehicles, perceived behavior [8], perception and adoption of new technologies and other studies; that is, research on electric vehicle product design is currently limited to technology or consumer attitudes.

In response to these limitations, some recent studies have tried to explore the appearance of electric vehicles from a design perspective [9–12]. However, research on electric vehicles is still limited to appearance styling or the future trends of electric vehicle styling; there has been no analysis of the relationship between consumer cognition and willingness to buy. In addition, in terms of research methods, research using the technology acceptance model has also been attempted [13], but there are limitations in expanding or improving existing research. Based on these research deficiencies, this study starts from the product design dimensions (i.e., aesthetics, function, and symbolism) in electric vehicle purchase intention, establishes an empirical analysis model, and analyzes the variables that affect electric vehicle purchase intentions. To this end, we develop an extended technology acceptance model (TAM) to explain the relationship between product design and technology acceptance based on new technologies, and to provide prediction results for electric vehicles through the identified predictors and estimation results. Development provides implications, and in terms of measurement, the impact of environmental awareness is added along with product design and cognitive acceptance in order to apply the extended model.

Finally, when studying the willingness to purchase electric vehicles (EVs), it is important to choose Chinese and Korean consumers as comparison objects. The Chinese market is growing rapidly, with strong policy support and subsidies, and local brands such as BYD and Xiaopeng are actively promoting market development. Although the Korean market is relatively small, the government is also actively promoting the popularization of EVs, and brands such as Hyundai and Kia have performed well in the global market. Comparing these two markets can reveal the impact of different policy environments on consumer purchase intentions, as well as reflecting differences in culture, economic level, and consumer behavior. In addition, the development of charging infrastructure, market size, and the influence of local brands are different in the two countries. Studying these differences can provide in-depth understanding of the complexity of the EV market, along with valuable insights for relevant policymakers and companies.

For companies, cross-cultural surveys can help them to better understand the needs and preferences of different cultural markets, so as to develop more effective marketing strategies and product designs. Cross-cultural differences in design tastes are common. Cultural background is one of the important determinants of consumers' perceptions of product design [14]. In order to explore consumers' purchase intentions and provide a basis for the differentiation strategies of enterprises, this study uses Chinese and Korean consumers as samples, explores the impact of product design dimensions on the purchase intentions of consumers in the two countries, and analyzes the differences in their purchase intentions for electric vehicles. Such research helps electric vehicle manufacturers in various countries to understand the characteristics of consumers in various countries, and it can also serve as a reference for other studies in the future.

2. Literature Review

2.1. Concept of Product Design

As one of the four Ps of marketing (product, price, place, and promotion), the importance of product design was proven long ago (e.g., Bloch,1995 [14]; Davis,(1989) [15]). Nussbaum's research (1988) [16] shows that when consumers can choose between two products with the same price and function, they will buy the product that they think is more attractive. Han (2021) [17] believes that product design trends should be understood from the perspective of social trends, consumer lifestyles, and work. The elements of product design are defined differently depending on the purpose of the research. For example, Kellaris (1993) [18] and Bloch, P.H (1995) [14] divide product design into elements such as shape, proportion, rhythm, proportion, material, color, reflectivity, decorativeness, and texture based on the consumer's product image; and in Sascha Mahlke's research (2007) [19], the user experience research was conducted by combining aesthetics, symbolism, and emotional user response with a user interaction experience method. Lee (2017) [20] studied consumers' purchase intentions through form (appearance) and feature design. Srinivasan et al. (2012) [21] proposed a customer-based product design framework based on customer experience design—the total product design concept (TPDC), which they defined as three elements of a product—function, aesthetics, and meaning—in order to understand the role and impact of the product.

However, the existing research has shortcomings in several respects. First, most studies have only focused on a specific field or design element, lacking a comprehensive perspective. Second, the mechanism of how product design specifically affects consumers' willingness to buy is still unclear. To make up for these shortcomings, Homburg et al. (2015) [22] and others have developed and validated a new scale to measure product design from three dimensions: aesthetics, function, and symbolism. In addition, they also studied the impact of these design dimensions on purchase intention, word of mouth, and willingness to pay. In summary, this study finally adopted the product design dimensions of aesthetics, function, and symbolism, as described by Homburg et al. (2015) [22], as measurement variables for our research.

2.2. Technology Acceptance Model

The technology acceptance model (TAM) is based on the model proposed by Davis (1989) [15]. It is a modified and developed model based on the causal relationship between attitude and intention to use, as proposed by Ajzen and Fishbein (1980) [23]—the theory of rational action (TRA).

Davis (1989) [15] pointed out through research that the "perceived usefulness (PU)" and "perceived ease of use (PEU)" of a particular technology affect the acceptance of new technologies. PU refers to the individual's belief that using a particular technology will improve their performance, while PEU refers to the individual's belief that using a particular technology is trouble-free. By studying the impact of a particular technology (external variable) on PU and PEU, one can understand the individual's acceptance of new technologies.

Later scholars established an extended TAM based on this model by adding external variables and other factors. For example, Venkatesh and Davis (2000) [24] added relevant external variables such as social influence and cognitive process to the TAM and proposed TAM2. Venkatesh and Bala (2008) [25] further refined perceived ease of use and perceived usefulness on the basis of TAM2, explored how intervention measures affect these factors, and proposed a more comprehensive model for user acceptance and use of technology: TAM3.

Subsequent studies introduced factors such as social influence and perceived risk when expanding the TAM. For example, Wang et al. (2013) [26] applied the TAM to study the influencing factors of new-energy vehicle purchase intentions; Hu et al. (2015) [27] combined the TAM with the perceived risk theory to study the willingness to purchase new-energy vehicles; Yin et al. (2019) [1] combined the TAM and perceived risk theory to analyze the impact of consumers' personality traits on their willingness to purchase new-energy vehicles when facing uncertainty. These studies verify that environmental awareness is an important variable affecting the acceptance of EVs.

In addition, Wang (2018) [28] explored the relationships between variables based on PEU, PU, and attitudes toward the intention to use electric motorcycles; Yuan et al. (2018) [29] confirmed through the TAM and rational behavior theory that users' environmental awareness (EA) has a significant positive impact on their attitude and willingness to use shared cars; Zhang (2022) [30] used the TAM as a theoretical framework based on variables such as PU and PEU, introduced extended variables such as social influence and environmental awareness, and established a willingness analysis model for the Mobility as a Service (MaaS) platform; He et al. (2018) [31] also showed that personality types such as personal innovation and environmental concern directly affect the willingness to purchase electric vehicles. Therefore, these previous studies can confirm that EA is an important variable in consumer acceptance of EVs.

However, the TAM also faces some challenges. First, the TAM focuses on the initial acceptance of technology's use, but there is less research on long-term usage intentions. Second, the model rarely considers the impact of social, cultural, and emotional factors on technology's acceptance.

2.3. Electric Vehicles (EVs)

Electric vehicles (EVs) can be divided into battery electric vehicles (BEVs), plug-in hybrid electric vehicles (PHEVs), fuel cell electric vehicles (FCEVs), hybrid electric vehicles (HEVs), etc., according to the driving system (see Figure 1) [12,13,32–35].

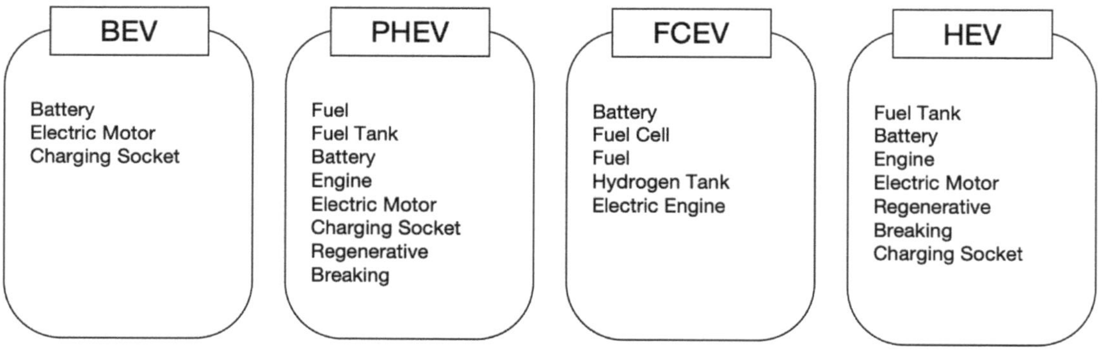

Figure 1. Electric vehicle classification.

1. Battery electric vehicle (BEV): Refers to a car that uses electricity to drive the motor and does not use any fuel, such as gasoline. The battery provides electricity to drive the motor, which drives the vehicle. A BEV is a vehicle that only needs a battery to drive the electric motor, and it requires fewer parts than an internal combustion engine vehicle (e.g., Tesla Model S; Nissan Leaf; Audi Bolt EV; BYD tang EV).
2. Plug-in hybrid electric vehicle (PHEV): A vehicle that can be charged and run on gasoline. The battery capacity is larger than that of ordinary hybrid vehicles. A PHEV is usually driven by electricity, but when the battery is exhausted, it switches to a

conventional internal combustion engine (e.g., Toyota Prius Prime; Chevrolet Volt; Mitsubishi Outlander PHEV; Hyundai oniq Plug-in Hybrid).

3. Fuel cell electric vehicle (FCEV): Uses environmentally friendly energy such as methanol and hydrogen to generate electricity, and relies on the converted electricity to drive the car, so it does not pollute the environment. However, the power system is driven by the chemical combination of hydrogen in the vehicle tank and oxygen in the air to generate electricity—not batteries (e.g., Toyota Mirai; Honda Clarity Fuel Cell; BMW i Hydrogen NEXT; Mercedes-Benz GLC F-CELL).
4. Hybrid electric vehicle (HEV): Does not support external charging. An HEV is a vehicle that combines an internal combustion engine and an electric motor, using an electric motor or a generator to run. It is equipped with two or more energy storage systems and energy conversion to allow the car to charge the electric motor while driving, and it rotates the transmission at the same time to make the gasoline engine and electric motor work (e.g., Toyota Prius; Honda Accord Hybrid).

2.4. Willingness to Buy (WTB)

Willingness to buy (WTB) refers to an individual's thoughts or plans for a specific action—that is, the consumer's intention to purchase a certain product [36]. Excellent design can differentiate products by creating and conveying value to customers, thereby improving customers' acquisition and retention of products [37,38].

Song et al. (2018) [7] studied the WTB and satisfaction of early adopters of electric vehicles based on satisfaction and anxiety about mileage, subjective knowledge, and important attributes. Lieven et al. (2011) [39] weighed personal priorities with social preferences and determined the WTB of individuals who were considered to be potential EV buyers through empirical research on German consumers. Kim et al. (2019) [40] studied the WTB of Chinese consumers for EVs based on the theory of planned behavior. Based on the above, Table 1 (Ashraf et al., 2020) [41] was compiled.

Table 1. Literature review.

No.	Research	Key Findings	Weaknesses	Suggested Solutions
1	Bloch (1995) [14]; Davis (1989) [15]	Importance of product design	Lack of comprehensive perspective	Introduce multi-dimensional design elements and comprehensive evaluation
2	Nussbaum (1988) [16]	Attractiveness affects purchase decisions	Research limited to specific conditions	Expand to different product categories and markets
3	Han (2021) [17]	Design should be understood from a lifestyle perspective	Lack of empirical research	Combine with empirical data verification
4	Homburg et al. (2015) [22]	Developed a new scale to measure product design	No involvement in electric vehicles	Apply to the electric vehicle field
5	Davis (1989) [15]	TAM explains technology acceptance	Ignoring long-term usage intentions	Introduce long-term use variables
6	Venkatesh and Davis (2000) [24]; Venkatesh and Bala (2008) [25]	Extended TAM	Lack of emotional factors	Add emotional and socio-cultural factors
7	Wang et al. (2013) [26]; Hu et al. (2015) [27];	Combining TAM with perceived risk to study new-energy vehicles	Mainly focused on initial acceptance	Study long-term use and market maturity
8	Song et al. (2018) [7]; Lieven et al. (2011) [39]; Kim et al. (2019) [40]	Factors affecting purchase intention	Immature market, insufficient research	Conduct more market and consumer perception research

Although existing studies have shown that the electric vehicle market continues to grow, the research and market are still not fully mature. There are currently few studies on consumer perceptions of electric vehicles, and there have not been many cases that studied the appearance design of electric vehicles in detail. Therefore, this study aims to explore the relationship between product design and consumer perceptions, in order to fill this research gap.

Through reviewing and analyzing the existing literature, this study explores the impact of product design on electric vehicle purchase intentions from a multi-dimensional perspective in order to make up for the shortcomings of existing research.

3. Conceptual Framework and Hypotheses

3.1. The Impact of Product Design on Technical Models and Environmental Awareness

3.1.1. The Impact of Aesthetics (Aes) on Technical Models and Environmental Awareness

Veryzer Jr, R. W. (1993) [42] believes that the aes response refers to a person's reaction to an object (e.g., based on their perception of the object). Aes is sometimes more important than technical features and is important in ensuring the success of a new product. Kotlw and Raih(1984) [43] also believe that good design can be transformed into high-quality product perceptions, bringing greater market share and higher profits. The connection between aes and technology acceptance is very important because aes design directly affects users' experience and acceptance. Elegant and intuitive interfaces not only improve perceived ease of use and make users feel that the operation is simple, they also enhance perceived usefulness and increase trust in the technology's function. In addition, aesthetic design can stimulate positive emotional responses, increasing users' satisfaction with the technology and long-term willingness to use it. An attractive visual and interactive design can also strengthen the brand image, further affecting users' acceptance of the technology. Therefore, aes not only affects the appearance of technology but also profoundly affects users' acceptance and use of technology, which merits further exploration in technology acceptance research [44,45].

Therefore, based on the above theoretical background, this study makes the following assumptions:

Hypothesis 1. *The higher the aesthetics dimension of product design, the greater the impact on PEU, PU, and EA.*

H1-1. *The aes dimension of product design has a significant positive impact on the PEU of EVs.*

H1-2. *The aes dimension of product design has a significant positive impact on the PU of EVs.*

H1-3. *The aes dimension of product design has a significant positive impact on the EA of EVs.*

3.1.2. The Impact of Function (Func) on Technical Models and Environmental Awareness

Functional design of products, in terms of practicality and functionality, involves product performance, reliability, and whether the product meets the basic needs of users. Fu et al. (2008) [46] proved through research on the functional requirements and intrinsic quality of automobiles that automobile technology plays a key role in automobile styling design. Chakrabarti and Bligh (2001) [47] believe that an ideal functional reasoning environment should have the following characteristics: able to support a variety of design tasks, whether routine or innovative; able to handle designs at any level of detail; and able to provide support between designs at different levels of detail. The functions of EVs have attracted much attention from researchers and users in the EV design process. Lv et al. (2014) [48] and Agarwal and Prasad (1998) [49] pointed out that the novelty of technology will affect practicality and ease of use, thereby affecting usage intention.

In addition, Agawal and Karhanna (2000) [50] showed that cognitive absorption significantly affects individuals' PU and PEU with respect to information technology, thereby affecting their intention to use it. Therefore, an individual's perceived technological innovativeness can be explained by cognitive absorption. Based on the discussion of these previous studies, this study proposes the following hypotheses on the role of func dimensions of product design in EV acceptance:

Hypothesis 2. *The higher the function (func) dimension of product design, the greater the impact on PEU, PU, and EA.*

H2-1. *The func dimension of product design has a significant positive impact on the PEU of EVs.*

H2-2. *The func dimension of product design has a significant positive impact on the PU of EVs.*

H2-3. *The func dimension of product design has a significant positive impact on the EA of EVs.*

3.1.3. The Impact of Symbolism (Sym) on Technical Models and Environmental Awareness

Symbolism (sym) refers to the symbolic meaning and cultural value of a product; it involves how products express the user's identity, status, and personal taste [42,51–54].

Xing (2022) [55] in order to promote the spread of Macao culture and the development of the tourist souvenir industry, took the symbolic meaning of cultural symbols as the starting point and explored effective methods for the design of Macao tourist souvenirs; Zhang et al. (2001) [56] pointed out that this is related to functional indicator symbols. Compared with packaging, the design and recognition of symbols on packaging are more complex. However, the connotative meaning conveyed by a symbol is usually broader and deeper than the denotative meaning of the symbol. Belk et al. (1989) [57] and Verganti (2008) [58] believe that sym is as important as func because consumers have a strong desire for meaning. In fact, sym can become the basis for consumers to experience personal value. Based on the above research, this study proposes the following hypotheses:

Hypothesis 3. *The higher the symbolism (sym) dimension of product design, the greater the impact on PEU, PU, and EA.*

H3-1. *The sym dimension of product design has a significant positive impact on the PEU of EVs.*

H3-2. *The sym dimension of product design has a significant positive impact on the PU of EVs.*

H3-3. *The sym dimension of product design has a significant positive impact on the EA of EVs.*

3.2. The Impact of the Technology Acceptance Model and Environmental Awareness on Willingness to Buy (WTB)

The technology acceptance model (TAM) has been widely used by many scholars at home and abroad to study the use of innovative technologies from different perspectives. The model proposes using PU and PEU as the main indicators for measuring technology acceptance behavior. PU refers to the individual's belief that using a specific application system can improve their work performance—i.e., users feel that using this innovative technology can bring convenience to their work or life—while PEU refers to the individual's belief that it is relatively easy to use or master a system or innovative technology [20].

The study of Jamal and Sharifuddin (2015) [59] shows that PU can positively promote the willingness to purchase products. Chitra (2007) [60] pointed out that consumers' EA has become an important factor affecting their willingness to pay. The stronger the consumer's EA, the higher the price that they are willing to pay for environmentally friendly products. Li (2001) [61] believes that efforts should be made to cultivate people's EA, so that they can clearly understand the interdependence between humanity and nature and humanity and the environment, and consciously pay attention to and actively participate in environmental protection. He et al. (2015) [62] demonstrated that consumers' PU and PEU with respect to innovative technologies of new-energy vehicles can significantly and positively affect their willingness to purchase such vehicles [1].

Based on these previous studies, we believe that PU, ease of use, and EA affect the WTB for EVs. Therefore, this paper proposes the following hypotheses and verifies them through empirical analysis:

Hypothesis 4. *The higher the PU, PEU, and EA, the greater the impact on the WTB.*

H4-1. *PEU has a significant positive impact on the WTB for EVs.*

H4-2. *PU has a significant positive impact on the WTB for EVs.*

H4-3. *EA has a significant positive impact on the WTB for EVs.*

3.3. Measurement Model

As mentioned above, this study is based on the extended TAM. The scale of the TAM is affected by multiple factors. Model complexity increases with the increase in the number of variables. More variables (such as user characteristics and environmental factors) will make the model larger, thereby increasing the amount of calculation and the analytical complexity. The size of the dataset is also critical. Large-scale datasets require more complex models to capture subtle differences, while simultaneously posing computing and storage challenges. Variable selection is equally important. Adding more external variables (such as social influences and personal characteristics) will make the model more complex, thereby affecting the effectiveness and efficiency of the model.

For the external variables, we selected Aes, Func, and Sym as product design variables, with PU, PEU, and EA as mediators and WTB as the dependent variable, to study EVs. In measuring perceived acceptance, this study not only included the influence of usefulness and ease of use on WTB, but also EA, because as people's awareness of environmental protection increases, it becomes necessary to confirm the difference between EA and perception in WTB. This study established the following research model by reviewing previous studies (see Figure 2).

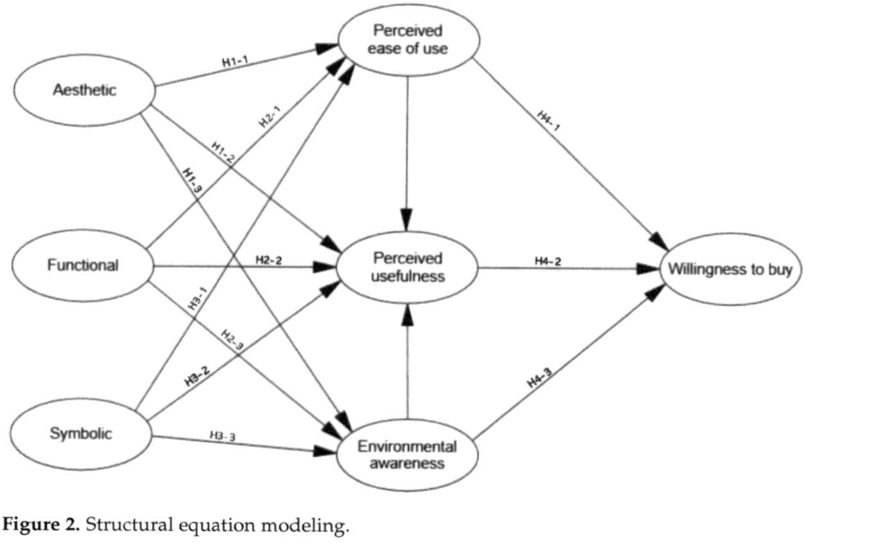

Figure 2. Structural equation modeling.

Based on the above assumptions, the expanded TAM model was established and used to analyze the purchase intentions of Chinese and Korean consumers toward EVs.

4. Research Methodology

Measures

The questionnaire survey of this study is designed to be divided into five parts. The first part is the demographic statistics (seven questions); in addition to the basic gender, age, and education level, it also includes the four user types classified by Lee (2017) [37] in the identification characteristics of the attractiveness of car appearance design (A: car enthusiasts, B: users who value car sym, C: users who value car functions, and D: users who are not interested in cars), in order to confirm the potential awareness of most consumers about cars. The second part measures the participants' attitudes toward the aes (four questions; question codes Q1, Q2, Q3, and Q4), func (three questions; Q5, Q6, and Q7), and sym (four questions; Q8, Q9, Q10, and Q11) of EV design. The questions are adapted from the works of Bloch, P. H. (1995) [14], Srinivasan et al., 2012 [21], Homburg et al., 2015 [22] and Gilal et al., 2018 [44]. The third part measures the participants' PEU (three; Q12, Q13, and Q14) and PU (three questions; Q15, Q16, and Q17) of EVs. The questions are adapted from the works of Davis [15,24,63–65] and Venkatesh et al. [24,25,64]. The fourth part measures the participants' attitudes toward EA (five questions; Q18, Q19, Q20, Q21, and Q22); the items are adapted from the works of Van et al. (2023) [66], Carley et al. (2013) [67], and Bunce et al. (2014) [68]. The fifth part measures the participants' willingness to purchase EVs (three questions; Q23, Q24, and Q25); the items are adapted from the works of Agarwal et al. (2000) [50], Zhang et al. (2011) [69], Park et al. (2020) [70], and Patara and Monroe (2008) [71] (see Table 2).

The second to fifth parts of the scale are measured by 7-point Likert scales, from high to low: 7 points = strongly agree; 6 points = somewhat agree; 5 points = agree; 4 points = neutral; 3 points = somewhat disagree; 2 points = disagree; 1 point = strongly disagree. After the scale design was completed, it was first reviewed by relevant experts and scholars, and then appropriately modified. Then, in order to distribute the questionnaire to Chinese and Korean consumers at the same time, the English questionnaire was translated.

Finally, in order to ensure measurement invariance, the final questionnaire was translated in two stages. First, all items of the questionnaire were translated from English to Chinese by two Chinese professors who were fluent in English, and from English to Korean by two Korean professors who were fluent in English. In the second stage, according to the double translation procedure, five other independent professional translators conducted back-translation checks on the Chinese and Korean questionnaires. After comparing the original English version and the translated versions of the questionnaire, and adding some minor changes to the Chinese version, the Chinese and Korean questionnaires were finalized.

To determine the validity of the questionnaire, a small-scale pre-survey was conducted in a group chat on a Korean software platform (KakaoTalk version 10.8) 42 people, with ages ranging from 20 to 65). The questionnaire was revised for the second time based on the survey results, and the final version of the questionnaire was determined. Before the final questionnaire data were collected, all participants, including Chinese and Korean consumers, were informed of the purpose of the study, the benefits and risks of participating in the study, and how their data would be used. All tests were conducted after obtaining the consent of the participants, and the questionnaires were completed anonymously and voluntarily.

Although the overall population bases of China and South Korea are very different, the representativeness of the samples can be ensured through appropriate sample selection and analysis methods. If the sample sizes from China and South Korea are more than 400 each, they may still be statistically valid. The key is to ensure the representativeness of the samples in terms of key characteristics such as gender, age, and income. Stratified sampling and weighted adjustment methods can be used to reflect the overall characteristics, and the representativeness of the samples can be confirmed through validation data and sensitivity analysis. Appropriate sample selection, weighted adjustment, and data verification can improve the reliability and validity of the research results.

Table 2. Measurement scales for variables.

<table>
<tr><th>No.</th><th>Variable Name</th><th>Code</th><th>Measurement Question</th><th>Reference Source</th></tr>
<tr><td rowspan="4">1</td><td rowspan="4">Aes</td><td>Q1</td><td>I think electric cars look good</td><td rowspan="11">Bloch, P. H. (1995); [14]
Srinivasan, R et al. (2012) [21];
Homburg, C et al. (2015) [22];
Gilal, N. G et al. (2018) [44];</td></tr>
<tr><td>Q2</td><td>I think electric car designs are attractive</td></tr>
<tr><td>Q3</td><td>I think electric car designs are eye-catching</td></tr>
<tr><td>Q4</td><td>I think electric car designs are impressive</td></tr>
<tr><td rowspan="3">2</td><td rowspan="3">Func</td><td>Q5</td><td>I think electric car functionality is important</td></tr>
<tr><td>Q6</td><td>I think electric car designs are safe</td></tr>
<tr><td>Q7</td><td>I think electric cars are very functional</td></tr>
<tr><td rowspan="4">3</td><td rowspan="4">Sym</td><td>Q8</td><td>I think electric cars can show my social status</td></tr>
<tr><td>Q9</td><td>I think electric cars can show that I value environmental awareness</td></tr>
<tr><td>Q10</td><td>I portray part of my lifestyle through electric cars</td></tr>
<tr><td>Q11</td><td>I want to define myself through electric cars</td></tr>
<tr><td rowspan="3">4</td><td rowspan="3">PEU</td><td>Q12</td><td>I think it is easy to use electric cars</td><td rowspan="6">Davis, F. D. (1987); (1989 [15]);
(1992) [63]; (2000) [24],
Venkatesh et al. (2003) [64];
(2008) [25];</td></tr>
<tr><td>Q13</td><td>I think electric cars are easy to learn to use</td></tr>
<tr><td>Q14</td><td>I find it easy to do what I want to do with electric cars</td></tr>
<tr><td rowspan="3">5</td><td rowspan="3">PU</td><td>Q15</td><td>I think using electric cars will improve my life</td></tr>
<tr><td>Q16</td><td>I think using electric cars will be useful</td></tr>
<tr><td>Q17</td><td>I think using electric cars is effective</td></tr>
<tr><td rowspan="5">6</td><td rowspan="5">EA</td><td>Q18</td><td>I think environmental issues are important</td><td rowspan="5">Bunce et al. (2014) [68];
Song et al. (2018) [7];
Van et al. (2023) [66];
Carley et al. (2013) [67];</td></tr>
<tr><td>Q19</td><td>I usually take practical actions to protect the environment (such as garbage sorting)</td></tr>
<tr><td>Q20</td><td>I usually use environmentally friendly products as much as possible</td></tr>
<tr><td>Q21</td><td>I think we should use environmentally friendly products</td></tr>
<tr><td>Q22</td><td>If it is an environmentally friendly product, I will buy it even if it is expensive</td></tr>
<tr><td rowspan="3">7</td><td rowspan="3">WTB</td><td>Q23</td><td>I plan to buy an electric car in the future</td><td rowspan="3">Wang et al. (2013) [26];
Agarwal et al. (2000) [50];
Zhang et al. (2011) [69];
Park et al. (2020) [70];
Patara and Monroe (2008) [71]</td></tr>
<tr><td>Q24</td><td>Next time I buy a new car, I will buy an electric car</td></tr>
<tr><td>Q25</td><td>I will recommend buying an electric car to people around me</td></tr>
</table>

5. Study

5.1. Chinese Consumers

5.1.1. Chinese Participants and Measures

This study used convenience sampling and conducted an online survey using the Chinese market research company "wjx.cn" from 11 April 2024 to 27 June 2024; wjx.cn (https://www.wjx.cn/, accessed from 11 April 2024 to 27 June 2024) is a popular online survey platform in China. Since its launch in 2006, users have published more than 274 million questionnaires and collected more than 21.307 billion questionnaires. The platform has maintained an annual growth rate of more than 100% and a market share of more than 60%. This study used the company to collect data from men and women aged 20 to over

60 in China in order to understand the WTB for electric vehicles. A total of 497 people were sent online surveys, of which 468 people's responses (excluding questionnaires with repeated answers or too short an answer time) were used for the final analysis. In terms of demographic characteristics of the survey subjects in the questionnaire collection for Chinese consumers, there were 257 males, accounting for 55%, and 211 females, accounting for 45%. In terms of age, 94 people (20%) were in their 20s, 174 people (37%) were in their 30s, and 109 people were in their 40s (23%); 14% (27 people) of the participants were over 50 years old, and 8% were over 60 years old. In terms of education level, the majority of the respondents were college graduates, accounting for 66% (307 people). In terms of the lifestyle of Chinese consumers, the proportions of respondents who thought that cars could show their social status and those who were interested in cars were similar, accounting for 127 and 125 people, respectively. In China, the majority of respondents owned electric cars, accounting for 239 people (51%), and the number of respondents who owned fuel cars was 156 (33%). For details, see Table 3.

Table 3. Demographic characteristics of the Chinese sample.

No.	Measure	Item	Count	%
1	Gender	Male	257	55
		Female	211	45
2	Age	20~29 years old	94	20
		30~39 years old	174	37
		40~49 years old	109	23
		50~59 years old	64	14
		<60 years old	27	6
3	Education level	High school	39	8
		Junior college	97	21
		University	307	66
		Graduate school	24	5
		Other	1	0
4	Lifestyle	I am attracted to the exterior of cars and I like cars	125	27
		I am not interested in cars	119	25
		I think the functionality of a car is important	97	21
		I think cars express my social status	127	27
5	Car ownership	I don't have a car	73	16
		I have a car with an internal combustion engine	156	33
		I have an electric car	239	51
6	Driving frequency	1 day or less per week	193	41
		2~3 days per week	178	38
		4~5 days per week	58	12
		More than 6 days per week	39	8

5.1.2. Analysis of Results

Reliability and Validity Test

This study first used SPSS27.0 for reliability and validity testing. When performing the KMO (Kaiser–Meyer–Olkin) test and Bartlett's test of sphericity, the KMO value was 0.902 (greater than 0.7), and Bartlett's test showed that the significance of sphericity was 0.000. The Cronbach's α test results of variables such as aes (measurement items AD1~AD4), func (measurement items FD1~FD3), sym (measurement items SD1~SD4), PEU (measurement items PEU1~PEU3), PU (measurement items PU1~PU3), and EA (measurement items EA1~EA5), and WTB (measurement items WTB1~WTB3) also showed that the Cronbach's α of each dimension of the questionnaire was greater than 0.7 [72]. At the same time, the CITC values of the questionnaires were all higher than 0.5, indicating that the correlation between the scales is good and suitable for further analysis (for details, see Table 4).

Table 4. Chinese proportion attributes.

KMO and Bartlett's Test				
KMO sampling suitability measure				0.902
Bartlett's test of sphericity		Approximate chi-squared Degrees of freedom Significance		6533.05 300 0.000
Reliability Analysis of the Questionnaire				
Variable Name	Measurement Item	Correction Total Correlation (CITC)	After Deleting the Item's Cronbach's α Value	Each Variable's Cronbach's α Value
Aes	AD1 AD2 AD3 AD4	0.759 0.771 0.803 0.781	0.879 0.874 0.863 0.871	0.901
Func	FD1 FD2 FD3	0.663 0.601 0.675	0.710 0.775 0.697	0.801
Sym	SD1 SD2 SD3 SD4	0.720 0.795 0.709 0.678	0.837 0.808 0.841 0.853	0.871
PEU	PEU1 PEU2 PEU3	0.684 0.659 0.694	0.752 0.777 0.742	0.824
PU	PU1 PU2 PU3	0.658 0.695 0.666	0.759 0.727 0.760	0.816
EA	EA1 EA2 EA3 EA4 EA5	0.694 0.765 0.735 0.710 0.813	0.884 0.868 0.875 0.880 0.857	0.896
WTB	WTB1 WTB2 WTB3	0.676 0.738 0.702	0.804 0.744 0.782	0.840

Next, the data were extracted for principal components to obtain the initial eigenvalue, variance contribution rate, cumulative variance contribution rate, etc. The results are shown in Table 5. According to the extraction principle that the eigenvalue is greater than 1, seven factors were extracted; the cumulative variance contribution rate was 74.12%, and the commonality corresponding to each item was above 0.5, indicating that the seven factors extracted from 25 questions (component numbers 1, 2, 3, 4, 5, . . ., 25) provide a relatively ideal explanation for the original data.

Finally, the Kaiser standardized maximum variance rotation method was used to rotate the factor loadings, and the results are shown in Table 6 (aes (measurement items AD1~AD4), func (measurement items FD1~FD3), sym (measurement items SD1~SD4), PEU (measurement items PEU1~PEU3), PU (measurement items PU1~PU3), EA (measurement items EA1~EA5), and WTB (measurement items WTB1~WTB3)). The loadings of each item on the corresponding factor are all over 0.5; color and bold are used in the table, so there is reason to believe that the scale of this study has good structural validity.

Table 5. Chinese variance explanation.

Component	Initial Eigenvalues			Sum of Squares of Rotating Loads			Commonality
	Total	Variance Percentage	Cumulative %	Total	Variance Percentage	Cumulative %	
1	8.402	33.606	33.606	3.566	14.262	14.262	0.754
2	2.637	10.549	44.156	3.156	12.625	26.888	0.764
3	2.232	8.929	53.084	2.969	11.874	38.762	0.800
4	1.733	6.932	60.016	2.278	9.111	47.873	0.783
5	1.279	5.117	65.133	2.269	9.076	56.949	0.730
6	1.186	4.744	69.876	2.189	8.758	65.707	0.670
7	1.060	4.241	74.117	2.103	8.410	74.117	0.749
8	0.579	2.315	76.433				0.729
9	0.535	2.142	78.575				0.800
10	0.497	1.987	80.561				0.729
11	0.440	1.760	82.322				0.676
12	0.432	1.728	84.049				0.716
13	0.403	1.611	85.660				0.757
14	0.398	1.591	87.251				0.748
15	0.384	1.538	88.789				0.750
16	0.361	1.443	90.232				0.716
17	0.347	1.390	91.621				0.773
18	0.324	1.295	92.916				0.661
19	0.293	1.172	94.088				0.759
20	0.283	1.131	95.219				0.707
21	0.265	1.058	96.277				0.677
22	0.253	1.010	97.288				0.795
23	0.247	0.987	98.275				0.735
24	0.235	0.939	99.214				0.791
25	0.197	0.786	100.00				0.757

Extraction method: principal component analysis.

Table 6. Load matrix after rotation.

Measures	Factor 1	Factor 2	Factor 3	Factor 4	Factor 5	Factor 6	Factor 7
AD1	0.107	**0.853**	0.032	0.038	0.045	0.070	0.072
AD2	0.099	**0.850**	0.074	0.042	0.126	0.030	0.091
AD3	0.137	**0.870**	0.114	0.024	0.047	0.027	0.088
AD4	0.140	**0.862**	−0.001	0.100	0.079	0.063	0.016
FD1	0.209	0.068	0.105	0.201	0.243	0.158	**0.739**
FD2	0.143	0.119	0.135	0.192	0.118	0.161	**0.735**
FD3	0.214	0.101	0.157	0.122	0.130	0.202	**0.772**
SD1	0.110	0.082	**0.828**	0.073	0.105	−0.020	0.091
SD2	0.166	0.052	**0.860**	0.107	0.086	0.066	0.084
SD3	0.097	0.089	**0.824**	0.001	0.097	0.148	0.034
SD4	0.203	−0.006	**0.767**	0.050	0.075	0.095	0.173
PU1	0.242	0.088	0.061	0.236	0.160	**0.736**	0.153
PU2	0.184	0.019	0.123	0.178	0.154	**0.781**	0.207
PU3	0.196	0.086	0.101	0.082	0.100	**0.808**	0.149
PEU1	0.171	0.074	0.124	**0.802**	0.144	0.104	0.160
PEU2	0.130	0.023	0.064	**0.779**	0.156	0.109	0.227
PEU3	0.131	0.105	0.029	**0.807**	0.170	0.240	0.084

Table 6. *Cont.*

Measures	Factor 1	Factor 2	Factor 3	Factor 4	Factor 5	Factor 6	Factor 7
EA1	**0.727**	0.085	0.150	0.018	0.237	0.152	0.154
EA2	**0.815**	0.139	0.125	0.163	0.152	0.100	0.031
EA3	**0.776**	0.151	0.126	0.105	0.081	0.177	0.129
EA4	**0.735**	0.178	0.161	0.102	0.054	0.133	0.220
EA5	**0.828**	0.065	0.144	0.164	0.120	0.161	0.136
WTB1	0.196	0.183	0.078	0.151	**0.782**	0.089	0.125
WTB2	0.184	0.077	0.141	0.164	**0.810**	0.160	0.149
WTB3	0.142	0.056	0.164	0.182	**0.784**	0.159	0.183

Rotation method: Kaiser normalized varimax. Rotation converged after six iterations.

In summary, the questionnaire used in this study has good validity and is suitable for further analysis.

In order to verify the validity of the questionnaire, confirmatory factor analysis was conducted on each structure. Confirmatory factor analysis can measure the correlation between latent variables in each model. This study divides the confirmatory factor analysis into product design variables such as aes, func, and sym; perceived variables such as PU, PEU, and EA; and WTB. The results of the analysis of the fitness of the model, using AMOS 26.0, showed that the chi-squared–degrees-of-freedom ratio of the confirmatory factor analysis model was 1.60, i.e., less than 3 (consistent with the standard value proposed by Bentler and Bonett (1980) [73] and Carmines and McIver (1981)) [74]. The RMSEA (root-mean-square error of approximation) was 0.04, i.e., less than 0.05 and within the good range [75]. The GFI (goodness-of-fit index) was 0.94, the AGFI (adjusted goodness-of-fit index) was 0.92, and the RFI (relative fit index) was 0.92, i.e., all greater than 0.8 (consistent with the recommended values of Browne and Cudeck (1992) [75], Hu and Bentler (1999) [76], Bentler (1990) [77], Byrne (1998) [78], and Doll et al. (1994)) [79]. The NFI (normed fit index) was 0.94, the IFI (incremental fit index) was 0.98, the TLI was 0.97, and the CFI (comparative fit index) was 0.98, i.e., all greater than 0.9 (consistent with the recommended values of Bentler and Bonett (1980)) [73]. As shown in Table 6, all fitting indices were within the critical range, so it can be considered that the confirmatory factor analysis model has a relatively good fit effect on the questionnaire data (see Table 7).

Table 7. Confirmatory factor analysis model fitness.

Fit Index		χ^2/df	RMSEA	GFI	AGFI	NFI	RFI	IFI	TLI	CFI
Evaluation standard	Ideal	<3	<0.05	>0.9	>0.9	>0.9	>0.9	>0.9	>0.9	>0.9
	Acceptable	<5	<0.08	>0.8	>0.8	>0.8	>0.8	>0.8	>0.8	>0.8
Structural model measurements		1.608	0.036	0.935	0.916	0.939	0.928	0.976	0.971	0.976

As shown in Table 8, the standardized loadings of the items corresponding to the seven variables (aes, func, sym, PEU, PU, EA, and WTB) were all greater than 0.5, indicating that each latent variable has high representativeness corresponding to the measurement item. According to the standard of Nunnally and Bernstein (1978) [80], when the composite reliability (CR) exceeds 0.8, and when the AVE (average variance extracted) exceeds 0.6 according to the standard of Bagozzi and Yi (1988) [81], the convergent validity is ideal. The results of this study show that the composite reliability (CR) is 0.88 for aes, 0.85 for func, 0.89 for sym, 0.85 for ease of use, 0.86 for usefulness, 0.92 for EA, and 0.84 for WTB—all exceeding 0.8—and the AVE is 0.65 for aes, 0.65 for func, 0.68 for sym, 0.65 for ease of use, 0.68 for usefulness, 0.7 for EA, and 0.63 for WTB. Since the average variance extracted (AVE) for each variable in the sample of Chinese consumers is greater than 0.6, and the composite reliability of each variable is greater than 0.8, it can be concluded that the convergent validity of the scale is ideal.

Finally, the test results of discriminant validity are presented. The values on the diagonal are the arithmetic square roots of the AVE of the seven variables: aes, func, sym, PEU, PU, EA, and WTB (color and bold are used in the table). The values of the lower triangular matrix are the Pearson correlation coefficients between the variables. Since the absolute value of the Pearson correlation coefficient between the variables is less than the arithmetic square root of the AVE, it can be inferred that the latent variables have good discrimination—that is, the discriminant validity of the scale used in this study is ideal (see Table 9).

Table 8. Convergent validity test.

No.	Variable	Measurement Item	Parameter Significance Estimate				Factor Loading	Item Reliability	Composite Reliability	Average Variance Extracted
			Unstd.	S.E.	*t*-Value	*p*	Std.	SMC	CR	AVE
1	Aes	AD4	1				0.833	0.694	0.881	0.652
		AD3	1.067	0.049	21.961	***	0.862	0.743		
		AD2	1.009	0.048	20.857	***	0.829	0.688		
		AD1	1.046	0.052	20.187	***	0.810	0.657		
2	Func	FD3	1				0.786	0.617	0.846	0.648
		FD2	0.894	0.063	14.278	***	0.694	0.481		
		FD1	1.032	0.064	16.102	***	0.796	0.633		
3	Sym	SD4	1				0.742	0.551	0.894	0.679
		SD3	1.097	0.067	16.346	***	0.777	0.604		
		SD2	1.134	0.062	18.220	***	0.881	0.776		
		SD1	1.052	0.064	16.367	***	0.778	0.606		
4	PEU	PEU1	1				0.787	0.619	0.848	0.654
		PEU2	1.005	0.065	15.559	***	0.753	0.567		
		PEU3	1.045	0.064	16.349	***	0.804	0.646		
5	PU	PU1	1				0.779	0.608	0.862	0.676
		PU2	0.990	0.060	16.375	***	0.806	0.649		
		PU3	1.095	0.071	15.369	***	0.745	0.554		
6	EA	EA5	1				0.871	0.759	0.919	0.695
		EA4	0.893	0.045	19.786	***	0.766	0.587		
		EA3	0.938	0.046	20.573	***	0.786	0.618		
		EA2	0.970	0.044	21.805	***	0.815	0.665		
		EA1	0.850	0.045	18.914	***	0.744	0.554		
7	WTB	WTB1	1				0.756	0.572	0.837	0.634
		WTB2	1.162	0.068	17.036	***	0.841	0.707		
		WTB3	1.192	0.072	16.495	***	0.801	0.642		

Note: *** indicate significance at the 1% levels.

Table 9. Correlation coefficient matrix and square roots of AVEs.

No.	Variable	1	2	3	4	5	6	7
1	Aes	**0.81**						
2	Func	0.408 ***	**0.81**					
3	Sym	0.400 ***	0.417 ***	**0.82**				
4	PU	0.446 ***	0.506 ***	0.486 ***	**0.82**			
5	PEU	0.340 ***	0.446 ***	0.344 ***	0.476 ***	**0.81**		
6	EA	0.357 ***	0.309 ***	0.443 ***	0.487 ***	0.412 ***	**0.83**	
7	WTB	0.395 ***	0.401 ***	0.463 ***	0.612 ***	0.395 ***	0.513 ***	**0.80**

Note: *** indicate significance at the 1% levels.

5.1.3. Structural Equation Modeling Results-Chinese Consumers

After conducting confirmatory factor analysis, AMOS 26.0 was used to construct a structural equation model. Through further processing, the path coefficients and significant coefficients between each latent variable were calculated to verify the hypotheses proposed in this study. The model path relationship diagram is shown in Figure 3.

The final model path fitness is shown in Table 10. Its chi-squared–freedom ratio is 1.7 < 3, which is within the ideal range of fitness; the RMSEA value is 0.04 < 0.08, which is within the acceptable range; the values of GFI, AGFI, and RFI are 0.93, 0.91, and 0.92, respectively—all greater than 0.8, and within the acceptable range; the values of NFI, IFI, TLI, and CFI are 0.94 and 0.97, which are also greater than 0.9, i.e., within the ideal range of fitness. In summary, it can be considered that the structural equation model has a relatively good fitting effect on the questionnaire data.

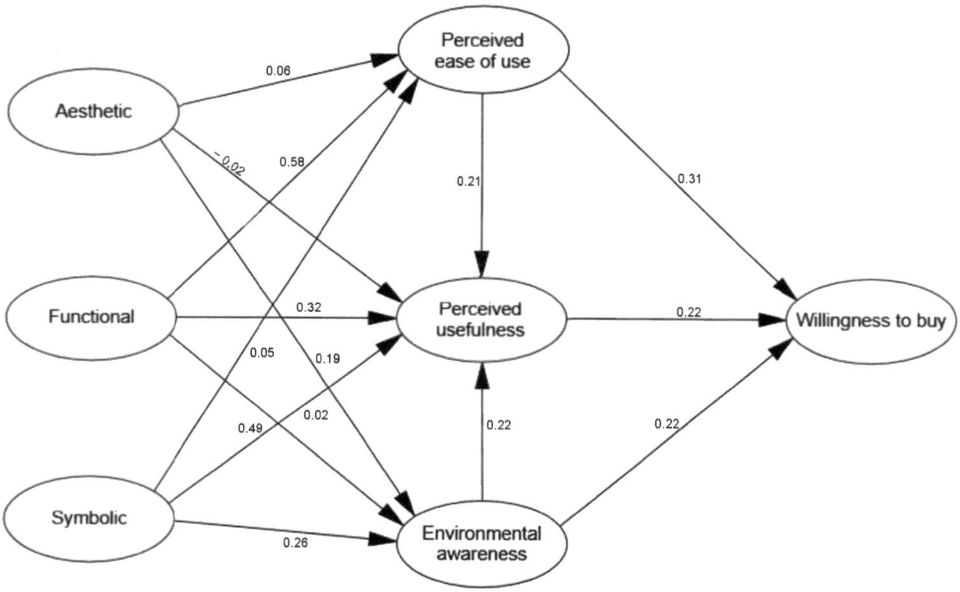

Figure 3. Results of structural equation modeling of Chinese consumers.

Table 10. The fitness of the Chinese consumer path model.

Fit Index		χ^2/df	RMSEA	GFI	AGFI	NFI	RFI	IFI	TLI	CFI
Evaluation standard	Ideal	<3	<0.05	>0.9	>0.9	>0.9	>0.9	>0.9	>0.9	>0.9
	Acceptable	<5	<0.08	>0.8	>0.8	>0.8	>0.8	>0.8	>0.8	>0.8
Structural model measurements		1.699	0.039	0.930	0.912	0.934	0.924	0.972	0.967	0.972

Relationship between Aesthetics (Aes) and Perceived Variables, Environmental Awareness (EA)

Hypothesis 1 is about the relationship between aes and perceived variables, EA. The results show that H1-1 that with the improvement of aes, the PEU of EV will increase (β = 0.05, T = 1.26) is not established, H1-2 that the PU of EV will increase (β = 0.04, T = −0.45) is not established, and only H1-3 that the EA of EV will increase (β = 0.05, T = 3.96) is established. Contrary to the research of Filieri and Lin (2017) [82], it can be seen that the aes of EV has not increased the purchasing intention of Chinese consumers but can actively improve consumers' EA. In other words, electric vehicles with excellent appearance cannot make Chinese consumers perceive ease of use or usefulness, but Chinese consumers can improve their EA through appearance, thereby protecting the environment.

Chinese Consumers: Relationship between Function (Func), Perceived Variables, and Environmental Awareness (EA)

From the hypothesis of the relationship between func and the perceived variable EA, H2-1 that the PEU of EVs will increase with the optimization of functions; β = 0.07, T = 8.56), H2-2 (that PU will increase; β = 0.07, T = 4.60), and H2-3 (that EA will increase; β = 0.06, T = 7.80) are all established. This result explains the results of Sovacool et al. (2019) [83], who found that Chinese consumers' willingness to accept the use of EVs is related to the performance characteristics of EVs. Additionally, like in Lieven et al.'s (2011) [39] study on the German market, func is a key criterion for consumers.

Relationship between Symbolism (Sym), Perceived Variables, and Environmental Awareness (EA)

H3-1 of the relationship between symbolic and perceived variables and environmental protection (the PEU of EVs will increase as the value of symbols increases; β = 0.06, T = 0.89) and H3-2 (the PU will increase; β = 0.05, T = 0.35) are not established. However, H3-3 (β = 0.06, T = 4.39)—that the EA of EVs will increase—is established, which can be seen as the Chinese government's restrictions on carbon emissions [84] and the strengthening of Chinese consumers' EA [63]. In addition, the results of this study are consistent with the findings of Liu et al. (2020) [85] and Heffner et al. (2007) [86], so it can be considered that the most important symbolic meaning of EVs is EA, followed by aes.

Chinese Consumers: The Relationship between Perceived Variables, Environmental Awareness (EA), and Willingness to Buy (WTB)

As with the findings of He et al. (2015) [62], our results show that H4-1 (the higher the PEU, the higher the intention to purchase EVs; β = 0.06, T = 5.22), H4-2 (the higher the PU, the higher the willingness to purchase EVs; β = 0.07, T = 3.08), and H4-3 (which states that the stronger the EA, the higher the WTB electric vehicles; β = 0.05, T = 4.29) are all established. This shows that PEU, PU, and EA can all have a positive impact on WTB.

Finally, consistent with the research results of Davis (1989) [15], PEU has a significant positive impact on PU (β = 0.06, T = 3.62), and the results prove that EA also has a significant impact on usefulness (β = 0.05, T = 4.29) (see Table 11).

Table 11. Results of hypothesis testing on Chinese consumers.

No.	Path Relationship			B	β	S.E.	C.R.	*p*-Value	Result
H1-1	Aes	→	PEU	0.061	0.063	0.048	1.261	0.207	Fail
H1-2		→	PU	−0.018	−0.021	0.041	−0.446	0.656	Fail
H1-3		→	EA	0.188	0.180	0.048	3.96	***	Pass
H2-1	Func	→	PEU	0.582	0.567	0.068	8.557	***	Pass
H2-2		→	PU	0.325	0.349	0.071	4.603	***	Pass
H2-3		→	EA	0.488	0.437	0.063	7.796	***	Pass
H3-1	Sym	→	PEU	0.053	0.048	0.060	0.886	0.376	Fail
H3-2		→	PU	0.018	0.018	0.051	0.353	0.724	Fail
H3-3		→	EA	0.264	0.217	0.061	4.392	***	Pass
H4-1	PEU	→	WTB	0.311	0.332	0.060	5.224	***	Pass
H4-2	PU	→		0.219	0.212	0.071	3.083	0.002	Pass
H4-3	EA	→		0.217	0.252	0.051	4.288	***	Pass

Note: *** indicate significance at the 1% levels.

5.1.4. Insights into Chinese Consumers

EVs are a new means of transportation that has been developed for environmental protection. Unlike existing fuel vehicles, they provide a new driving experience by integrating new technologies. It is expected that the EV market will continue to grow in the future and become the main mode of transportation. As consumers' perceptions of electric vehicles change, this study uses Davis's TAM to apply the aes, func, and sym of product design to the TAM, along with EA, and explores the impact of product design dimensions on the willingness to purchase EVs.

The results of the analysis of Chinese consumers in this study are summarized as follows. First, in the product design dimension, aes was found to have no effect on PU and PEU, but it had a direct effect on EA. This suggests that the aes of electric vehicles can affect Chinese consumers' EA, thereby triggering WTB.

Second, in the product design dimension, func has a significant impact on all perceived variables and EA. As existing research mainly focuses on technology [87,88], the new technology of EVs will have a direct impact on WTB. Additionally, as reported by Clarkson et al. (2013) [89], consumers' knowledge can affect the novel consumption that they seek. As consumers' knowledge and experience increase, it can be inferred that their demand for EVs will also change accordingly.

Finally, in terms of the symbolic dimension of product design, although it has no direct impact on PEU and PU, it does affect EA. For Chinese consumers, PEU and EA both positively affect PU and directly affect WTB.

5.2. Korean Consumers

5.2.1. Korean Participants and Measures

This study also adopted the convenience sampling method. From 7 April to 29 June 2024, an online survey was conducted on 426 Korean men and women aged 20 to 60 years, using Google's questionnaire website "Google Forms". The responses of 409 people (excluding invalid questionnaires) were used for the final analysis. In terms of the demographic characteristics of the survey subjects, 229 were male, accounting for 56%, and 180 were female, accounting for 44%. In terms of age, 129 were in their 20s, accounting for 32%; 119 were in their 30s, accounting for 29%; 103 were in their 40s, accounting for 25%; and 3% of the participants were over 50 years old. In terms of education level, the majority of respondents were university graduates, accounting for 274 people, or 67%. In terms of the lifestyle of Korean consumers, the majority of respondents were interested in cars, accounting for 53%, followed by respondents who thought that cars' functions are the most important, accounting for 105 people, or 26%. Among Korean consumers, the majority of respondents owned internal combustion cars, accounting for 50% (204 people) (see Table 12).

Table 12. Demographic characteristics of the Korean sample.

No.	Measure	Item	Count	%
1	Gender	Male	229	56
		Female	180	44
2	Age	20~29 years old	129	32
		30~39 years old	119	29
		40~49 years old	103	25
		50~59 years old	47	11
		<60 years old	11	3
3	Education level	High school	52	13
		Junior college	47	12
		University	274	67
		Graduate school	34	8
		Other	2	1

Table 12. *Cont.*

No.	Measure	Item	Count	%
4	Lifestyle	I am attracted to the exterior of cars and I like cars	217	53
		I am not interested in cars	48	12
		I think the functionality of a car is important	105	26
		I think cars express my social status	39	10
5	Car ownership	I don't have a car	166	41
		I have a car with an internal combustion engine	204	50
		I have an electric car	39	10
6	Driving frequency	1 day or less per week	176	43
		2~3 days a week	84	21
		4~5 days a week	76	19
		More than 6 days a week	73	18

5.2.2. Analysis of Results

Reliability and Validity Testing

As with the reliability and validity testing for Chinese consumers, the reliability and validity analysis for Korean consumers also first used SPSS 27.0. The result was that the KMO value was 0.928 (greater than 0.7), and Bartlett's test of sphericity was significant, at 0.001. The Cronbach's α of each dimension of the questionnaire (aes, func, sym, PEU, PU, EA, and WTB) was greater than 0.7, indicating that the correlation between the scales of Korean consumers is good and suitable for further research. For further analysis, see Table 13.

Table 13. Korean proportion attributes.

KMO and Bartlett's Test				
KMO sampling suitability measure			0.928	
Bartlett's test of sphericity	6037.694 300 0.001		6533.05 300 0.000	
Reliability Analysis of the Questionnaire				
Variable Name	Measurement Item	Correction Total Correlation (CITC)	After Deleting the Item's Cronbach's α Value	Each Variable's Cronbach's α Value
Aes	AD1	0.717	0.852	0.878
	AD2	0.744	0.842	
	AD3	0.738	0.844	
	AD4	0.753	0.838	
Func	FD1	0.699	0.76	0.832
	FD2	0.703	0.756	
	FD3	0.673	0.786	
Sym	SD1	0.737	0.832	0.872
	SD2	0.726	0.837	
	SD3	0.702	0.847	
	SD4	0.742	0.830	
PEU	PEU1	0.721	0.765	0.842
	PEU2	0.721	0.765	
	PEU3	0.677	0.808	
PU	PU1	0.706	0.787	0.844
	PU2	0.696	0.796	
	PU3	0.728	0.765	

Table 13. *Cont.*

KMO and Bartlett's Test				
EA	EA1	0.766	0.875	0.900
	EA2	0.755	0.877	
	EA3	0.774	0.873	
	EA4	0.738	0.881	
	EA5	0.724	0.884	
WTB	WTB1	0.725	0.794	0.853
	WTB2	0.708	0.810	
	WTB3	0.739	0.781	

Next, the principal component extraction of the Korean consumer data was performed to obtain the initial eigenvalues, variance contribution rates, cumulative variance contribution rates, etc. The results are shown in Table 14. According to the extraction principle that the eigenvalue is greater than 1, seven factors were extracted, the cumulative variance contribution rate was 74.67%, and the commonality of each item was above 0.5, indicating that the seven factors extracted from the 25 questions provided a relatively ideal explanation for the original data.

Table 14. Korean variance explanation.

Component	Initial Eigenvalues			Sum of Squares of Rotating Loads			Commonality
	Total	Variance Percentage	Cumulative %	Total	Variance Percentage	Cumulative %	
1	10.173	40.692	40.692	3.651	14.606	14.606	0.726
2	1.979	7.918	48.610	3.009	12.034	26.640	0.748
3	1.526	6.105	54.714	2.976	11.906	38.545	0.730
4	1.503	6.010	60.724	2.301	9.206	47.751	0.757
5	1.271	5.083	65.807	2.287	9.149	56.900	0.744
6	1.193	4.773	70.580	2.281	9.124	66.024	0.773
7	1.022	4.087	74.667	2.161	8.644	74.667	0.719
8	0.542	2.169	76.836				0.732
9	0.479	1.916	78.752				0.723
10	0.454	1.814	80.566				0.705
11	0.446	1.785	82.351				0.758
12	0.429	1.715	84.066				0.789
13	0.395	1.579	85.644				0.792
14	0.382	1.527	87.172				0.730
15	0.362	1.449	88.621				0.763
16	0.354	1.418	90.039				0.753
17	0.337	1.346	91.385				0.780
18	0.333	1.333	92.718				0.739
19	0.311	1.245	93.964				0.740
20	0.29	1.159	95.122				0.747
21	0.273	1.093	96.216				0.691
22	0.264	1.055	97.271				0.691
23	0.25	1.001	98.271				0.788
24	0.218	0.872	99.143				0.758
25	0.214	0.857	100.00				0.792

Extraction method: principal component analysis.

Finally, the Kaiser standardized maximum variance rotation method was used to rotate the factor loadings, and the results are shown in Table 15. The loadings of each item on the corresponding factor are all over 0.5. Color and bold are used in the table, so there is reason to believe that the scale of this study has good structural validity.

Table 15. Load matrix after rotation.

Measures	Factor 1	Factor 2	Factor 3	Factor 4	Factor 5	Factor 6	Factor 7
AD1	0.221	0.184	**0.766**	0.059	0.047	0.189	0.121
AD2	0.190	0.114	**0.788**	0.103	0.146	0.087	0.193
AD3	0.218	0.174	**0.749**	0.134	0.173	0.110	0.176
AD4	0.214	0.082	**0.795**	0.139	0.141	0.122	0.134
FD1	0.228	0.180	0.157	0.157	**0.751**	0.169	0.131
FD2	0.180	0.124	0.157	0.156	**0.808**	0.124	0.090
FD3	0.144	0.185	0.136	0.205	**0.748**	0.190	0.095
SD1	0.155	**0.780**	0.146	0.143	0.095	0.146	0.165
SD2	0.152	**0.779**	0.152	0.178	0.089	0.098	0.142
SD3	0.170	**0.767**	0.080	0.106	0.196	0.078	0.160
SD4	0.132	**0.798**	0.146	0.151	0.133	0.196	0.049
PU1	0.153	0.184	0.148	0.160	0.100	**0.808**	0.144
PU2	0.141	0.208	0.169	0.151	0.114	**0.808**	0.104
PU3	0.164	0.092	0.138	0.095	0.304	**0.741**	0.158
PEU1	0.191	0.141	0.268	0.151	0.105	0.166	**0.757**
PEU2	0.258	0.210	0.157	0.179	0.108	0.122	**0.748**
PEU3	0.260	0.184	0.208	0.119	0.131	0.152	**0.762**
EA1	**0.792**	0.138	0.157	0.132	0.159	0.103	0.121
EA2	**0.787**	0.073	0.199	0.174	0.059	0.156	0.130
EA3	**0.776**	0.201	0.212	0.100	0.127	0.095	0.157
EA4	**0.726**	0.154	0.192	0.115	0.175	0.164	0.182
EA5	**0.761**	0.146	0.163	0.068	0.157	0.058	0.178
WTB1	0.198	0.219	0.127	**0.798**	0.099	0.160	0.109
WTB2	0.130	0.151	0.159	**0.779**	0.174	0.159	0.173
WTB3	0.154	0.192	0.105	**0.791**	0.262	0.093	0.133

Rotation method: Kaiser normalization with varimax. The rotation converged after six iterations.

In summary, the questionnaire used by Korean consumers has good validity and is suitable for further analysis.

Confirmatory Factor Analysis

Similar to the analysis of the results of Chinese consumers, in order to verify the validity of the questionnaire, confirmatory factor analysis was conducted on each structure. The results of the analysis of the fitness of the model using AMOS 26.0 showed that the χ^2/df value of the confirmatory factor analysis model was 1.34 (less than 3), the RMSEA was 0.03 (less than 0.05), the GFI was 0.94, the AGFI was 0.92, and the RFI was 0.94 (all greater than 0.8). Additionally, the NFI was 0.95, the IFI was 0.99, the TLI was 0.98, and the CFI was 0.99, i.e., all greater than 0.9 (Table 16). All of the fitting indicators were within the critical range, so it can be considered that the questionnaire data have a relatively good fitting effect.

Table 16. Confirmatory factor analysis model fitness.

Fit Index		χ^2/df	RMSEA	GFI	AGFI	NFI	RFI	IFI	TLI	CFI
Evaluation standard	Ideal	<3	<0.05	>0.9	>0.9	>0.9	>0.9	>0.9	>0.9	>0.9
	Acceptable	<5	<0.08	>0.8	>0.8	>0.8	>0.8	>0.8	>0.8	>0.8
Structural model measurements		1.339	0.029	0.938	0.921	0.945	0.935	0.985	0.983	0.985

Similar to the analysis of the results for Chinese consumers, the standardized loads of each item corresponding to the seven variables aes, func, sym, PEU, PU, EA, and WTB of Korean consumers were measured. The results showed that the composite reliability (CR) was 0.88 for aes, 0.83 for func, 0.87 for sym, 0.84 for ease of use, 0.84 for usefulness, 0.9 for EA, and 0.85 for WTB—all exceeding 0.8. The AVE was 0.65 for aes, 0.62 for func, 0.63 for sym, 0.64 for ease of use, usefulness, and EA, and 0.66 for WTB—all exceeding 0.6 (see Table 17). Therefore, it can be inferred that the convergent validity of the scale for Korean consumers is ideal.

Finally, the discriminant validity test results for the Korean consumer scale are presented. The values on the diagonal are the arithmetic square roots of the AVE of the seven variables: aes, func, sym, PEU, PU, EA, and WTB (color and bold are used in the table). As with the Chinese consumer scale, the discriminant validity of the Korean consumer scale is ideal (see Table 18).

Table 17. Convergent validity test.

Variable	**Measurement Item**	**Parameter Significance Estimate**				**Factor Loading**	**Item Reliability**	**Composite Reliability**	**Average Variance Extracted**
		Unstd.	**S.E.**	***t*-Value**	***p***	**Std.**	**SMC**	**CR**	**AVE**
Aes	AD4	1				0.813	0.661	0.879	0.645
	AD3	0.997	0.056	17.91	***	0.814	0.662		
	AD2	0.905	0.051	17.739	***	0.807	0.652		
	AD1	0.906	0.053	17.126	***	0.777	0.604		
Func	FD3	1				0.77	0.593	0.832	0.624
	FD2	0.994	0.065	15.291	***	0.786	0.618		
	FD1	1.051	0.069	15.333	***	0.812	0.659		
Sym	SD4	1				0.813	0.661	0.872	0.631
	SD3	0.952	0.058	16.365	***	0.763	0.583		
	SD2	0.93	0.054	17.073	***	0.791	0.626		
	SD1	0.967	0.054	17.836	***	0.811	0.658		
PEU	PEU1	1				0.816	0.665	0.842	0.641
	PEU2	0.992	0.058	17.02	***	0.818	0.668		
	PEU3	0.919	0.059	15.668	***	0.768	0.589		
PU	PU1	1				0.795	0.632	0.844	0.644
	PU2	0.946	0.059	16.097	***	0.787	0.62		
	PU3	1.005	0.059	17.023	***	0.826	0.682		
EA	EA5	1				0.793	0.589	0.900	0.643
	EA4	1.047	0.057	18.265	***	0.832	0.629		
	EA3	0.986	0.056	17.473	***	0.799	0.693		
	EA2	1.017	0.057	17.965	***	0.818	0.639		
	EA1	0.953	0.057	16.724	***	0.768	0.668		
WTB	WTB1	1			1	0.812	0.659	0.854	0.660
	WTB2	0.955	0.057	16.614	0.955	0.792	0.627		
	WTB3	1.012	0.058	17.516	1.012	0.834	0.695		

Note: *** indicate significance at the 1% levels.

Table 18. Correlation coefficient matrix and square roots of AVEs.

No.	**Variable**	**1**	**2**	**3**	**4**	**5**	**6**	**7**
1	Aes	**0.803**						
2	Func	0.523 ***	**0.790**					
3	Sym	0.486 ***	0.527 ***	**0.795**				
4	PU	0.511 ***	0.576 ***	0.526 ***	**0.801**			
5	PEU	0.628 ***	0.512 ***	0.554 ***	0.547 ***	**0.803**		
6	EA	0.610 ***	0.556 ***	0.504 ***	0.493 ***	0.639 ***	**0.802**	
7	WTB	0.472 ***	0.604 ***	0.558 ***	0.519 ***	0.549 ***	0.503 ***	**0.813**

Note: *** indicate significance at the 1% levels.

5.2.3. Structural Equation Modeling Results

The same applies to the scale for Korean consumers. After confirmatory factor analysis, AMOS 26.0 was used to construct a structural equation model. Through further processing, the path coefficients and significant coefficients between each latent variable were calculated to verify the hypotheses proposed in this study. The model path relationship diagram is shown in Figure 4.

Table 19 shows the final model path fitness for Korean consumers, with a chi-squared–degree-of-freedom ratio of 1.47 (<3), RMSEA value of 0.03 (<0.08), a GFI value of 0.93, AGFI value of 0.91, and RFI value of 0.92 (all greater than 0.8) and NFI, IFI, TLI, and CFI values of 0.94 and 0.98 (all greater than 0.9). Therefore, it can be considered that the structural equation model for Korean consumers has a good fitting effect on the questionnaire data.

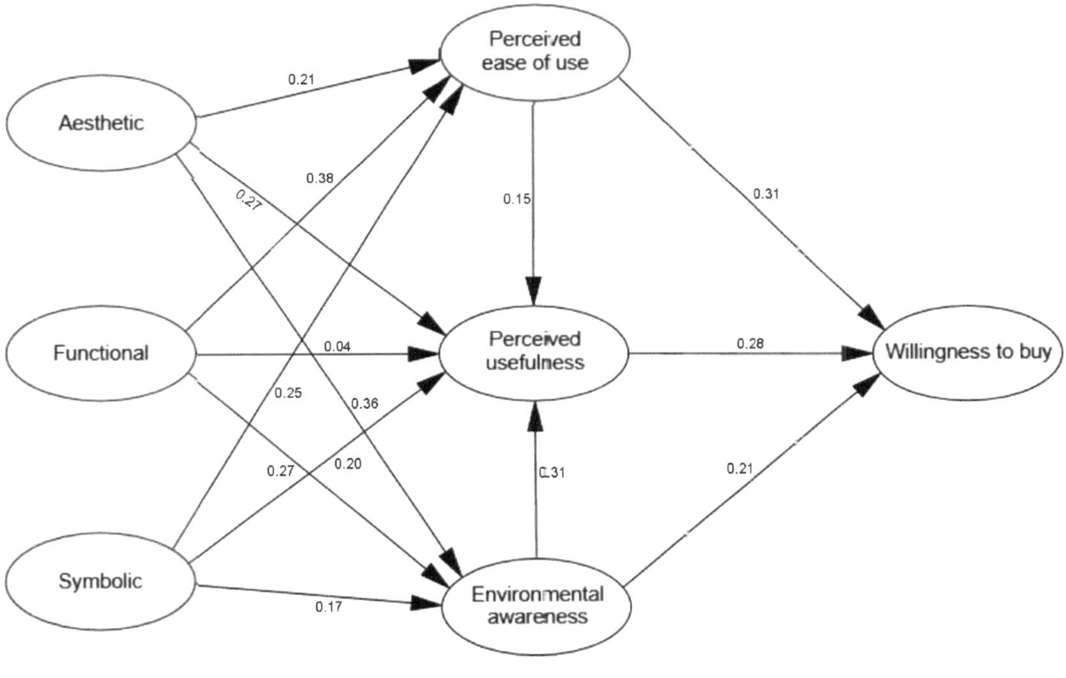

Figure 4. Results of structural equation modeling for Korean consumers.

Table 19. The fitness of the Korean consumer path model.

Fit Index		χ^2/df	RMSEA	GFI	AGFI	NFI	RFI	IFI	TLI	CFI
Evaluation standard	Ideal	<3	<0.05	>0.9	>0.9	>0.9	>0.9	>0.9	>0.9	>0.9
	Acceptable	<5	<0.08	>0.8	>0.8	>0.8	>0.8	>0.8	>0.8	>0.8
Structural model measurements		1.470	0.034	0.932	0.914	0.939	0.929	0.980	0.976	0.979

Relationship between Aesthetics (Aes), Perceived Variables, and Environmental Awareness (EA)

The research results between Korean consumers' aes, perceived variables, and EA show that H1-1 (aes affects the PEU of EVs; β = 0.06, T = 3.47), H1-2 (aes affects the PU of EVs; β = 0.07, T = 4.20), and H1-3 (aes affects the EA of electric vehicles; β = 0.06, T = 6.43) are all established. The results of this study show that, among Korean consumer groups, aesthetic factors have a significant impact on the PEU, PU, and EA of electric vehicles. Specifically, aes can not only enhance consumers' perceptions of the ease of use and practicality of electric vehicles but also enhance their EA. Therefore, for Korean consumers, aesthetic factors play an important role in selecting and accepting EV technology, and this result reveals how aesthetic values influence Korean consumers' attitudes toward and adoption of new technologies in the context of Korean culture. This is contrary to the results of Jung et al. (2021) [90], who found that Korean consumers attach great importance to the future value of the car when buying a car but do not pay much attention to the aes of the EV.

Relationship between Function (Func), Perceived Variables, and Environmental Awareness (EA)

The survey results of Korean consumers show that H2-1 and H2-2, concerning the PEU of electric vehicles ($\beta = 0.07$, $T = 5.25$) and EA ($\beta = 0.06$, $T = 4.32$), respectively, are established, but H2-3 (that the PU of electric vehicles will increase) is not established ($\beta = 0.08$, $T = 0.57$). This means that Korean consumers tend to think that EVs are easier to use and more environmentally friendly. However, the PU can only be improved after a test drive [91].

Relationship between Symbolism (Sym), Perceived Variables, and Environmental Awareness (EA)

From the perspective of the relationship between symbolic and perceived variables and environmental protection, H3-1 ($\beta = 0.06$, $T = 3.19$), H3-2 ($\beta = 0.05$, $T = 3.18$), and H3-3 ($\beta = 0.06$, $T = 2.81$) are all valid for Korean consumers. In other words, sym can improve Korean consumers' PEU and PU, and it can also improve EA. This is consistent with the findings of Beak et al. (2020) [92] that Korean consumers have a relatively high willingness to pay for the carbon dioxide emission reduction rate of EVs.

The Relationship between Perceived Variables, Environmental Awareness (EA), and Willingness to Buy (WTB)

The research results on H4-1, H4-2, and H4-3 clearly show that, for the Korean consumers in this study, the effects of PEU, PU, and EA on the intention to purchase EVs are all positive. The higher the consumer's PEU for electric vehicles ($\beta = 0.07$, $T = 4.67$), the stronger the intention to purchase EVs. In other words, Korean consumers are more likely to choose EVs that they find easy to drive and operate. PU was also shown to have a positive impact on WTB ($\beta = 0.08$, $T = 3.68$). This proves that Korean consumers' perceptions of the functionality of EVs, such as their aes, func, and sym, will significantly influence their purchase decisions. The positive impact of EA on WTB has also been verified ($\beta = 0.07$, $T = 2.81$). This shows that concern and recognition of environmental protection issues will prompt Korean consumers to choose EVs.

In summary, in the EV market, PEU, PU, and EA have a significant impact on improving Korean consumers' WTB electric vehicles. Therefore, these factors should be paid attention to and strengthened in marketing and product development to meet consumers' needs and promote the popularity and market growth of EVs.

Finally, Korean consumers, much like Chinese consumers, were consistent with the findings of Davis (1989) [15]. PEU has a significant positive impact on PU ($\beta = 0.07$, $T = 2.18$), and the results prove that Korean consumers' EA also has a significant positive impact on perceived usefulness ($\beta = 0.07$, $T = 4.41$) (see Table 20).

Table 20. Results of hypothesis testing on Korean consumers.

No.	Path Relationship			B	β	S.E.	C.R.	p-Value	Result
H1-1	Aes	→	PEU	0.207	0.212	0.060	3.472	***	Pass
H1-2		→	PU	0.274	0.271	0.065	4.199	***	Pass
H1-3		→	EA	0.355	0.383	0.055	6.425	***	Pass
H2-1	Func	→	PEU	0.379	0.351	0.072	5.245	***	Pass
H2-2		→	PU	0.043	0.039	0.075	0.57	0.569	Fail
H2-3		→	EA	0.274	0.269	0.063	4.315	***	Pass
H3-1	Sym	→	PEU	0.255	0.252	0.064	4.057	***	Pass
H3-2		→	PU	0.200	0.192	0.063	3.186	0.001	Pass
H3-3		→	EA	0.175	0.183	0.055	3.186	0.001	Pass
H4-1	PEU	→	WTB	0.309	0.302	0.066	4.673	***	Pass
H4-2	PU	→		0.277	0.279	0.075	3.677	***	Pass
H4-3	EA	→		0.208	0.192	0.074	2.814	0.005	Pass

Note: *** indicate significance at the 1% levels.

5.2.4. Insights into Korean Consumers

Similar to the analysis of Chinese consumption, the aesthetics, function, symbolism, and environmental awareness of product design were applied to Davis's TAM to explore the impact of product design dimensions on the intention to purchase electric vehicles. However, unlike the analysis results for Chinese consumers, in the product design dimension of Korean consumers, aesthetics was found to have a direct impact on perceived usefulness, perceived ease of use, and environmental awareness. This shows that the aesthetics of electric vehicles can affect all perceived attitudes and environmental awareness, thereby triggering purchase intentions. Second, in the product design dimension, function has a significant impact on both perceived ease of use and environmental awareness; however, it has no effect on perceived usefulness. As reported by Kim et al. (2019) [91], for Korean consumers, perceived usefulness can only be improved after a test drive. Finally, in the symbolic dimension of product design, contrary to the results for Chinese consumers—that it has no direct effect on perceived ease of use and perceived usefulness, but affects environmental awareness—symbols have a positive effect on both perceived attitude and environmental awareness for Korean consumers. Additionally, like Chinese consumers, for Korean consumers, perceived ease of use and environmental awareness both positively affect perceived usefulness and directly affect purchase intentions.

6. Discussion

This study used the TAM to explore the factors that affect the intention to purchase EVs in the product design dimension of China (the world's largest EV market) and South Korea (an emerging EV market). The results show that, for Chinese consumers, H1-1, H1-2, H3-1, and H3-2 are not established. It can be concluded that Chinese consumers do not attach much importance to the aes and sym of EVs. The study of Korean consumers showed that the product design dimensions of aes, func, and sym have an impact on PEU, PU, and EA and can ultimately affect WTB; however, only H2-2 is not established, indicating that Korean consumers have a relatively low PU when accepting new technologies.

However, product preferences are influenced by cultural and social factors [93]. Cultural values are widely accepted beliefs, norms, and codes of conduct that influence people's thinking, decision-making, and behavior. History, religion, education, economic development, and social structure all influence cultural values, and these values affect the acceptance of specific cultural styles [94].

Cultural values directly affect the market position of EVs. For example, in Northern Europe, EVs are popular due to the emphasis on environmental protection and sustainable development, and governments have also introduced incentives. In Silicon Valley, United States, EVs are more acceptable due to the high acceptance of technology; in conservative cultures, promoting EVs requires more time and education. Companies need to understand these cultural differences and develop appropriate market strategies.

Hofstede's research (2001) [95] also explains the differences in results in different countries. China is a highly collectivist society, and consumers pay more attention to function rather than appearance when popularizing EVs under the government's promotion. South Korea scores high in uncertainty avoidance, and consumers have a strong avoidance of technological innovation. Therefore, the Korean market needs to reduce the uncertainty brought by new technologies by providing pre-driving experiences, and companies should reduce consumer concerns and promote the acceptance of new technologies through effective market communication and product education.

It should be added that, from a marketing management perspective, consumers' willingness to buy EVs is also affected by other options, such as cars with oil and diesel engines. Therefore, when studying EV purchasing behavior, one should not only consider product design, perceived acceptance and environmental awareness, but also consider the competition from other types of vehicles on the market, which differ from EVs in fuel, pricing, comfort, and aesthetic features and can provide consumers with diverse choices.

Finally, just as the 2024 Paris Summer Olympics will take "sustainable development" as a core concept, when exploring the factors affecting EV purchase intentions, we need to consider not only product design dimensions but also broader behavioral motivations and attitudes. In recent years, there have been an increasing number of studies on clean energy technologies, which reveal consumers' behavioral motivations and attitudes towards clean energy technologies in their daily lives.

For example, Kyriakopoulos (2022) [96] recently provided a comprehensive overview of the research on energy community management policies, economic aspects, technologies, and models, exploring the application of clean energy technologies therein. The study emphasized the social acceptance and support for clean energy technologies, which is very important for understanding the context of EV purchasing behavior.

7. Managerial Implications

The results of this study have several important implications for theory and practice.

First, this study applied product design to the TAM and investigated the purchase intentions of Chinese and Korean consumers.

Second, the research findings provided by this study can help automakers and marketers gain a deeper understanding of the characteristics of Chinese and Korean electric vehicle consumers. For example, Chinese consumers pay more attention to the new technology and functions of products than to aes and sym. This means that, when targeting the Chinese market, automakers should focus on promoting the advanced technological features and functional advantages of electric vehicles in order to attract the interest and purchasing desire of Chinese consumers.

Third, Korean consumers are more interested in aes and sym than in functions. This suggests that, in the Korean market, automakers should emphasize the aes and sym in the design of electric vehicles in order to cater to the preferences of Korean consumers. This can include promoting the fashionable design, brand image, and sym of EVs as environmental and social responsibilities.

Fourth, another effective communication strategy is to emphasize the contribution of electric vehicles to environmental protection, because both Chinese and Korean consumers believe that EA is an important factor in using electric vehicles. Automakers and marketers can enhance the appeal of electric vehicles and meet consumers' demand for environmental protection by promoting their low emissions and sustainability. As the Korean EV market matures, environmental concerns and reputation issues may become more important. Therefore, automakers should continue to pay attention to and respond to Korean consumers' concerns about environmental protection and social responsibility. This will not only enhance the brand image but also increase consumers' recognition and trust in EVs.

In summary, the findings of this study provide automakers with a basis for formulating differentiated marketing strategies for different markets. By understanding the unique needs and preferences of Chinese and Korean consumers, EVs can be promoted more effectively, and market share can be increased.

8. Limitations and Future Research Directions

This study has some limitations and provides multiple directions for future research. The following is a discussion of these limitations and future research directions.

8.1. Limitations

First, this study mainly used the technology acceptance model (TAM) to explore the impact of product design dimensions on the intention to purchase EVs among Chinese and Korean consumers, so the results may be limited to product design dimensions. This study ignored other consumer needs and dimensions, such as sustainability and technological innovation. Sustainability affects consumers' environmental awareness and social responsibility, while technological innovation meets the demand for new functions and high technology. These dimensions interact with aes, func, and sym and may jointly affect consumers' purchase decisions.

Second, this study did not analyze consumer differences such as gender and age. Future research could be extended to explore gender differences (such as males and females) and age differences, in order to investigate whether these differences affect the role of aes, func, and sym design on the intention to purchase EVs. For example, male consumers may be more interested in technological innovation and performance, whereas female consumers may pay more attention to safety and environmental protection; young consumers may be more open to new technologies, whereas older consumers may pay more attention to the reliability and long-term value of vehicles.

In addition, this study is also lacking in theoretical research depth. Only partial theories using the TAM may overlook factors such as social influence, personal attitudes, or external environment. Future research could incorporate the theory of planned behavior (TPB) or the unified theory of behavior (UTAUT) to build a more comprehensive theoretical model, in order to provide a deeper understanding and more accurate prediction of consumer acceptance of EVs.

In addition, the results of this research are based only on data from Chinese and Korean consumers, which may not be applicable to consumers in other countries. The cultural background and market environment of each country are different, and future research should consider the impacts of cultural, regional, and subcultural differences on consumer behavior. For example, the difference in the degree of modernization between eastern and inland China, South Korea, and other regions, as well as ethnic minority culture, age generation differences, etc., may significantly affect consumer purchasing behavior and preferences.

Possible biases in this study include social desirability bias, memory bias, and inconsistent responses, which may affect the accuracy and reliability of the data. Therefore, future research should consider how to reduce these biases in order to improve the validity of the research results.

8.2. Future Research Directions

- Multidimensional analysis: Future research should consider the impacts of sustainability and technological innovation on the willingness to purchase EVs and explore the interactive relationships between these dimensions and aes, func, and sym.
- Gender and age differences: In-depth exploration of the impact of gender and age on the willingness to buy EVs and analysis of the differences in the needs of different groups in aes, func, and sym design.
- Expansion of the theoretical framework: Combining other theories in the TAM or integrating the theory of planned behavior (TPB) and the unified theory of behavior (UTAUT) to build a more comprehensive theoretical model, in order to enhance the explanatory power of consumers' willingness to buy EVs.
- Cultural differences research: Expand to other countries and regions to analyze the impacts of cultural background, economic development, degree of modernization, and ethnic minority culture on consumer behavior.
- Dynamic data collection: A longitudinal research design should be adopted to obtain dynamic data by tracking the attitude and behavior changes of the same group over a long period of time, especially under new technologies and market changes.
- Application of biometric technology: Citing the research results of biometric technology, explore its potential impact on EV consumer decision-making. For example, the use of advanced biometric technology may enhance the safety and personalization of EVs and attract more consumers who pay attention to these features [97].
- As the electric vehicle market develops rapidly, and as consumer preferences and technological advances continue to change, manufacturers should continue to innovate, invest in R&D, cooperate with technology companies, conduct dynamic market analysis, flexibly produce, optimize the supply chain, adjust brand positioning, strengthen digital marketing, and provide comprehensive charging and after-sales services in order to maintain market relevance and meet changing needs.

In short, future research should focus on making up for the shortcomings of existing research and, through multidimensional theoretical discussions and data analysis, gaining a deeper understanding of consumers' willingness to purchase EVs, so as to provide more valuable insights and suggestions for the further development of the EV market.

Author Contributions: Z.S.: conceptualization, data curation, formal analysis, funding acquisition, investigation, methodology, project administration, software, validation, writing—original draft, writing—review and editing. B.L.: conceptualization, data curation, formal analysis, funding acquisition, investigation, methodology, project administration, resources, supervision, validation, writing—original draft, writing—review and editing. All authors have read and agreed to the published version of the manuscript.

Funding: This research received no external funding.

Data Availability Statement: Restrictions apply to the availability of these data. Data were obtained from [google] and are available [at https://drive.google.com/drive/u/0/home] with the permission of [google].

Acknowledgments: The original contributions presented in the study are included in the article; further inquiries can be directed to the corresponding authors, all individuals included consented to the acknowledgement.

Conflicts of Interest: The authors declare no conflicts of interest.

References

1. Yin, J.; Zhang, Z.; Liao, G.; Ge, X. Research on consumers' willingness to purchase new energy vehicles based on technology acceptance model and perceived risk theory. *Forecasting* **2019**, *38*, 83–89.
2. Statista, Electric Vehicles: Market Data & Analysis. Available online: https://www.statista.com/study/103895/electric-vehicles-report/ (accessed on 18 April 2024).
3. Yong, J.Y.; Ramachandaramurthy, V.K.; Tan, K.M.; Mithulananthan, N. A review on the state-of-the-art technologies of electric vehicle, its impacts and prospects. Renew. *Sustain. Energy Rev.* **2015**, *49*, 365–385. [CrossRef]
4. Hertzke, P.; Müller, N.; Schenk, S. *Dynamics in the Global Electric-Vehicle Market*; McKinsey Center for Future Mobility: New York, NY, USA, 2017.
5. Crabtree, G. The coming electric vehicle transformation. *Science* **2019**, *366*, 422–424. [CrossRef] [PubMed]
6. Han, J.-S.; Lee, J.-H. Market segmentation analysis on electric vehicle purchasing behavior in Seoul. *J. Korean Transp. Soc.* **2018**, *36*, 129–140. [CrossRef]
7. Song, M.-R.; Joo, W.-J.; Lim, M.-J. Analysis of consumer characteristics in the Korean electric vehicle market. *J. Manag. Stud.* **2018**, *47*, 271–306.
8. Huang, X.; Ge, J. Electric vehicle development in Beijing: An analysis of consumer purchase intention. *J. Clean. Prod.* **2019**, *216*, 361–372. [CrossRef]
9. Byun, Y. A study on the factors that change the formative design of self-driving electric vehicles—Focusing on changes in front design due to the use of self-driving systems. *Form. Media Stud.* **2023**, *26*, 271–279.
10. Button, Q.E. *The Influence of Exterior Design Attributes on Consumer Preference for Electric Vehicles*; Oklahoma State University: Stillwater, OK, USA, 2019.
11. Jo, W.-C. Low-speed electric vehicle (NEV) exterior design research for a sustainable society. *Korean Soc. Sci. Arts Converg.* **2012**, *10*, 211–221.
12. Wang, Y. Research on the Exterior Design of Household Electric Vehicles. Master's Thesis, Yanshan University, Qinhuangdao, China, 2015.
13. Lim, J.W.; Park, J.S. A study on the structural relationship between factors influencing the purchase of electric vehicles—Focusing on the TAM model, recognition of electric vehicle attributes, and subjective knowledge. *Manag. Stud.* **2022**, *37*, 165–181.
14. Bloch, P.H. Seeking the ideal form: Product design and consumer response. *J. Mark.* **1995**, *59*, 16–29. [CrossRef]
15. Davis, F.D. Perceived Usefulness, Perceived Ease of Use, and User Acceptance of Information Technology. *MIS Q.* **1989**, *13*, 319–340. [CrossRef]
16. Nussbaum, B. Smart Design. *Business Week*, 11 April 1988; pp. 102–117.
17. Han, G.-J. CMF research based on changes in product design trends. *Form. Des. Res.* **2021**, *24*, 207–221.
18. Kellaris, J.J.; Kent, R.J. An exploratory investigation of responses elicited by music varying in tempo, tonality, and texture. *J. Consum. Psychol.* **1993**, *2*, 381–401. [CrossRef]
19. Mahlke, S. User experience: Usability, aesthetics and emotions in human-technology interaction. In Proceedings of the towards UX Manifesto, COST294-MAUSE Affiliated Workshop, Lancaster, UK, 3 September 2007; p. 26.
20. Lee, S.; Johnson, Z.S. The effect of new product design and innovation on South Korean consumer's willingness to buy. *Asia Pac. J. Mark. Logist.* **2017**, *29*, 98–113. [CrossRef]
21. Srinivasan, R.; Lilien, G.L.; Rangaswamy, A.; Pingitore, G.M.; Seldin, D. The Total Product Design Concept and an Application to the Auto Market. *J. Prod. Innov. Manag.* **2012**, *29*, 3–20. [CrossRef]
22. Homburg, C.; Schwemmle, M.; Kuehnl, C. New Product Design: Concept, Measurement, and Consequences. *J. Mark.* **2015**, *79*, 41–56. [CrossRef]
23. Ajzen, I.; Martin, F. Understaning attitudes and predicting social behavior. In *TRA: Theory of Reasoned Action*; Prentice Hall: Englewood Cliffs, NJ, USA, 1980.
24. Venkatesh, V.; Davis, F.D. A Theoretical Extension of the Technology Acceptance Model: Four Longitudinal Field Studies. *Manag. Sci.* **2000**, *46*, 186–204. [CrossRef]
25. Venkatesh, V.; Bala, H. Technology Acceptance Model 3 and a Research Agenda on Interventions. *Decis. Sci.* **2008**, *39*, 273–315. [CrossRef]
26. Wang, Y.; Wang, Q. Factors affecting Beijing residents' buying behavior of new energy vehicle: An integration of technology acceptance model and theory of planned behavior. *Chin. J. Manag. Sci.* **2013**, *21*, 691–698.
27. Hu, L.; Tang, Y.; Mao, Y. Factors influencing the use of small electric vehicles based on the technology acceptance model. *Sci. Technol. Manag. Res.* **2015**, *35*, 122–125.
28. Wang, Z. Research on the Purchase Intention of Environmentally Friendly Electric Motorcycles-TAM Model Perspective. Master's Thesis, Tamkang University, Taipei, Taiwan, 2018; pp. 1–83.
29. Yuan, X.; Wang, A. Research on the factors influencing the willingness to use shared cars based on TAM. *J. Wuhan Univ. Technol. (Inf. Manag. Eng. Ed.)* **2018**, *40*, 434–438.
30. Zhang, K.; Zhao, S. Research on MaaS usage intention based on TAM extended model. *Transp. Res.* **2022**, *8*, 120.
31. He, X.; Zhan, W.; Hu, Y. Consumer purchase intention of electric vehicles in China: The roles of perception and personality. *J. Clean. Prod.* **2018**, *204*, 1060–1069. [CrossRef]

32. Zhang, Y. BYD "Han" New Energy Vehicle Styling Design Based on Brand Genes. Master's Thesis, Kunming University of Science and Technology, Kunming, China, 2017.
33. Ajanovic, A. The future of electric vehicles: Prospects and impediments. *WIREs Energy Environ.* **2015**, *4*, 521–536. [CrossRef]
34. Sun, Z.; Park, K. A Study on the Preference of Electric Vehicle Front Design Elements. *Appl. Sci.* **2024**, *14*, 3262. [CrossRef]
35. ElectricForAll, Basics of Electric Vehicle. Available online: https://www.electricforall.org/ko/ev-basics/ (accessed on 14 January 2024).
36. Feng, J.; Mu, W.; Fu, Z. A review of research on consumer purchasing intention. *Mod. Manag. Sci.* **2006**, *11*, 7–9.
37. Lee, M.-Y. Perception characteristics of the attractiveness of automobile exterior design—Proposal of a recognition model based on user characteristics. *Basic Plast. Stud.* **2017**, *18*, 291–305.
38. Gilal, F.G.; Zhang, J.; Gilal, R.G.; Gilal, R.G.; Gilal, N.G. Supply chain management practices and product development: A moderated mediation model of supply chain responsiveness, organization structure, and research and development. *J. Adv. Manuf. Syst.* **2017**, *16*, 35–56. [CrossRef]
39. Lieven, T.; Mühlmeier, S.; Henkel, S.; Waller, J.F. Who will buy electric cars? An empirical study in Germany. *Transp. Res. Part D Transp. Environ.* **2011**, *16*, 236–243. [CrossRef]
40. Kim, H.; Kim, H. A study on Chinese consumers' purchase intention factors for electric vehicles. *Distrib. Sci. Res.* **2019**, *17*, 85–90.
41. Ashraf, S.; Alfandi, O.; Ahmad, A.; Khattak, A.M.; Hayat, B.; Kim, K.H.; Ullah, A. Bodacious-Instance Coverage Mechanism for Wireless Sensor Network. *Wirel. Commun. Mob. Comput.* **2020**, *2020*, 8833767. [CrossRef]
42. Veryzer, R.W., Jr. Aesthetic response and the influence of design principles on product preferences. In *Advances in Consumer Research*; Association for Consumer Research: Provo, UT, USA, 1993; Volume 20.
43. Kotler, P.; Rath, G.A. Design: A powerful but neglected strategic tool. *J. Bus. Strat.* **1984**, *5*, 16–21. [CrossRef]
44. Gilal, N.G.; Zhang, J.; Gilal, F.G. Linking product design to consumer behavior: The moderating role of consumption experience. *Psychol. Res. Behav. Manag.* **2018**, *11*, 169–185. [CrossRef] [PubMed]
45. Kim, H.W.; Kankanhalli, A. Investigating user resistance to information systems implementation: A status quo bias perspective. *MIS Q.* **2009**, *33*, 567–582. [CrossRef]
46. Fu, L.; Fu, L. Research on technical factors of automobile styling. *Packag. Eng.* **2008**, *29*, 167–169.
47. Chakrabarti, A.; Bligh, T.P. A scheme for functional reasoning in conceptual design. *Des. Stud.* **2001**, *22*, 493–517. [CrossRef]
48. Lü, S.; Wang, X.; Ni, H.; Yuan, Y. Functions and research progress of electric vehicle energy management system. *Power Sources Technol.* **2014**, *38*, 386–389.
49. Agarwal, R.; Prasad, J. A Conceptual and Operational Definition of Personal Innovativeness in the Domain of Information Technology. *Inf. Syst. Res.* **1998**, *9*, 204–215. [CrossRef]
50. Agarwal, R.; Karahanna, E. Time Flies When You're Having Fun: Cognitive Absorption and Beliefs about Information Technology Usage. *MIS Q.* **2000**, *24*, 665–694. [CrossRef]
51. Belk, R.W. Possessions and the Extended Self. *J. Consum. Res.* **1988**, *15*, 139–168. [CrossRef]
52. Cheah, I.; Phau, I.; Chong, C.; Shimul, A.S. Antecedents and outcomes of brand prominence on willingness to buy luxury brands. *J. Fash. Mark. Manag. Int. J.* **2015**, *19*, 402–415. [CrossRef]
53. Creusen, M.E.H.; Schoormans, J.P.L. The Different Roles of Product Appearance in Consumer Choice. *J. Prod. Innov. Manag.* **2004**, *22*, 63–81. [CrossRef]
54. Seva, R.R.; Helander, M.G. The influence of cellular phone attributes on users' affective experiences: A cultural comparison. *Int. J. Ind. Ergon.* **2009**, *39*, 341–346. [CrossRef]
55. Xing, Y. Research on the design of Macau tourist souvenirs from the perspective of cultural symbols. *Packag. Eng. Art Ed.* **2022**, *43*, 392–400.
56. Zhang, L.; Liu, G. Application of Connotative Semantics in Product Identification. *J. Wuxi Univ. Light Ind. (Soc. Sci. Ed.)* **2001**, *413*, 406–408, 413. [CrossRef]
57. Belk, R.W.; Wallendorf, M.; Sherry, J.F., Jr. The sacred and the profane in consumer behavior: Theodicy on the odyssey. *J. Consum. Res.* **1989**, *16*, 1–38. [CrossRef]
58. Verganti, R. Design, Meanings, and Radical Innovation: A Metamodel and a Research Agenda. *J. Prod. Innov. Manag.* **2008**, *25*, 436–456. [CrossRef]
59. Jamal, A.; Sharifuddin, J. Perceived value and perceived usefulness of halal labeling: The role of religion and culture. *J. Bus. Res.* **2015**, *68*, 933–941. [CrossRef]
60. Chitra, K. In search of the green consumers: A perceptual study. *J. Serv. Res.* **2007**, *7*, 173–191.
61. Li, N. Environmental awareness and environmental behavior. *Xuehai* **2001**, *1*, 120–124.
62. He, W.; He, R. An empirical analysis of the factors affecting the diffusion of new energy vehicles into the public market: Based on TAM-IDT theory. *J. Dalian Univ. Technol. Soc. Sci. Ed.* **2015**, *36*, 28–33.
63. Davis, F.D.; Bagozzi, R.P.; Warshaw, P.R. Extrinsic and Intrinsic Motivation to Use Computers in the Workplace. *J. Appl. Soc. Psychol.* **1992**, *22*, 1111–1132. [CrossRef]
64. Venkatesh, V.; Morris, M.G.; Davis, G.B.; Davis, F.D. User acceptance of information technology: Toward a unified view. *MIS Q.* **2003**, *27*, 425–478. [CrossRef]

65. Davis, F.D. User Acceptance of Information Systems: The Technology Acceptance Model (TAM). 1987, pp. 1–36. Available online: https://deepblue.lib.umich.edu/bitstream/handle/2027.42/35547/b1409190.0001.001.pdf (accessed on 17 January 2024.).
66. Van Tonder, E.; Fullerton, S.; De Beer, L.; Saunders, S. Social and personal factors influencing green customer citizenship behaviours: The role of subjective norm, internal values and attitudes. *J. Retail. Consum. Serv.* **2023**, *71*, 103190. [CrossRef]
67. Carley, S.; Krause, R.M.; Lane, B.W.; Graham, J.D. Intent to purchase a plug-in electric vehicle: A survey of early impressions in large US cites. *Transp. Res. Part D Transp. Environ.* **2013**, *18*, 39–45. [CrossRef]
68. Bunce, L.; Harris, M.; Burgess, M. Charge up then charge out? Drivers' perceptions and experiences of electric vehicles in the UK. *Transp. Res. Part A Policy Pract.* **2014**, *59*, 278–287. [CrossRef]
69. Zhang, Y.; Yu, Y.; Zou, B. Analyzing public awareness and acceptance of alternative fuel vehicles in China: The case of EV. *Energy Policy* **2011**, *39*, 7015–7024. [CrossRef]
70. Park, C.-S.; Park, H.-S. A study on the influencing factors of self-driving car acceptance and purchase intention. *Logist. Res.* **2020**, *28*, 13–27.
71. Patara, Y.; Monroe, K. Using cue consistency theory to investigate potential effects of the country-of-origin labeling act. In Proceedings of the Society for Consumer Psychology 2008 Winter Conference, New Orleans, LA, USA, 21–23 February 2008; pp. 39–43.
72. Donnici, G.; Frizziero, L.; Liverani, A.; Buscaroli, G.; Raimondo, L.; Saponaro, E.; Venditti, G. A New Car Concept Developed with Stylistic Design Engineering (SDE). *Inventions* **2020**, *5*, 30. [CrossRef]
73. Bentler, P.M.; Bonett, D.G. Significance tests and goodness of fit in the analysis of covariance structures. *Psychol. Bull.* **1980**, *88*, 588–606. [CrossRef]
74. Carmines, E.G.; McIver, J.P. Analyzing models with unobserved variables: Analysis of covariance structures. In *Social Measurement: Current Issues*; Bohrnstedt, G.W., Borgatta, E.F., Eds.; Sage: Beverly Hills, CA, USA, 1981; pp. 65–115.
75. Browne, M.W.; Cudeck, R. Alternative ways of assessing model fit. *Sociol. Methods Res.* **1992**, *21*, 230–258. [CrossRef]
76. Hu, L.T.; Bentler, P.M. Cutoff criteria for fit indexes in covariance structure analysis: Conventional criteria versus new alternatives. *Struct. Equ. Model. Multidiscip. J.* **1999**, *6*, 1–55. [CrossRef]
77. Bentler, P.M. Comparative Fit Indexes in Structural Models. *Psychol. Bull.* **1990**, *107*, 238–246. [CrossRef] [PubMed]
78. Byrne, B.M. *Structural Equation Modeling with LISREL, PRELIS, and SIMPLIS: Basic Concepts, Applications, and Programming*; Lawrence Erlbaum Associates, Publishers: Mahwah, NJ, USA, 1998.
79. Doll, W.J.; Xia, W.; Torkzadeh, G. A confirmatory factor analysis of the end-user computing satisfaction instrument. *MIS Q.* **1994**, *18*, 453–461. [CrossRef]
80. Nunnally, J.; Bernstein, I. *Bernstein Psychometric Theory*; McGraw-Hill: New York, NY, USA, 1978.
81. Bagozzi, R.P.; Yi, Y. On the evaluation of structural equation models. *J. Acad. Mark. Sci.* **1988**, *16*, 74–94. [CrossRef]
82. Filieri, R.; Lin, Z. The role of aesthetic, cultural, utilitarian and branding factors in young Chinese consumers re-purchase intention of smartphone brands. *Comput. Hum. Behav.* **2017**, *67*, 139–150. [CrossRef]
83. Sovacool, B.K.; Abrahamse, W.; Zhang, L.; Ren, J. Pleasure or profit? Surveying the purchasing intentions of potential electric vehicle adopters in China. *Transp. Res. Part A Policy Pract.* **2019**, *124*, 69–81. [CrossRef]
84. Xiong, Z.; Zhang, P.; Guo, N. The impact of carbon tax and consumer environmental awareness on carbon emissions in the supply chain. *Syst. Eng. Theory Pract.* **2014**, *34*, 2245–2252.
85. Liu, R.; Ding, Z.; Wang, Y.; Jiang, X.; Jiang, X.; Sun, W.; Wang, D.; Mou, Y.; Liu, M. The relationship between symbolic meanings and adoption intention of electric vehicles in China: The moderating effects of consumer self-identity and face consciousness. *J. Clean. Prod.* **2020**, *288*, 125116. [CrossRef]
86. Heffner, R.R.; Kurani, K.S.; Turrentine, T.S. Symbolism in California's early market for hybrid electric vehicles. *Transp. Res. Part D Transp. Environ.* **2007**, *12*, 396–413. [CrossRef]
87. Poullikkas, A. Sustainable options for electric vehicle technologies. *Renew. Sustain. Energy Rev.* **2015**, *41*, 1277–1287. [CrossRef]
88. Catenacci, M.; Verdolini, E.; Bosetti, V.; Fiorese, G. Going electric: Expert survey on the future of battery technologies for electric vehicles. *Energy Policy* **2013**, *61*, 403–413. [CrossRef]
89. Clarkson, J.J.; Janiszewski, C.; Cinelli, M.D. The Desire for Consumption Knowledge. *J. Consum. Res.* **2013**, *39*, 1313–1329. [CrossRef]
90. Jung, J.; Yeo, S.; Lee, Y.; Moon, S.; Lee, D.-J. Factors affecting consumers' preferences for electric vehicle: A Korean case. *Res. Transp. Bus. Manag.* **2021**, *41*, 100666. [CrossRef]
91. Kim, J.-H.; Kim, H.-J.; Yoo, S.-H. Willingness to pay for fuel-cell electric vehicles in South Korea. *Energy* **2019**, *174*, 497–502. [CrossRef]
92. Beak, Y.; Kim, K.; Maeng, K.; Cho, Y. Is the environment-friendly factor attractive to customers when purchasing electric vehicles? Evidence from South Korea. *Bus. Strat. Environ.* **2020**, *29*, 996–1006. [CrossRef]
93. McCracken, G. Culture and Consumption: A Theoret ical Account of tbe Structure and Movement of Cultural Meaning of Consumer Goods. *J. Consum. Res.* **1986**, *13*, 71–84. [CrossRef]
94. Kron, J. *Home-Psych: The Social Psychology of Home and Decoration*; Clarkson N Potter: New York, NY, USA, 1983.
95. Hofstede, G. *Culture's Consequences: Comparing Values, Behaviors, Institutions, and Organizations across Nations*; Sage: Beverly Hills, CA, USA, 2001.

96. Kyriakopoulos, G.L. Energy Communities Overview: Managerial Policies, Economic Aspects, Technologies, and Models. *J. Risk Financ. Manag.* **2022**, *15*, 521. [CrossRef]
97. Ashraf, S.; Saleem, S.; Ahmed, T.; Aslam, Z.; Shuaeeb, M. Iris and Foot based Sustainable Biometric Identification Approach. In Proceedings of the 2020 International Conference on Software, Telecommunications and Computer Networks (SoftCOM), Split, Croatia, 17–19 September 2020.

vehicles

MDPI

Article

Using Multimodal Large Language Models (MLLMs) for Automated Detection of Traffic Safety-Critical Events

Mohammad Abu Tami [1], Huthaifa I. Ashqar [2,3,*], Mohammed Elhenawy [4,5], Sebastien Glaser [4] and Andry Rakotonirainy [4]

1 Natural, Engineering and Technology Sciences Department, Arab American University, Jenin P.O Box 240, Palestine; m.abutami@student.aaup.edu
2 Civil Engineering Department, Arab American University, Jenin P.O Box 240, Palestine
3 Artificial Intelligence Program, Fu Foundation School of Engineering and Applied Science, Columbia University, New York, NY 10027, USA
4 CARRS-Q, Queensland University of Technology, Kelvin Grove, QLD 4059, Australia; mohammed.elhenawy@qut.edu.au (M.E.); sebastien.glaser@qut.edu.au (S.G.); r.andry@qut.edu.au (A.R.)
5 Centre for Data Science, Queensland University of Technology, Kelvin Grove, QLD 4059, Australia
* Correspondence: huthaifa.ashqar@aaup.edu

Abstract: Traditional approaches to safety event analysis in autonomous systems have relied on complex machine and deep learning models and extensive datasets for high accuracy and reliability. However, the emerge of multimodal large language models (MLLMs) offers a novel approach by integrating textual, visual, and audio modalities. Our framework leverages the logical and visual reasoning power of MLLMs, directing their output through object-level question–answer (QA) prompts to ensure accurate, reliable, and actionable insights for investigating safety-critical event detection and analysis. By incorporating models like Gemini-Pro-Vision 1.5, we aim to automate safety-critical event detection and analysis along with mitigating common issues such as hallucinations in MLLM outputs. The results demonstrate the framework's potential in different in-context learning (ICT) settings such as zero-shot and few-shot learning methods. Furthermore, we investigate other settings such as self-ensemble learning and a varying number of frames. The results show that a few-shot learning model consistently outperformed other learning models, achieving the highest overall accuracy of about 79%. The comparative analysis with previous studies on visual reasoning revealed that previous models showed moderate performance in driving safety tasks, while our proposed model significantly outperformed them. To the best of our knowledge, our proposed MLLM model stands out as the first of its kind, capable of handling multiple tasks for each safety-critical event. It can identify risky scenarios, classify diverse scenes, determine car directions, categorize agents, and recommend the appropriate actions, setting a new standard in safety-critical event management. This study shows the significance of MLLMs in advancing the analysis of naturalistic driving videos to improve safety-critical event detection and understanding the interactions in complex environments.

Keywords: multimodal large language models (MLLMs); safety-critical events; in-context learning (ICL); self-ensemble learning; object-level question–answers (QAs)

Citation: Abu Tami, M.; Ashqar, H.I.; Elhenawy, M.; Glaser, S.; Rakotonirainy, A. Using Multimodal Large Language Models (MLLMs) for Automated Detection of Traffic Safety-Critical Events. *Vehicles* **2024**, *6*, 1571–1590. https://doi.org/10.3390/vehicles6030074

Academic Editor: Elżbieta Macioszek

Received: 5 August 2024
Revised: 30 August 2024
Accepted: 1 September 2024
Published: 2 September 2024

1. Introduction

The advent and development of autonomous driving technologies have marked a significant transformation in the automotive industry, reshaping how vehicles interact with their environment and with each other. The concept of autonomous driving has evolved over several decades, beginning with basic cruise control systems in the 1950s [1,2] and progressing to the sophisticated connected and automated vehicles (CAVs) of today [3,4]. The integration of advanced sensors, machine learning algorithms, and communication technologies has enabled vehicles to perform complex tasks such as navigation, obstacle avoidance, and decision-making with minimal human intervention [5].

As the capabilities of autonomous vehicles have expanded, so too have the challenges associated with ensuring their safety and reliability. Traffic safety has always been a critical concern in the development of autonomous vehicles, as these systems must be able to respond to a wide range of dynamic and unpredictable situations on the road. This is particularly important in the context of safety-critical events, such as sudden changes in traffic patterns, unexpected obstacles, and potential collisions. Additionally, traffic safety is influenced not only by the type of vehicle—whether conventional or autonomous—but also by the volume of traffic, as higher traffic volumes increase the probability of unsafe conditions [6]. The ability of autonomous vehicles to detect, analyze, and respond to these events in real time is crucial for preventing accidents and ensuring the safety of all road users [7].

The recent advancement breakthrough in large language models (LLMs) has revealed the potential usage in the complex challenging environment of analyzing driving videos. Many researchers have investigated the potential of utilizing LLMs in analyzing driving videos through textual representations [8–11]. With the advancement of MLLMs, a new merger has been reached with the power reasoning of LLMs in the different modalities of text, image, and audio [12,13].

Critical-safety event analysis is considered one of the complex and critical environments that could benefit from the new MLLM breakthrough. While full critical-safety event detection might still be a far reach, MLLM could advance the understanding of the dynamic variety of road transportation through providing textual analysis of the visual representation of the environment and the different agents in it, then using this analysis to provide a direct, concise, and actionable early warning to the ego-driver in the case of any potential hazards.

This capability of MLLMs to synthesize information across multiple modalities—such as visual cues from driving videos, environmental sounds, and contextual data—opens new avenues for enhancing driver assistance systems. By integrating textual analysis with real-time video and audio inputs, MLLMs can facilitate more accurate and context-aware interpretations of driving scenarios, which are crucial for preventing accidents and improving road safety. The ability to generate natural language descriptions or warnings based on complex visual and auditory inputs allows for a more intuitive interface between the technology and the driver, which has the ability to reduce cognitive load and improve reaction times in critical situations.

Moreover, the adaptability of MLLMs to learn from diverse data sources, including different driving environments and conditions, enhances their robustness and reliability. As MLLMs continue to evolve, their role in the domain of autonomous driving and driver assistance is expected to expand, offering more sophisticated solutions for anticipating and mitigating safety risks. The integration of these models into real-world applications could mark a significant step forward in achieving safer and more efficient transportation systems. In this context, the development of MLLMs presents an exciting opportunity to revolutionize the field of critical-safety event detection in driving. As the technology matures, the potential for creating systems that can not only detect but also predict, prevent, and recommend about critical-safety events becomes increasingly feasible, moving us closer to a future where road transportation is not only smarter but also significantly safer.

2. Related Works

Before the era of MLLM, researchers in safety event analysis relied on developing a complex machine learning model from the ground up, utilizing thousands of annotated datasets to achieve high accuracy and reliability. For instance, the authors in [14] proposed a supervised encoder–decoder model where a pre-trained ResNet-101 was used as encoder to extract the visual and flow features of 17k distinct ego-car dash cam scenarios, and a neural image caption generation structure [15] as a decoder to predict the caption of the street frames while attaining the extracted features from the encoder part.

The study by Zhenjie et. al. LLM4Drive [16] reviews the integration of large language models (LLMs) in autonomous driving systems, highlighting their potential to enhance decision-making, perception, and interaction through advanced reasoning and contextual understanding. The survey categorizes current research into planning, perception, question answering, and generation, addressing the challenges of transparency, scalability, and real-world application. It underscores the need for robust datasets and interpretable models to build trust and improve system reliability in autonomous driving.

Cui et al. explores the integration of LLMs and vision foundation models (VFMs) in enhancing autonomous driving systems [17]. The work covers the historical evolution from early sensor-based approaches to advanced deep learning techniques that improve perception, planning, and decision-making. The paper also reviews existing multimodal tools and datasets like KITTI [18] and nuScenes [19] A study by Chen et al. [20] proposed a pre-training method that aligns numeric vector modalities with LLM (GPT3.5) representations, improving the system's ability to interpret driving scenarios, answer questions, and make decisions. Furthermore, the study titled "DriveMLM" [21] introduces an LLM-based autonomous driving (AD) framework that aligns multi-modal LLMs with behavioral planning states, enabling closed-loop autonomous driving in realistic simulators. It bridges the gap between language decisions and vehicle control commands by standardizing decision states according to the off-the-shelf motion planning module. On another hand, the "Drive As you Speak" paper [22] presents an approach to enabling human-like interaction with large language models in autonomous vehicles. It leverages LLMs to understand and respond to human commands, demonstrating the potential of LLMs in creating more intuitive and user-friendly autonomous driving experiences. Moreover, AccidentGPT [23] introduces a multi-modal model for traffic accident analysis, which was capable of reconstructing crash processes and providing comprehensive reports.

Recent advancements also explored the integration of sensor data and real-time processing using LLMs to enhance autonomous driving capabilities. A study by Zhang et al. [24] examined the integration of LLMs with LiDAR and radar data to improve object detection and tracking. Similarly, a study by Singh et al. [25] highlighted the use of LLMs in predicting pedestrian behavior by analyzing both visual signals and contextual information, which increased the reliability of autonomous systems in urban settings. Moreover, another study by Lopez et al. [26] focused on utilizing LLMs to interpret driver motions and voice commands, which might facilitate a more natural interaction between the driver and the vehicle. Furthermore, a study by Kim et al. [27] explored the analysis of live video feeds from dashboard cameras, which enabled the early detection of potential hazards such as sudden lane changes or road obstacles. This approach allowed for timely warnings and interventions, which has the potential to enhance the safety of autonomous driving systems.

A study by Hussien et al. [28] also highlighted the potential of integrating LLMs with knowledge graphs and retrieval-augmented generation (RAG)-based explainable frameworks to provide explainable predictions of road user behaviors, which is crucial for developing safe automated driving systems. This study underscored the importance of explainability in the potential deployment of LLMs in critical environments like autonomous driving.

In addition to LLMs, contributions have been made in the domain of cooperative control of CAVs. The study by Liang et al [29] explores a multi-agent system (MAS) architecture designed to facilitate the cooperative control of CAVs. This hierarchical architecture enables vehicles to collaborate effectively in complex traffic environments, sharing information and making collective decisions that enhance safety and efficiency. The study emphasizes the importance of cooperation among autonomous vehicles, particularly in scenarios where rapid decision-making and coordination are critical to preventing accidents and ensuring smooth traffic flow.

Despite the promising developments in using MLLMs for autonomous driving and intelligent transportation systems, a significant gap remains in the application of these models for safety-critical event analysis. Existing studies focused on enhancing autonomous driving capabilities through improved perception and decision-making processes without specifically addressing the unique challenges posed by safety-critical situations. This gap shows the need for a specialized approach that leverages the multimodal capabilities of LLMs to directly address the complexity of safety-critical events in driving scenarios and provide more explainable information and recommendations, which is very important for taking the right safety countermeasures. Current methods mostly rely on complex machine and deep learning models and extensive annotated datasets, which are not always feasible or scalable in real-world applications. There is a need for an easy-to-implement, scalable, and explainable framework that can automate the extraction of visual representations from raw video feeds and utilize object-level question–answer (QA) prompts to guide MLLMs in generating actionable insights for hazard detection and response.

This study aims to bridge this gap by introducing an MLLM framework specifically designed for the analysis and interpretation of safety-critical events. By integrating the different modalities of texts and images, our framework seeks to provide a more holistic and scalable view of dynamic driving environments. Furthermore, our approach emphasizes the automation of extracting visual representation from the raw video and feeding it to an MLLM with the creation of object-level QA prompts to guide the MLLM's analysis, which focuses on generating actionable insights for safety-critical event detection and response. This study introduces a novel application of MLLMs in a domain where precision and reliable decision-making are essential, marking a significant step forward in the development of safer driving.

3. Preliminary

This section introduces the dataset utilized for evaluating the proposed framework and the Gemini model, which form the foundation for the experimental work in this study.

3.1. Dataset

Creating a dataset from driving videos that integrates language for visual understanding is a challenging task. This process is a resource-extensive task that requires trained human annotators for optimal accuracy and reliability. In addition, the variety and complexity of driving scenarios require a dataset rich in visual scenes. The dataset needs to cover a variety ranging from simple driving directions to complex situations involving pedestrians, other vehicles, and road signs.

Many researchers have either enhanced existing datasets with textual information [30–33] or developed new ones from scratch [14,34]. Notable among these are the DRAMA datasets [14]. DRAMA focuses on driving hazards and related objects, featuring video and object-level inquiries. This dataset supports visual captioning with free-form language descriptions and accommodates both closed and open-ended questions, making it essential for assessing various visual captioning skills in driving contexts. In addition, the vast variety found in DRAMA scenarios makes it a uniquely comprehensive resource for investigating and evaluating MLLM models on complex driving situations.

Considering these factors, this study selected the DRAMA dataset to utilize the ground truth label to report this paper's experimental results. DRAMA's detailed focus on hazard detection and its comprehensive framework for handling natural language queries make it exceptionally suitable for pushing forward research in safety-critical event analysis.

The DRAMA dataset includes multiple levels of human-labeled question–answer (QA) pairs. These include base questions regarding whether a risk exists in the scene, with Yes/No answers; scene classification into urban road, intersection, and narrow lane categories; questions about the direction of the ego-car, with options such as straight, right, and left; questions about potential hazard-causing agents in the scene, such as vehicles, pedestrians, cyclists, and infrastructure; and finally, questions about recommended actions for the ego-car driver based on the scene analysis, with eight possible actions, including stop, slow down, be aware, follow the vehicle ahead, carefully maneuver, start moving, accelerate, and yield. Figure 1 illustrates the distribution of each QA used in the studies, where 300 distinct videos ranging from 2 to 5 s were employed to examine the effectiveness of MLLMs in detecting traffic safety-critical events.

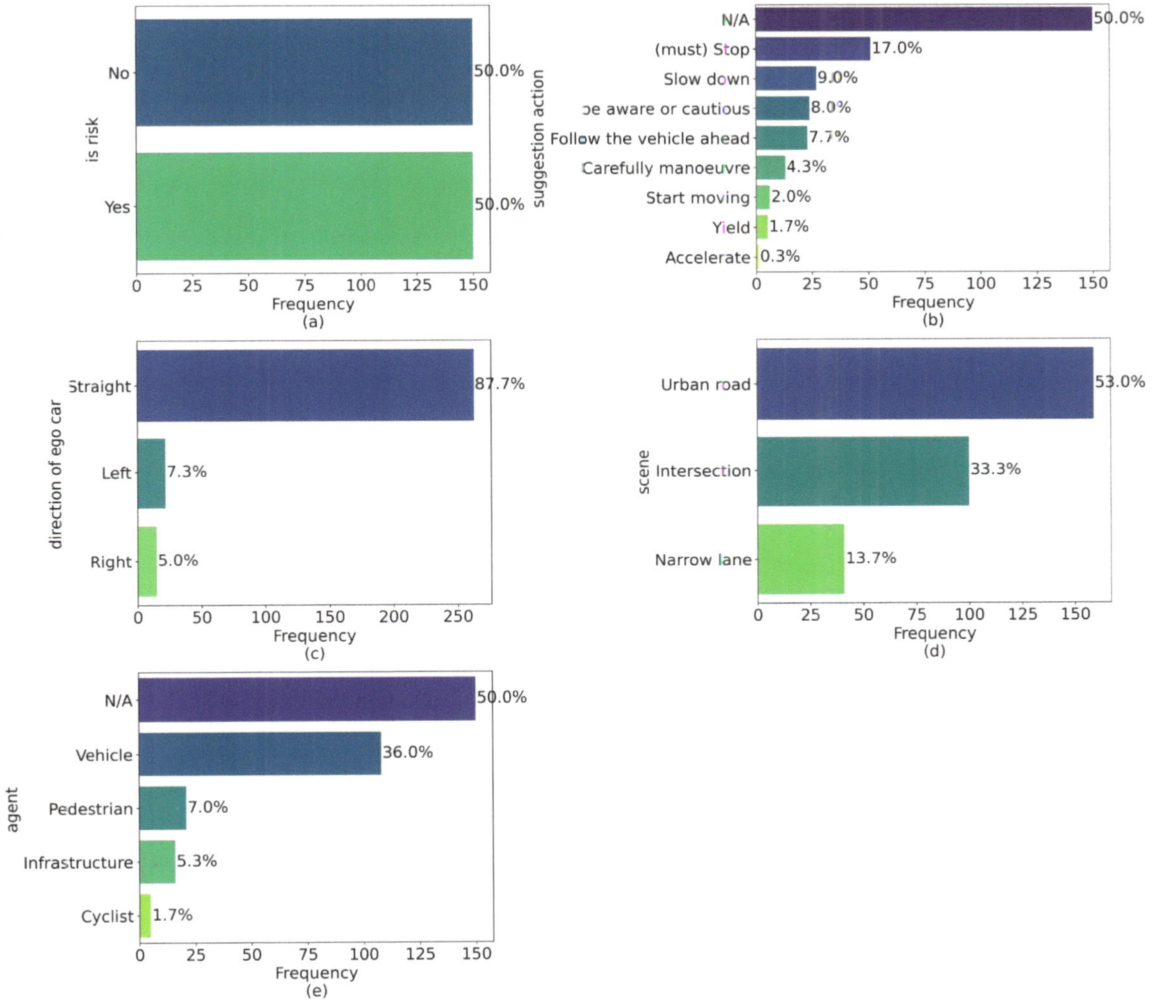

Figure 1. Distribution of QA categories in the DRAMA dataset for traffic safety-critical event detection including (**a**) is risk, (**b**) suggested action, (**c**) direction of ego car, (**d**) scene description, and (**e**) agent type.

3.2. Gemini MLLM

The framework for detecting safety-critical events from driving videos in this study utilizes the Gemini-Pro-Vision 1.5 MLLM [35]. This model was chosen for its advanced capabilities in logical and visual reasoning, particularly in identifying potential hazards across diverse traffic scenarios.

Gemini 1.5 is designed to process and integrate information across multiple modalities—text, images, and video—making it highly effective for tasks that require a deep understanding of both visual and textual data. This is crucial for driving scenarios where the model needs to interpret video frames and respond to natural language prompts simultaneously. One of the most striking features of the Gemini 1.5 model is its ability to handle a context window of up to

1 million tokens, which is significantly larger than most other models. This allows Gemini to process large amounts of data in a single pass.

4. Methodology

We conducted multiple experiments to investigate the capability and logical and visual reasoning power of MLLMs in identifying potential hazards across diverse traffic scenarios. To guide our investigation, we formulated the following research questions (RQs):

- RQ1: How effective are MLLMs at identifying traffic hazards using in-context learning (ICL) with zero-shot and few-shot learning approaches?
- RQ2: Does the number of frames used impact the accuracy of hazard detection in traffic scenarios?
- RQ3: What is the impact of self-ensembling techniques on the reliability and robustness of MLLMs in detecting critical traffic safety events?

The employed methods range from ICL with zero-shot and few-shot learning to varying the number of frames, utilizing textual context alongside visual frames, and implementing self-ensembling techniques. The subsequent sections present the proposed framework and its operational flow for detecting critical traffic safety events, followed by an overview of the different methodologies employed and the implemented prompt design.

4.1. Framework

The framework illustrated in Figure 2 is designed for detecting safety-critical events from driving video extracted from car dash cams, utilizing a multi-stage QA approach with an MLLM, specifically, Gemini-pro-vision 1.5. The process initiates with frame extraction, where the system automatically collects video frames from the ego-vehicle's camera at regular intervals (i.e., every second). These frames are subjected to the hazard detection phase, where the model assesses the scene for potential dangers.

Upon identifying a hazard, the framework employs a tripartite categorization strategy to probe the nature of the threat further, using "What," "Which," and "Where" queries to reveal the object-level details. In the "What" phase, the MLLM classifies the entities detected by the camera, differentiating among agents like pedestrians, vehicles, or infrastructure elements. The "Which" stage involves the MLLM identifying specific features and attributes of these agents, such as pedestrian appearance, vehicle make and model, or infrastructure type, providing vital contextual insights for decision-making.

The final "Where" phase tasks the MLLM with determining the spatial location and distance of the hazard agents from the ego-car, including their position on the road, proximity to the vehicle, and movement direction. This spatial information is critical for the ego-car system to make a safer navigation decision. We tested the model across different dimensions to evaluate model performance in various tasks for each safety-critical event, including identifying risky scenarios, classifying different scenes, determining car direction, classifying agents, and suggesting correct actions.

The framework addresses traffic safety-critical events through a thorough analysis of interactions and road environments in three folds. First, the framework recognizes and evaluates scenarios where the interaction between the ego-vehicle and other road users (i.e., vehicles, pedestrians, and cyclists) or infrastructure may result in safety-critical events. These events include sudden stops, lane changes, or crossing pedestrians that could lead to hazardous situations if not managed correctly. Second, the model identifies and localizes risks within the driving environment, determining the exact location and potential impact of hazardous objects. It assesses the relative position of these objects, such as vehicles cutting into the lane or pedestrians crossing unexpectedly, which are crucial for proactive hazard mitigation. Third, the framework adapts to various road types, such as wide roads, intersections, and narrow streets, each presenting unique challenges. For example, intersections are flagged as particularly high risk due to the convergence of multiple traffic flows, necessitating precise detection and decision-making capabilities from the system.

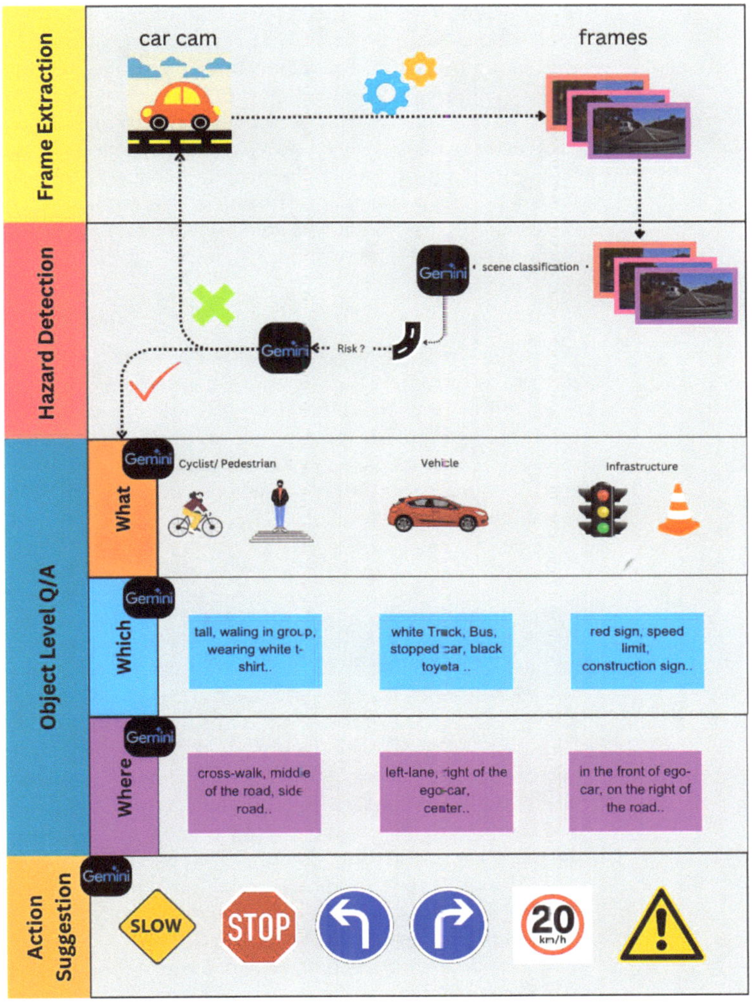

Figure 2. Automated multi-stage hazard detection framework for safety-critical events using MLLMs.

4.2. Analysis Methods

We incorporated different methods to enhance the detection of safety-critical events. These methods were experimented with to enable the system to focus on the most relevant information, thereby optimizing processing speed and accuracy in detecting safety-critical events.

One approach employed was the sliding window frame capture technique, which systematically captures subsets of video frames by defining a window that slides over the video timeline. This window captures a specific range of frames from t_i to t_{i+n}, where t_i represents the initial frame in the window, and n is the number of frames included in each window. The mathematical representation of this method can be expressed in Equation (1). This method allows for the dynamic adjustment of the window size based on the specific requirements of the analysis, which allows the framework to balance data completeness and processing efficiency.

$$Window(t_i) = \{Frame(t_i), Frame(t_{i+1}), \ldots, Frame(t_{i+n-1})\} \tag{1}$$

In-context learning (ICL) was also integrated into the framework to enhance the predictive capabilities of MLLMs by providing them with relevant examples during the inference process. This technique is particularly effective in scenarios where annotated data are rare or the model has to adapt to new situations while in progress. In our framework, we explored two primary settings of in-context learning: zero-shot and few-shot learning. Zero-shot learning allows the model to make predictions about safety-critical events without having seen any prior examples specific to the task. The model relies completely on its base knowledge and reasoning capabilities. This method might be beneficial for its ability to generalize across diverse and unforeseen scenarios. The zero-shot learning process can be mathematically described in Equation (2).

$$Prediction = MLLM(Prompt, Frames) \tag{2}$$

where $Frames$ is a sequence of specific frames and $Prompt$ is a general question or instruction provided to the MLLM to guide the analysis.

On the other hand, few-shot learning allows the model to be exposed to a small number of annotated examples relevant to the task before making predictions. This kind of learning enables the model to quickly adapt and improve its accuracy by leveraging specific patterns and features observed in the examples seen. The few-shot learning process is presented in Equation (3).

$$Prediction = MLLM(Prompt, Examples, Frames) \tag{3}$$

where $Examples$ is the annotated observations used to fine-tune the model's reasoning.

Label-augmented learning (LAL) was another method employed, providing context for the MLLM about how the data were originally annotated. This method helps the MLLM to understand the labeling scheme and the specific characteristics that were considered during the annotation process. By incorporating this context, the model can align its outputs more closely with the annotated data, thereby improving accuracy and consistency.

Image-augmented learning (IAL) involved applying various image augmentation techniques to the images before they were fed into the MLLM for safety-critical event detection QA. These augmented images aim to direct the MLLM to different areas within the language distribution it relies on for generating responses, as illustrated in Figure 3.

By introducing augmented images in the prompt, the MLLM can start at various points within the data distribution, which influences the diversity of local sampling results.

Subsequently, the outcomes from different model sampling processes are aggregated using a top-k voting mechanism to determine the outcome response. This approach aims to aid the model in producing textual responses that more accurately represent the scene under query.

Self-ensemble learning is another strategy used to boost the performance of our framework. It involves generating multiple predictions from the MLLM using slightly different contexts or parameters, such as model temperature, and then combining these predictions to obtain a more robust and accurate result. This approach reduces the likelihood of errors and increases the reliability of hazard detection. The self-ensemble process can be described mathematically Equation (4).

$$Predection = Top - k\left(\{MLLM(Prompt, Frames)\}_{k=1}^{N}\right) \tag{4}$$

where N is the number of individual predictions. The $top - k$ voting mechanism selects the k most frequent predictions among the N generated predictions, which enhances the overall performance by focusing on the most consistently identified outcomes.

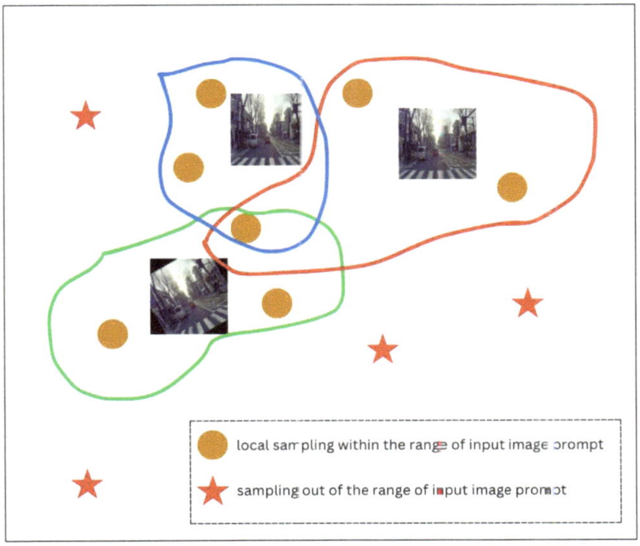

Figure 3. Conceptual 2-D diagram of augmented image prompting. The key idea of using different augmentation for the same scene under investigation is to direct the model to different places in the language distribution, which could help the model with generating more textual representation of the scene when generating a response through local sampling. The different colored areas showed the an example of how image augmentation can be done.

4.3. Prompt Design

The design of the prompt is pivotal in guiding the MLLM to accurately evaluate and respond to safety-critical events in driving scenarios. This prompt was designed to ensure a structured and systematic analysis of the input frames, thereby minimizing the risk of hallucination and enhancing the reliability of the MLLM's outputs, as seen in Figure 4. The structure of the prompt is intended to break down the evaluation process into clear, logical steps, which helps the MLLM to focus on specific aspects of the scene sequentially.

The prompt design benefits the MLLM by ensuring a structured analysis that breaks down the evaluation into discrete steps, allowing the model to focus on one aspect at a time and reducing cognitive overload. By using predefined categories and limited response options, it controls the model's output, minimizing the risk of hallucinations. Each step builds on the previous one, providing a holistic and context-aware understanding of the scene, which improves decision-making. The final step of suggesting actionable recommendations ensures that the analysis is not just descriptive but also prescriptive, offering clear and practical guidance for safe driving.

In summary, the prompt design is tailored to enhance the MLLM's ability to accurately and reliably detect safety-critical events from driving video frames. Its structured approach, combined with controlled output options, significantly mitigates the risk of hallucination, ensuring that the model's responses are both relevant and actionable.

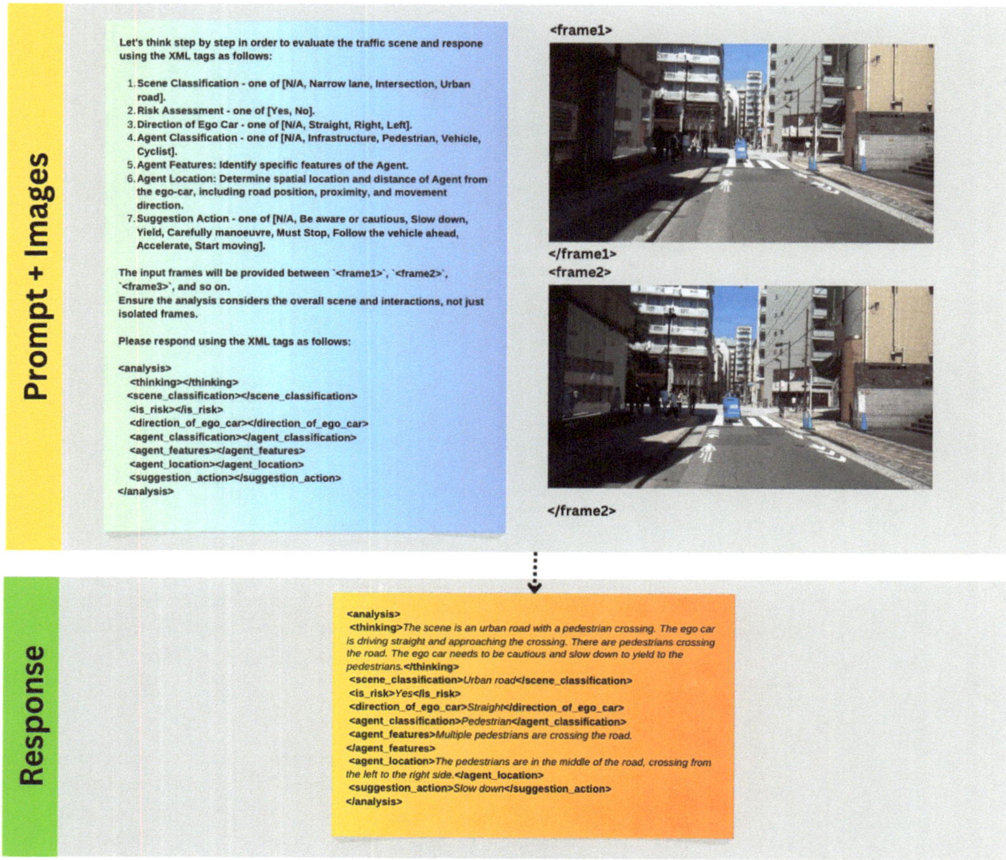

Figure 4. Example of textual prompt with two-frame scene with the corresponding response from Gemini.

5. Results

The work presented in this paper investigated the potential of leveraging the capabilities of MLLMs in analyzing safety-critical event scenarios using multi-modal data integration and dynamic contextual data reduction for guiding the model's output. The prediction illustrated in Figure 5 showcases the proficiency of Gemini-Pro-Vision 1.5 in zero-shot learning scenarios.

To understand the effectiveness of the proposed framework, a series of experiments was carried out utilizing Gemini-Pro-Vision 1.5. The results as shown in Table 1 are analyzed across various frames and in-context learning settings, including zero-shot and few-shot learning, as well as additional strategies like self-ensemble learning and image-augmented learning. We tested the model across different dimensions to evaluate model performance in various tasks for each safety-critical event, including identifying risky scenarios, classifying different scenes, determining car direction, classifying agents, and suggesting correct actions.

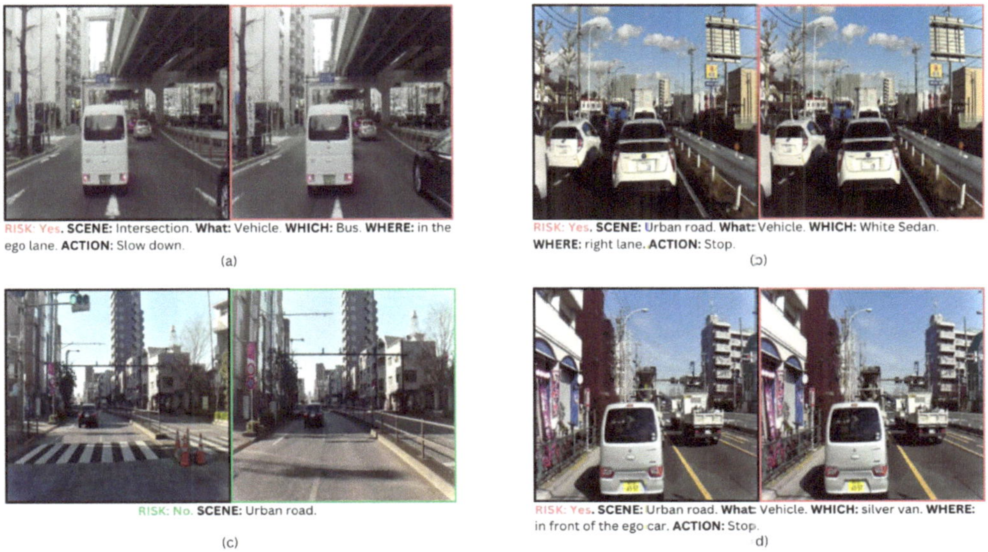

Figure 5. Output from Gemini-Pro-Vision 1.5 analysis with sliding window ($n = 2$). Gemini predicted (**a**), (**b**), and (**d**) as critical-safety events, while (**c**) is not.

Table 1. Comparative performance analysis of QA frameworks across different methods.

Method	Frame [1]/Candidates [2]	Is Risk %	Scene %	Direction of Car %	Agent %	Suggestion Action %	Overall %
Zero-shot	1	68	64	87	56	37	62.4
Zero-shot	2	51	66	89	46	39	58.2
Zero-shot	3	52	66	86	48	38	58
Zero-shot	4	47	72	86	44	35	56.8
Few-shot	1	72	73	87	57	39	65.6
Few-shot	3	76	76	87	59	40	67.6
Few-shot	5	76	76	87	59	40	67.6
Few-shot	7	75	80	89	63	45	70.4
Few-shot	10	79	81	90	64	44	71.6
Self-ensemble	3	69	65	87	56	38	63
Self-ensemble	5	71	67	89	55	39	64.2
Self-ensemble	7	70	67	88	55	39	63.8
Self-ensemble	9	66	66	88	54	38	62.4
LAL	-	68	66	83	48	34	59.8
IAL	-	67	60	80	55	33	59

[1] For zero- and few-shot learning. [2] For self-ensemble learning.

5.1. Zero-Shot Learning Results

Zero-shot learning demonstrated a variable performance profile across different metrics and frame counts, as seen in Figure 6. Initially, a single frame yielded an overall accuracy of 62.4%, with notable performance in detecting the direction of the car (87%) and scene classification (64%). However, as the number of frames increased, overall performance slightly decreased, reaching 56.8% with four frames. The decrease in performance with additional frames suggests potential trade-offs between the depth of context provided and the model's ability to generalize without prior task-specific examples. The impact of frame count on metrics like agent classification and suggested actions also reflects the model's challenge in maintaining accuracy across varying contexts.

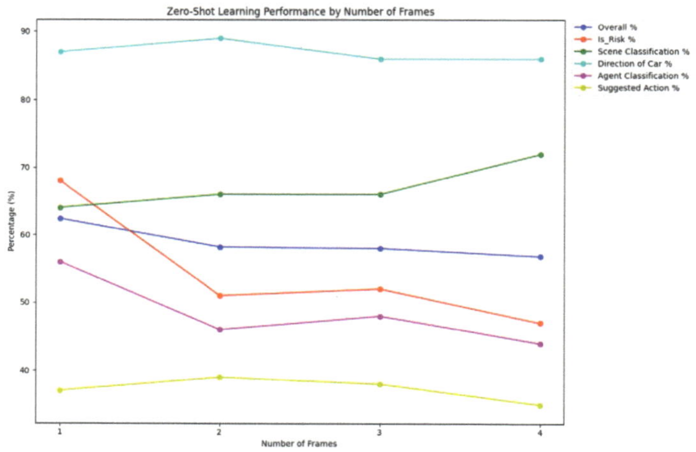

Figure 6. Zero-shot learning performance across different numbers of frames.

5.2. Few-Shot Learning Results

Few-shot learning demonstrated a clear trend of improvement with an increasing number of shots. The performance improved progressively from 1-shot to 10-shot scenarios, with the highest overall percentage (71.6%) achieved with 10 shots. This improvement, as seen in Figure 7, was evident across all metrics, particularly in scene classification and direction of cars, where the highest values were observed with 10-shot learning. The consistency in performance metrics with five-shot and seven-shot suggests that a moderate number of examples already offers substantial benefits.

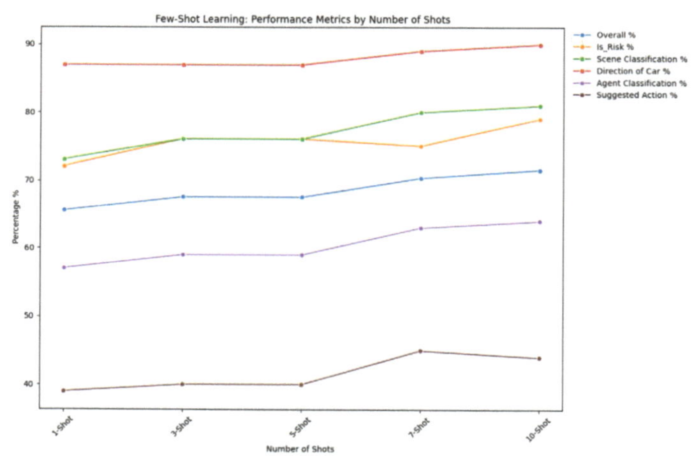

Figure 7. A few-shot learning performance across different numbers of examples.

When comparing zero-shot methods (including one frame, two frames, three frames, and four frames) to few-shot methods, it is evident that few-shot learning consistently outperforms zero-shot learning, as seen in Figure 8. The bar plots highlight that, with zero-shot methods achieving lower percentages across all metrics. For instance, the "is_risk %" metric showed a significant improvement from 68% in the one-frame zero-shot method to 79% in the 10-shot method. Similarly, "scene classification %" saw an increase from 64% with 1 frame to 81% with 10 shots. The comparison underscores the robustness of few-shot learning in improving model performance across various metrics, showcasing its superiority over zero-shot learning approaches.

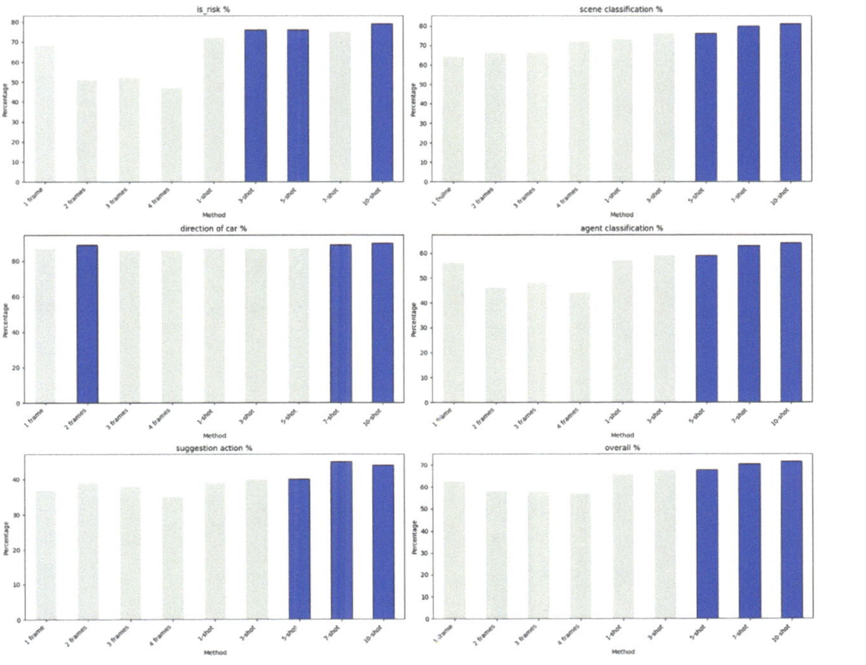

Figure 8. Comparison of zero-shot and few-shot methods across various metrics (top 3 highlighted).

5.3. Self-Ensemble Learning Results

Self-ensemble learning provided a relatively stable performance, with slight improvements as the number of candidates increased. The five-candidate configuration yielded the highest overall percentage (64.2%), as illustrated in Figure 9, showing that aggregating predictions from multiple candidates helped enhance performance. Although the improvements in metrics such as is_risk and scene classification were not drastic, the approach demonstrated increased reliability in hazard detection.

When comparing zero-shot (1-frame) methods to self-ensemble methods (3, 5, 7, 9 candidates), as in Figure 10, it is clear that self-ensemble methods generally offered better performance. The bar plots show that self-ensemble methods frequently surpassed the zero-shot (1-frame) approach. For instance, the "scene classification %" and "direction of car %" metrics showed noticeable improvements with self-ensemble methods. The three-candidate and five-candidate configurations consistently performed well across these metrics.

The use of self-ensemble methods enhanced the overall metric, which meant a more balanced and robust model performance. The highest overall in self-ensemble methods (64.2% with 5 candidates) still outperformed the zero-shot (1-frame) approach (62.4%). This trend is consistent across other metrics, such as "agent classification %" and "suggestion action %," where the self-ensemble methods exhibited a slight edge.

While the improvements in individual metrics like "is_risk %" and "scene classification %" were modest, the aggregated gains across all metrics suggest that self-ensemble learning provides a more reliable and effective approach than the zero-shot (1-frame) method. This highlights the value of leveraging multiple candidate predictions to improve the robustness and accuracy of the model's performance across diverse evaluation criteria.

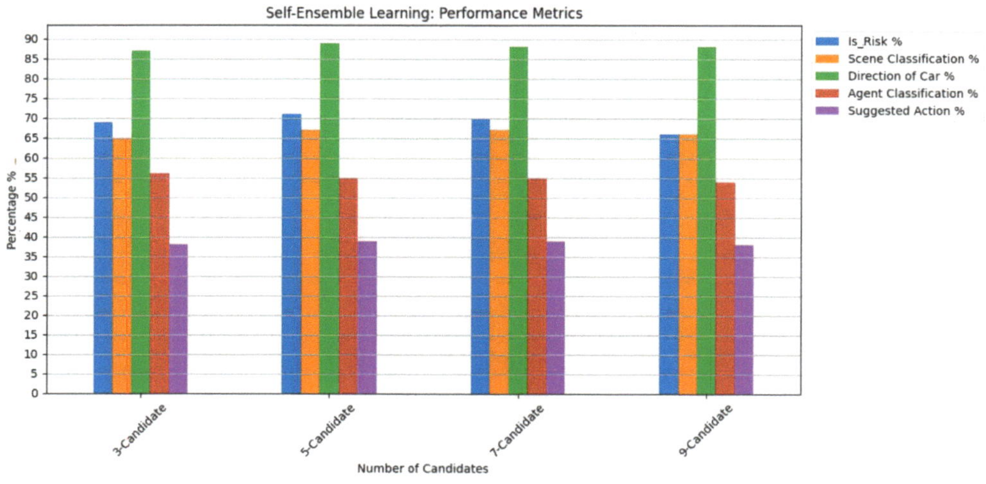

Figure 9. Self-ensemble learning across different number of candidates with top-k voting.

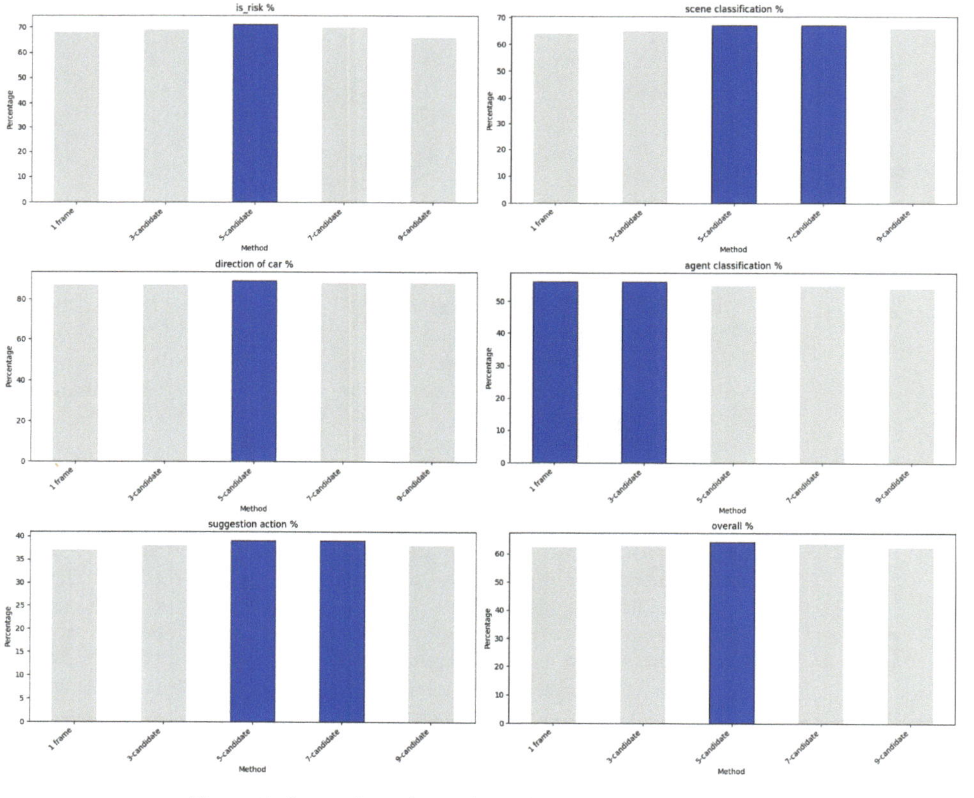

Figure 10. Comparison of zero-shot (1-frame) and self-ensemble methods across various metrics (top bar highlighted).

5.4. Image-Augmented Learning Results

Image-augmented learning with the top-k method resulted in lower overall performance compared to other methods, as seen in Figure 11, with an overall percentage of

59.0%. The image augmentation approach appeared to have a mixed impact, providing a moderate enhancement in some metrics but falling short in overall accuracy and suggested action classification. This suggests that while image augmentation introduces variability, its effect on overall performance needs further refinement and evaluation.

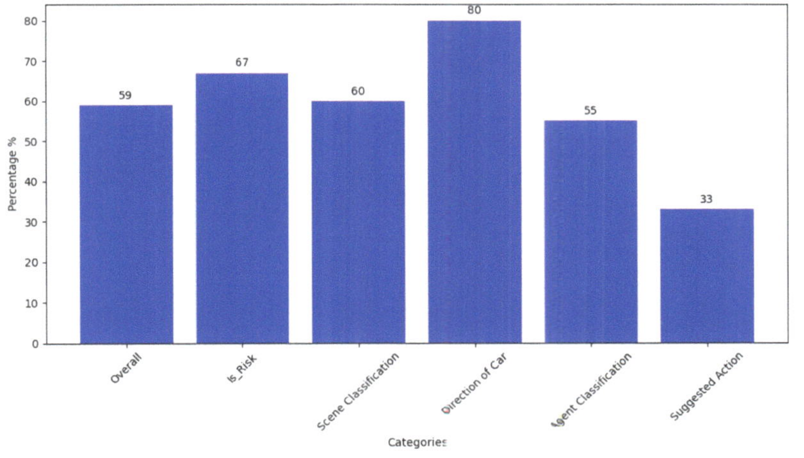

Figure 11. Image-augmented learning performance with top-k voting.

When comparing the zero-shot (1-frame) method to the image-augmented method across different metrics, as in Figure 12, it becomes evident that each approach has its strengths and weaknesses. The zero-shot (1-frame) method achieved a higher "is_risk %" (68%) compared to the image-augmented method (67%). Similarly, in "scene classification %," the zero-shot method performed better (64%) than the image-augmented method (60%).

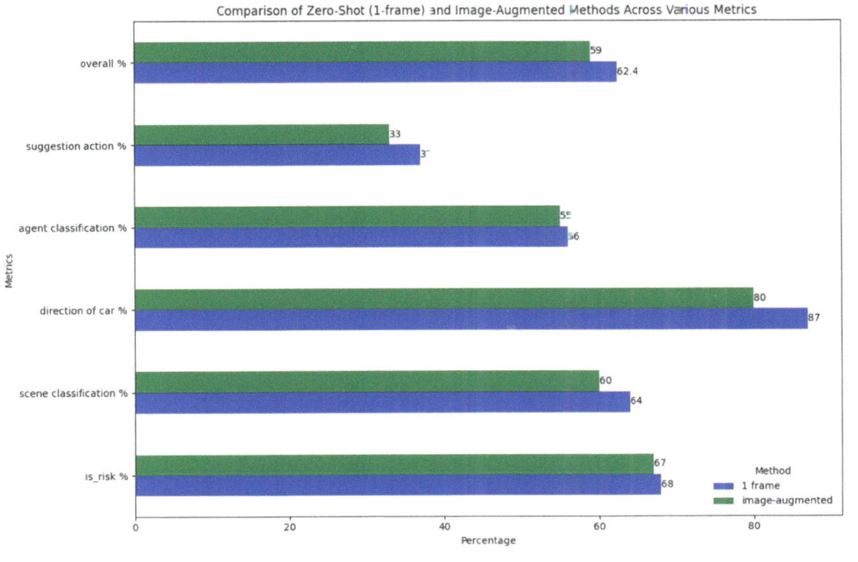

Figure 12. Comparison of zero-shot (1-frame) and image-augmented methods across various metrics.

Few-shot learning, as shown in Figure 13, consistently outperformed other methodologies across most metrics, with a notable improvement in overall performance as the number of shots increased. Zero-shot learning, while useful, showed decreased performance with an increasing number of frames, indicating that it may benefit from being combined with other methods for optimal results. Self-ensemble learning provided a modest increase in performance and stability, particularly in is_risk and scene classification metrics. Image-augmented learning, although innovative, showed less effectiveness compared to the other methods, suggesting that further exploration and refinement of augmentation techniques are necessary. These results highlight the potential of MLLMs in automated traffic safety event detection and offer insights into optimizing their use for various safety-critical scenarios.

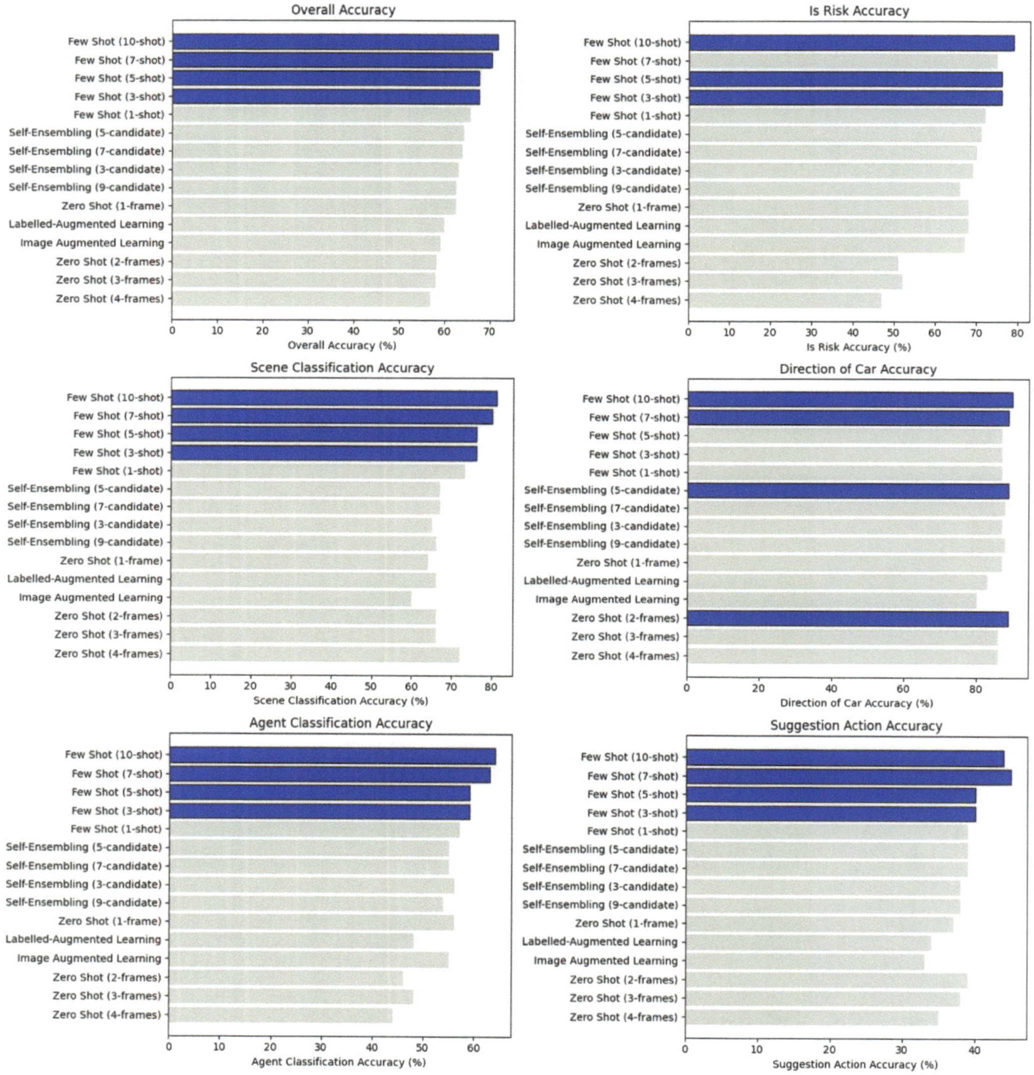

Figure 13. Overall performance comparison across different learning methods. The highlighted bars showed the highest accuracy from each category.

6. Discussion

Across all methods, few-shot learning stands out as the most effective approach for improving overall accuracy and performance in various metrics. The ability to leverage annotated examples allows for significant enhancements in scene classification, direction of the car, and agent classification. This aligns with the general observation that models benefit from specific, contextually relevant examples to improve their predictive capabilities.

Self-ensemble learning provides a robust alternative by stabilizing performance across different candidate predictions, showcasing its strength in minimizing errors and achieving consistent results. This approach is particularly useful in scenarios where model outputs can be uncertain or variable.

Zero-shot learning, while valuable for its generalization capabilities, shows limitations when handling varying frame contexts and specific hazard scenarios. The decrease in performance with additional frames indicates a need for more sophisticated methods to balance context depth with generalization.

Image-augmented learning, although effective in enhancing specific metrics, does not match the overall accuracy of few-shot or self-ensemble learning methods. This suggests that while image augmentation can improve certain aspects of model performance, it may not provide a comprehensive solution for all types of safety-critical event detection.

The results obtained from the proposed framework compared with other baselines that utilize visual-language QA for driving safety are presented in Table 2.

Table 2. Comparative performance analysis of QA frameworks across different baselines.

Method	Dataset	Accuracy
LLaVA-1.5 [36]	VQA-v2 [37]	38.5
Cube-LLM [38]	Talk2Car [30]	38.5
SimpleLLM4AD [39]	DriveLM-nuScenes [32]	66.5
Our Proposed Model	DRAMA	79

The comparative performance analysis in the table highlights the differences in how various visual-language QA frameworks perform in the context of driving safety tasks. Each method was tested on different datasets, and the results reveal significant variations in accuracy, reflecting the strengths and limitations of each approach. LLaVA-1.5 is a model that represents an advanced multimodal approach to integrating a vision encoder with an LLM fine-tuned using the VQA-v2 dataset. The model achieved a moderate accuracy of 38.5% in the driving safety context, suggesting that LLaVA-1.5 is not capable of handling all visual-language tasks well. Similarly, Cube-LLM, which was tested using the Talk2Car dataset, also achieved an accuracy of about 38.5%. The moderate performance of both models indicates that they might struggle with the real-time command interpretation in dynamic and mixed driving environments. In the case of SimpleLLM4AD, when tested on the DriveLM-nuScenes dataset, it achieved a significantly higher accuracy of about 66.5%. This suggests that SimpleLLM4AD is better optimized for driving-related tasks and is more able to understand the challenging scenarios that are closer to real-world driving conditions. However, our proposed model outperformed all the abovementioned methods, with an accuracy of about 79%. Our proposed model appears to be able to understand different driving scenarios and extract the contextual information necessary to excel in visual-language tasks related to driving safety. In addition, our MLLM model can also perform various tasks for each safety-critical event, including identifying risky scenarios, classifying different scenes, determining car direction, classifying agents, and suggesting correct actions, which, to the best of our knowledge, is the first model to do so. This performance shows the importance of domain-specific fine-tuning and training. This allows the model to better understand and respond to the unique challenges presented in autonomous driving.

7. Conclusions

The findings underscore the potential of MLLMs to advance the automated analysis of driving videos for traffic safety. The performance of different learning methods highlights the importance of choosing appropriate techniques based on specific detection requirements and available resources. Few-shot learning offers a promising avenue for improving hazard detection accuracy and adaptability in real-world scenarios. The few-shot model consistently outperformed other learning techniques, achieving the highest overall accuracy (about 67%), "is_risk" accuracy (78%), scene classification accuracy (65%), direction of car accuracy (82%), and agent classification accuracy (68%). This demonstrates its superior effectiveness across various tasks.

Future research should explore the integration of these methodologies to leverage their complementary strengths. Combining few-shot learning with self-ensemble or image-augmented techniques might provide a balanced approach that enhances overall performance while addressing the limitations observed in individual methods. Additionally, fine-tuning MLLMs on task-specific data is a crucial area for future investigation. Fine-tuning could enhance model performance by adapting the pre-trained models to the nuances of safety-critical event detection, thus improving accuracy and reliability. Further exploration into optimizing frame selection and processing strategies could help refine model accuracy and efficiency.

Moreover, we plan to incorporate RAG flow in future work. This approach would enable the model to dynamically retrieve and apply relevant information, such as implicit traffic rules, during inference. Incorporating RAG could further enhance the model's capability in handling complex traffic safety scenarios, making it more robust in detecting and managing safety-critical events. This addition to the future work underscores our commitment to advancing the effectiveness of MLLMs in autonomous driving systems.

Although the DRAMA dataset was constructed in limited geographical locations, our proposed MLLM model was tested using different scenarios, which included a variety of road scenes, such as urban and rural roads, narrow lanes, and intersections. Our proposed framework has demonstrated promising results across these different road conditions, indicating its potential robustness to be scaled and generalized in other geographical locations. Validating the framework's ability to detect safety-critical traffic events in diverse geographical contexts is indeed a crucial step, and we plan to incorporate this in our future research.

This study demonstrates the value of MLLMs in traffic safety applications and provides a foundation for further exploration and development of automated hazard detection systems. The insights gained from this research can guide the design of more effective and reliable safety-critical event detection frameworks in autonomous driving systems.

Author Contributions: Conceptualization, M.A.T., H.I.A. and M.E.; methodology, M.A.T., H.I.A. and M.E.; software, M.A.T.; formal analysis, M.A.T.; investigation, M.A.T.; resources, M.A.T.; data curation, M.A.T.; writing—original draft preparation, M.A.T.; writing—review and editing, H.I.A., M.E., S.G. and A.R.; visualization, M.A.T.; supervision, H.I.A. and M.E.; project administration, S.G. and A.R. All authors have read and agreed to the published version of the manuscript.

Funding: This research received no external funding.

Data Availability Statement: The data used in this study can be requested for research work from this link: https://usa.honda-ri.com/drama (accessed on 22 February 2024).

Conflicts of Interest: The authors declare no conflicts of interest.

References

1. Beiker, S. History and status of automated driving in the united states. In *Road Vehicle Automation*; Springer: Cham, Switzerland, 2014; pp. 61–70.
2. Ashqar, H.I.; Alhadidi, T.I.; Elhenawy, M.; Jaradat, S. Factors affecting crash severity in Roundabouts: A comprehensive analysis in the Jordanian context. *Transp. Eng.* **2024**, *17*, 100261. [CrossRef]
3. Eskandarian, A.; Wu, C.; Sun, C. Research advances and challenges of autonomous and connected ground vehicles. *IEEE Trans. Intell. Transp. Syst.* **2019**, *22*, 683–711. [CrossRef]

4. Butt, F.A.; Chattha, J.N.; Ahmad, J.; Zia, M.U.; Rizwan, M.; Naqvi, I.H. On the integration of enabling wireless technologies and sensor fusion for next-generation connected and autonomous vehicles. *IEEE Access* **2022**, *10*, 14643–14668. [CrossRef]
5. Bathla, G.; Bhadane, K.; Singh, R.K.; Kumar, R.; Aluvalu, R.; Krishnamurthi, R.; Kumar, A.; Thakur, R.N.; Basheer, S. Autonomous vehicles and intelligent automation: Applications, challenges, and opportunities. *Mob. Inf. Syst.* **2022**, *2022*, 7632892. [CrossRef]
6. Macioszek, E. Analysis of the Volume of Passengers and Cargo in Rail and Road Transport in Poland in 2009–2019. *Sci. J. Silesian Univ. Technology. Ser. Transp.* **2021**, *113*, 133–143. [CrossRef]
7. Faisal, A.; Kamruzzaman, M.; Yigitcanlar, T.; Currie, G. Understanding autonomous vehicles. *J. Transp. Land Use* **2019**, *12*, 45–72.
8. Raiaan, M.A.K.; Mukta, M.S.H.; Fatema, K.; Fahad, N.M.; Sakib, S.; Mim, M.M.J.; Ahmad, J.; Ali, M.E.; Azam, S. A Review on Large Language Models: Architectures, Applications, Taxonomies, Open Issues and Challenges. *IEEE Access* **2024**, *12*, 26839–26874. [CrossRef]
9. Bai, Y.; Wu, D.; Liu, Y.; Jia, F.; Mao, W.; Zhang, Z.; Zhao, Y.; Shen, J.; Wei, X.; Wang, T.; et al. Is a 3D-Tokenized LLM the Key to Reliable Autonomous Driving? *arXiv* **2024**, arXiv:2405.18361.
10. Cui, C.; Ma, Y.; Cao, X.; Ye, W.; Wang, Z. Receive, Reason, and React: Drive as You Say, With Large Language Models in Autonomous Vehicles. *IEEE Intell. Transp. Syst. Mag.* **2024**, *16*, 81–94. [CrossRef]
11. Prabhod, K.J. Advanced Techniques in Reinforcement Learning and Deep Learning for Autonomous Vehicle Navigation: Integrating Large Language Models for Real-Time Decision Making. *J. AI-Assist. Sci. Discov.* **2023**, *3*, 1–20.
12. Masri, S.; Ashqar, H.I.; Elhenawy, M. Leveraging Large Language Models (LLMs) for Traffic Management at Urban Intersections: The Case of Mixed Traffic Scenarios. *arXiv* **2024**, arXiv:2408.00948.
13. Zeng, A.; Attarian, M.; Ichter, B.; Choromanski, K.; Wong, A.; Welker, S.; Tombari, F.; Purohit, A.; Ryoo, M.; Sindhwani, V.; et al. Socratic Models: Composing Zero-Shot Multimodal Reasoning with Language. *arXiv* **2022**, arXiv:2204.00598.
14. Malla, S.; Choi, C.; Dwivedi, I.; Choi, J.H.; Li, J. DRAMA: Joint Risk Localization and Captioning in Driving. In Proceedings of the IEEE/CVF Winter Conference on Applications of Computer Vision, Waikoloa, HI, USA, 3–8 January 2022; pp. 1043–1052.
15. Ashqar, H.I.; Alhadidi, T.I.; Elhenawy, M.; Khanfar, N.O. The Use of Multimodal Large Language Models to Detect Objects from Thermal Images: Transportation Applications. *arXiv* **2024**, arXiv:2406.13898.
16. Elhenawy, M.; Abutahoun, A.; Alhadidi, T.I.; Jaber, A.; Ashqar, H.I.; Jaradat, S.; Abdelhay, A.; Glaser, S.; Rakotonirainy, A. Visual Reasoning and Multi-Agent Approach in Multimodal Large Language Models (MLLMs): Solving TSP and mTSP Combinatorial Challenges. *Mach. Learn. Knowl. Extr.* **2024**, *6*, 1894–1920. [CrossRef]
17. Cui, C.; Ma, Y.; Cao, X.; Ye, W.; Zhou, Y.; Liang, K.; Chen, J.; Lu, J.; Yang, Z.; Liao, K.D.; et al. A survey on multimodal large language models for autonomous driving. In Proceedings of the IEEE/CVF Winter Conference on Applications of Computer Vision, Waikoloa, HI, USA, 3–8 January 2024; pp. 958–979.
18. Geiger, A.; Lenz, P.; Stiller, C.; Urtasun, R. Vision meets robotics: The KITTI dataset. *Int. J. Robot. Res.* **2013**, *32*, 1231–1237. [CrossRef]
19. Caesar, H.; Bankiti, V.; Lang, A.H.; Vora, S.; Liong, V.E.; Xu, Q.; Krishnan, A.; Pan, Y.; Baldan, G.; Beijbom, O. nuscenes: A multimodal dataset for autonomous driving. In Proceedings of the IEEE/CVF Conference on Computer Vision and Pattern Recognition, Seattle, WA, USA, 13–19 June 2020; pp. 11621–11631.
20. Chen, L.; Sinavski, O.; Hünermann, J.; Karnsund, A.; Willmott, A.J.; Birch, D.; Maund, D.; Shotton, J. Driving with llms: Fusing object- level vector modality for explainable autonomous driving. *arXiv* **2023**. [CrossRef]
21. Wang, W.; Xie, J.; Hu, C.; Zou, H.; Fan, J.; Tong, W.; Wen, Y.; Wu, S.; Deng, H.; Li, Z.; et al. Drivemlm: Aligning multi-modal large language models with behavioral planning states for autonomous driving. *arXiv* **2023**. [CrossRef]
22. Cui, C.; Ma, Y.; Cao, X.; Ye, W.; Wang, Z. Drive as you speak: Enabling human-like interaction with large language models in autonomous vehicles. In Proceedings of the IEEE/CVF Winter Conference on Applications of Computer Vision, Waikoloa, HI, USA, 2–7 January 2023; pp. 902–909. [CrossRef]
23. Wang, L.; Ren, Y.; Jiang, H.; Cai, P.; Fu, D.; Wang, T.; Cui, Z.; Yu, H.; Wang, X.; Zhou, H.; et al. AccidentGPT: A V2X Environmental Perception Multi-modal Large Model for Accident Analysis and Prevention. In *2024 IEEE Intelligent Vehicles Symposium (IV)*; IEEE: New York, NY, USA, 2024; pp. 472–477.
24. Zhang, A.H.; Vora, S.; Caesar, H.; Zhou, L.; Yang, J.; Beijbom, O. Deep LiDAR-Radar-Visual Fusion for Object Detection in Urban Environments. *Remote Sens.* **2023**, *14*, 12697–12705.
25. Singh, M.; Lopez, E.; Antunes, M.; Gomez-Huélamo, C.; de la Peña, J.; and Bergasa, L.M. Towards LiDAR and RADAR Fusion for Object Detection and Multi-object Tracking in CARLA Simulator. *SpringerLink* **2023**, *14*, 710–715.
26. Lopez, E.; Singh, M.; Gomez-Huélamo, C.; de la Peña, J.; Antunes, M.; Bergasa, L.M. Real-time Object Detection Using LiDAR and Camera Fusion for Autonomous Driving. *Sci. Rep.* **2023**, *14*, 58443–58469.
27. Kim, S.; Julier, S.J.; Uhlmann, J.K. Smartmot: Exploiting the Fusion of HD Maps and Multi-object Tracking for Real-time Scene Understanding in Intelligent Vehicles Applications. *IEEE Intell. Veh. Symp.* **2023**, *14*, 710–715.
28. Hussien, M.M.; Melo, A.N.; Ballardini, A.L.; Maldonado, C.S.; Izquierdo, R.; Sotelo, M.Á. RAG-based Explainable Prediction of Road Users Behaviors for Automated Driving using Knowledge Graphs and Large Language Models. *arXiv* **2024**, arXiv:2405.00449.
29. Liang, J.; Li, Y.; Yin, G.; Xu, L.; Lu, Y.; Feng, J.; Shen, T.; Cai, G. A MAS-based hierarchical architecture for the cooperation control of connected and automated vehicles. *IEEE Trans. Veh. Technol.* **2022**, *72*, 1559–1573. [CrossRef]

30. Deruyttere, T.; Vandenhende, S.; Grujicic, D.; van Gool, L.; Moens, M.-F. Talk2Car: Taking Control of Your Self-Driving Car. In Proceedings of the 2019 Conference on Empirical Methods in Natural Language Processing and the 9th International Joint Conference on Natural Language Processing (EMNLP-IJCNLP), Hong Kong, China, 3–7 November 2019; pp. 2088–2098 [CrossRef]
31. Qian, T.; Chen, J.; Zhuo, L.; Jiao, Y.; Jiang, Y.G. Nuscenes-qa: A multi-modal visual question answering benchmark for autonomous driving scenario. In Proceedings of the AAAI Conference on Artificial Intelligence, Vancouver, BC, Canada, 20–27 February 2024; Volume 38, pp. 4542–4550.
32. Sima, C.; Renz, K.; Chitta, K.; Chen, L.; Zhang, H.; Xie, C.; Luo, P.; Geiger, A.; Li, H. Drivelm: Driving with graph visual question answering. *arXiv* **2023**, arXiv:2312.14150.
33. Wu, D.; Han, W.; Wang, T.; Liu, Y.; Zhang, X.; Shen, J. Language prompt for autonomous driving. *arXiv* **2023**, arXiv:2309.04379.
34. Kim, J.; Rohrbach, A.; Darrell, T.; Canny, J.; Akata, Z. Textual explanations for self-driving vehicles. In Proceedings of the European Conference on Computer Vision (ECCV), Munich, Germany, 8–14 September 2018; pp. 563–578.
35. Team, G.; Anil, R.; Borgeaud, S.; Wu, Y.; Alayrac, J.B.; Yu, J.; Soricut, R.; Schalkwyk, J.; Dai, A.M.; Hauth, A.; et al. Gemini: A family of highly capable multimodal models. *arXiv* **2023**, arXiv:2312.11805.
36. Liu, H.; Li, C.; Li, Y.; Lee, Y.J. Improved baselines with visual instruction tuning. In Proceedings of the IEEE/CVF Conference on Computer Vision and Pattern Recognition, Seattle, WA, USA, 17–21 June 2024; pp. 26296–26306.
37. Goyal, Y.; Khot, T.; Summers-Stay, D.; Batra, D.; Parikh, D. Making the v in vqa matter: Elevating the role of image understanding in visual question answering. In Proceedings of the IEEE Conference on Computer Vision and Pattern Recognition, Honolulu, HI, USA, 21–26 July 2017; pp. 6904–6913.
38. Cho, J.H.; Ivanovic, B.; Cao, Y.; Schmerling, E.; Wang, Y.; Weng, X.; Li, B.; You, Y.; Krähenbühl, P.; Wang, Y.; et al. Language-Image Models with 3D Understanding. *arXiv* **2024**, arXiv:2405.03685.
39. Zheng, P.; Zhao, Y.; Gong, Z.; Zhu, H.; Wu, S. SimpleLLM4AD: An End-to-End Vision-Language Model with Graph Visual Question Answering for Autonomous Driving. *arXiv* **2024**, arXiv:2407.21293.

vehicles

Article

Design, Topology Optimization, Manufacturing and Testing of a Brake Caliper MADE of Scalmalloy® for Formula SAE Race Cars

Luca Vecchiato, Federico Capraro and Giovanni Meneghetti *

Department of Industrial Engineering, University of Padova, Via Venezia 1, 35131 Padova, Italy; luca.vecchiato@unipd.it (L.V.)
* Correspondence: giovanni.meneghetti@unipd.it; Tel.: +39-049-8276751; Fax: +39-049-8276785

Abstract: This paper details the conceptualization, design, topology optimization, manufacturing, and validation of a hydraulic brake caliper for Formula SAE race cars made of Scalmalloy®, an innovative Al-Mg-Sc alloy which was never adopted before to manufacture a brake caliper. A monoblock fixed caliper with opposing pistons was developed, focusing on reducing mass for a fixed braking force. The design process began with a theoretical analysis to establish braking force and pressure requirements, followed by preliminary design and topology optimization. The caliper was then manufactured using laser powder bed fusion (LPBF). Comprehensive experimental validation, including testing with static and rotating brake discs on an inertial dynamometer, confirmed the expected caliper's performance, with the results showing strong alignment with finite element analysis estimations. In particular, strain and displacement measurements showed excellent correlation with numerical estimates, validating the design's accuracy and effectiveness.

Keywords: brake caliper; topology optimization; finite element analysis; additive manufacturing; brake test bench

Citation: Vecchiato, L.; Capraro, F.; Meneghetti, G. Design, Topology Optimization, Manufacturing and Testing of a Brake Caliper MADE of Scalmalloy® for Formula SAE Race Cars. *Vehicles* **2024**, *6*, 1591–1612. https://doi.org/10.3390/vehicles6030075

Academic Editors: Mohammed Chadli and Ralf Stetter

Received: 12 July 2024
Revised: 10 August 2024
Accepted: 2 September 2024
Published: 4 September 2024

1. Introduction

In FSAE race cars [1], braking systems must deliver high deceleration while minimizing unsprung mass. Therefore, the design and optimization of these systems require careful consideration of factors to balance performance, weight, and durability [2–12]. Hydraulic braking systems are commonly employed in race cars, where a hydraulic actuator presses brake pads with lining material against the brake disc, generating the necessary friction to decelerate the vehicle [13]. Among the system's components, the brake caliper plays a critical role in converting hydraulic pressure into mechanical force, enabling effective braking. The caliper must endure high braking forces and thermal loads without compromising structural integrity or performance. Ensuring its durability and reliability under severe racing conditions is crucial, and traditional brake calipers, often designed for mass-produced vehicles, are generally oversized, i.e., heavier and not optimized for the unique requirements of a FSAE race cars.

This study focuses on the design of a monoblock fixed caliper with opposing pistons, specifically tailored for FSAE race cars. In the literature, other authors have addressed similar topics. For example, Phad et al. [9] as well as Ravi Kumar et al. [14] focused on the design of a monoblock floating caliper in Al 7075 for FSAE cars, using traditional manufacturing methods. In their work, the authors provided a detailed description of the role of many parameters and conducted numerical analyses to verify the proposed caliper geometry. Similarly, the authors in [4,8] tackled the design of a monoblock fixed caliper with opposing pistons for FSAE cars in Al 7075, presenting a traditional design with numerical verifications of the realized model. Regarding traditional manufacturing methods, Ugemuge and Das [15] addressed the design, structural analysis using Finite Element method, manufacturing, and testing of a two-piece fixed caliper with two opposing pistons made of Al 7075-T6 alloy. Compared to a monoblock fixed caliper, a two-piece fixed caliper (constructed from two parts connected by bolted joints) offers easier manufacturing by eliminating undercuts and the need for specialized tooling. However, it can result in a heavier component due to the presence of flanges and bolts, and it typically has a lower braking performance due to increased caliper compliance [12,15].

Subsequently, Ugemuge and Das [16] addressed the redesign of the bridge in a four-piston fixed caliper made of Ti6Al4V alloy by using topology optimization and finite element analyses, focusing only on the numerical aspects of the design. Also, Meyer and Barnes [17] used topology optimization to perform a preliminary design of a four-piston monoblock fixed caliper. The objective of their work was to demonstrate that, in some cases, topology optimization can justify the use of high-performance, costly alloys in components produced via additive manufacturing. To do this, Meyer and Barnes [17] considered various alloys suitable for 3D printing, including AlSi10Mg, Ti6Al4V, and Scalmalloy®, an aerospace Al-Mg-Sc alloy with high specific strength. Interestingly, the authors showed that, under identical load cases and constraints in the topology optimization, Scalmalloy® enables the production of a lighter and less expensive caliper compared to AlSi10Mg or Ti6Al4V [17]. However, it is important to note that their investigation was limited to numerical analyses, and they did not proceed with the physical printing or experimental testing of the caliper. Instead, Farias et al. [6] combined topology optimization with additive manufacturing (AM) technologies to develop and fabricate a four-piston monoblock fixed caliper made of AlSi10Mg. Their work aimed to optimize an existing design, resulting in a lighter design with higher stiffness and better thermal dissipation capabilities. Similarly, Tyflopoulos et al. [7] used topology optimization and AM to redesign a commercial caliper, achieving a 3D-printed caliper made of Ti6Al4V that was 40% lighter than the original. Unfortunately, no experimental validation was reported in either [6] or [7]. Finally, Sergent et al. [10,12] presented the complete design of an optimized monoblock fixed caliper, not specifically tailored for FSAE vehicles, and made of an aluminum–lithium alloy. The authors also presented some experimental testing, mainly devoted to verifying the effectiveness of the proposed solution, by using both dial gauge measurement and pressure sensitive paper.

Instead, the present study offers one of the first examples of comprehensive perspective on the process of concept development, design, fabrication, and testing of a brake caliper made of an innovative Al-Mg-Sc (Scalmalloy®) alloy from scratch. This study starts by analyzing the vehicle dynamic to define the design loads of the brake caliper and it continues by discussing the preliminary design phase, which aims at defining the size of the caliper as well as its main geometrical feature. Then, it explores the use of topology optimization and laser powder bed fusion (LPBF) additive manufacturing to design and realize a high-performance, lightweight brake caliper made of an Al-Mg-Sc (Scalmalloy®) alloy, which was never been adopted to manufacture a brake caliper. Eventually, a comprehensive experimental validation is presented based mainly on strain gauge measurements and including testing at the dynamometer with static and rotating brake discs to verify the caliper's performance and the reliability of the finite element simulations.

2. Design of the Braking Caliper

2.1. Maximum Braking Force

To accurately design a brake caliper and the entire braking system, it is crucial to understand the various factors affecting the required braking force. For this purpose, refer to Figure 1, which describes the sub-assembly consisting of a single wheel and the brake disc during the braking phase.

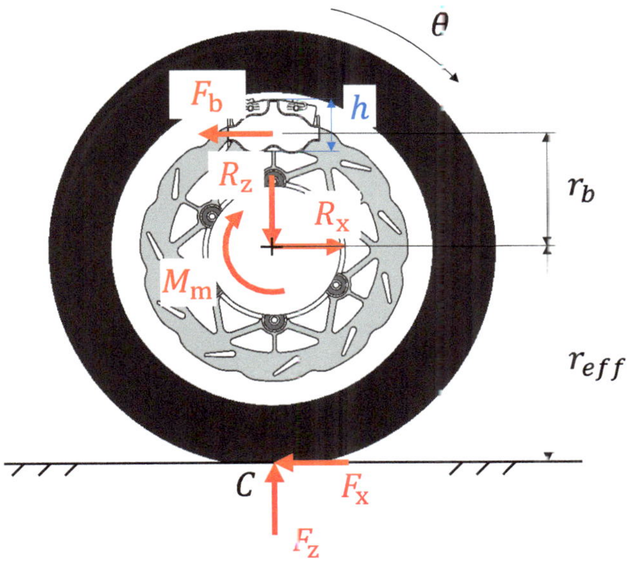

Figure 1. Schematic of a wheel and brake disc assembly.

Regardless of whether the front or rear wheel group is considered, R_x and R_z are the longitudinal and vertical forces exchanged between the wheel and the hub, the axis of which is located at a distance equal to the effective rolling radius r_{eff} from the ground. M_m is the torque transmitted by the drive shaft (positive if it comes from the engine and negative if relevant to engine braking or regenerative braking), while F_b, positioned at a distance r_b from the wheel's axis, is the resultant of the friction forces generated by the contact of the brake pads with the rotating brake disc. Lastly, F_x and F_z are the external forces exchanged between the tires and the ground through contact point C in the longitudinal and vertical directions, respectively. Note that this simplified schematic deliberately neglects aerodynamic loads and the weight force of the wheel assembly, assuming negligible rolling resistance (i.e., the center of pressure C of the contact forces is assumed to have no offset from the rotation axis). According to Figure 1, the spin dynamic of a wheel assembly is described by the following expression:

$$r_{eff}F_x - r_bF_b + M_m = I_w\ddot{\theta} \quad (1)$$

where I_w is the moment of inertia of the considered system. Rearranging Equation (1), the following expression is obtained:

$$F_b = \frac{1}{r_b}\left[r_{eff}F_x + M_m - I_w\ddot{\theta}\right] \quad (2)$$

Equation (2) provides several interesting considerations and insights regarding the design of a brake caliper. When designing a brake system for a racing car, the objective is to maximize performance by creating a lightweight system capable of generating very high decelerations. This approach aims to delay the initiation of braking as much as possible, thereby reducing the duration of the braking phase. In general, achieving high deceleration $\ddot{\theta}$ (which is negative during braking) necessitates a substantial braking force F_b. Typically, this results in relatively large and heavy brake calipers and associated components. However, this requirement conflicts with the well-established necessity to minimize unsprung mass, which includes the mass of the brake caliper. Therefore, it is essential to find a good compromise to reduce the braking force F_b and, at the same time, to maximize the deceleration $\ddot{\theta}$ as much as possible.

In light of this, it is easier to interpret the role of each term in Equation (2) and to understand why the inertial component $I_w\ddot{\theta}$, although usually negligible, has a detrimental effect (remember that $\ddot{\theta} < 0$). Secondly, Equation (2) suggests that it is preferable to maximize the average radius r_b of the brake disc. However, increasing r_b undesirably raises both the moment of inertia and the (unsprung) mass of the disc. Similarly, a positive driving torque ($M_m > 0$), although typically absent during braking, would require a higher braking force for the same angular deceleration $\ddot{\theta}$. Conversely, a negative torque $M_m < 0$, either generated by the engine brake or through regenerative braking, is obviously advantageous for the lightweight design of the brake caliper. Finally, F_x, which describes the resultant of friction forces between the tyre and the asphalt, reaches its maximum value at the point of incipient wheel slip, immediately prior to wheel lock-up. At this instant, its value equals the static friction force, expressed as follows:

$$\max\{\mathrm{F_x}\} = \mu_x \max\{F_z\} \tag{3}$$

where μ_x is the static longitudinal friction coefficient between the tires and (dry) asphalt (in this work, $\mu_x \approx 1.7 - 1.8$). To estimate the value of F_z, it is useful to refer to the diagram in Figure 2, where the longitudinal model of a race car is shown assuming it is symmetric and running on a perfectly flat track under constant acceleration/braking, inducing symmetric loads in the pitch plane (perfectly straight accelerations and braking).

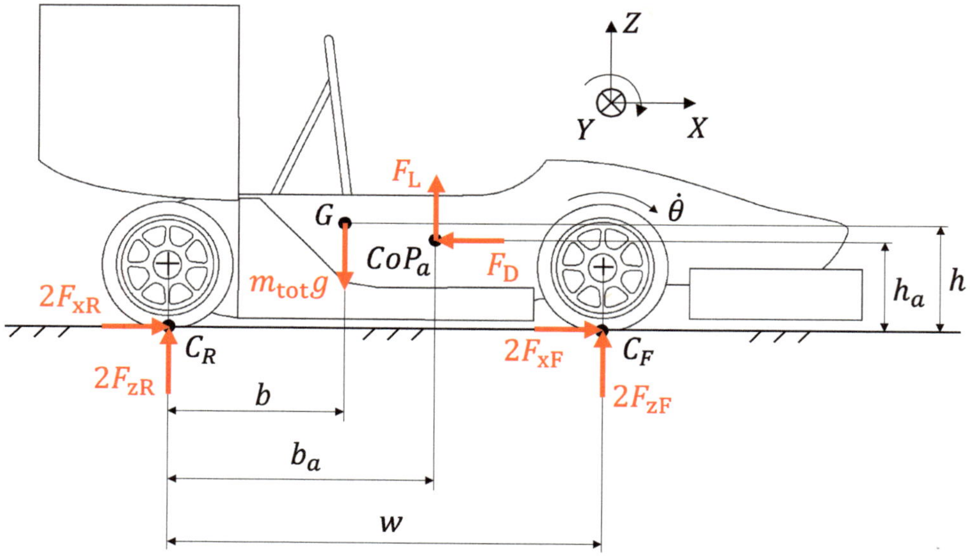

Figure 2. Schematic of a race car in the pitch plane.

In this model, four characteristics points are used to describe the system, i.e., the global center of mass G, the contact points of the rear and front wheels C_R and C_F, respectively, and the aerodynamic center of pressure CoP_a. A concentrated mass m_{tot} is applied to point G to schematize both the driver and the car inertia, while the resultant aerodynamic lift F_L and drag forces F_D are applied at point CoP_a. With reference to this schematic, the following quasi-steady-state equations of motion can be written as follows [18]:

$$\begin{cases} 2F_{xR} + 2F_{xF} - F_D = m_{tot}\ddot{x}_G \\ 2F_{zR} + 2F_{zF} + F_L - m_{tot}g = 0 \\ b2F_{zR} - (w-b)2F_{zF} - h(2F_{xR} + 2F_{xF}) + (h-h_a)F_D - (b_a - b)F_L = I_F\ddot{\theta}_F + I_R\ddot{\theta}_R \pm I_e\ddot{\theta}_e \end{cases} \quad (4)$$

where $\ddot{x}_G$ is the longitudinal acceleration of the vehicle, and I_F and I_R are the moments of inertia of the front and the rear axles, respectively, while I_e represents that of all rotating parts belonging to the engine. Finally, $\ddot{\theta}_F$, $\ddot{\theta}_R$ e $\ddot{\theta}_e$ are their corresponding angular accelerations. Note that F_x and F_z for both front and rear axles are multiplied by a factor of 2 because of the symmetry of the problem.

Then, rearranging these equations, an explicit formulation of both F_{zF} and F_{zR} can be defined as follows:

$$\begin{aligned} F_{zF} &= \tfrac{1}{2}\left\{ m_{tot}g\tfrac{b}{w} - m_{tot}\ddot{x}_G\tfrac{h}{w} - F_D\tfrac{h_a}{w} - F_L\tfrac{b_a}{w} - \tfrac{I_F}{w}\ddot{\theta}_F - \tfrac{I_R}{w}\ddot{\theta}_R \mp \tfrac{I_e}{w}\ddot{\theta}_e \right\} \\ F_{zR} &= \tfrac{1}{2}\left\{ m_{tot}g\tfrac{w-b}{w} + m_{tot}\ddot{x}_G\tfrac{h}{w} + F_D\tfrac{h_a}{w} - F_L\tfrac{w-b_a}{w} + \tfrac{I_F}{w}\ddot{\theta}_F + \tfrac{I_R}{w}\ddot{\theta}_R \pm \tfrac{I_e}{w}\ddot{\theta}_e \right\} \end{aligned} \quad (5)$$

where the first term is related to the static distribution of mass, while the others are responsible for load transfer, them being non-zero only when the vehicle is in motion. More specifically, the second is the longitudinal inertial terms, the third and the fourth are aerodynamic contributions, and the last three terms are related to the inertia of the rotating parts. It should be noted that the two aerodynamic terms contribute differently to the load transfer. Indeed, the drag component F_D is always positive ($F_D > 0$), while the lift component F_L is generally negative ($F_D < 0$), as it actually is a downforce. Equation (5) suggests the highest vertical load during the brake, i.e., the highest braking force F_b, is that occurring at the front wheel (Equation (5)). Then, considering the front wheel and by ignoring for simplicity the contribution due to aerodynamic and to the inertia of the rotating parts, the relationship is significantly simplified and can be written as follows:

$$\max\{F_{zF}\} = \frac{1}{2}\left[m_{tot}g\frac{b}{w} - m_{tot}\min\{\ddot{x}_G\}\frac{h}{w}\right] \quad (6)$$

Assuming that, during the braking phase, the engine torque is zero and that the inertial component of the wheel assembly is negligible compared to the other forces, the following expression can be found by inserting Equation (6) into Equation (3) and then into Equation (2):

$$\max\{F_{b,F}\} = \frac{1}{2}\frac{r_{eff}}{r_b}\,\mu_x\, m_{tot}\, g\frac{b}{w}\left(1 - \frac{h}{b}\min\left\{\frac{\ddot{x}_G}{g}\right\}\right) \quad (7)$$

Equation (7) allows for an estimation of the maximum braking force required by the braking system and represents one of the key parameters in the sizing and design of the brake caliper. In the case under consideration, the parameters of the FSAE vehicle permit an estimation of a maximum braking force of approximately $\max\{F_{b,F}\} \approx 5200$ N, corresponding to a braking torque of approximately $\max\{M_b\} = r_b\max\{F_{b,F}\} \approx 570$ Nm with $r_b = 109$ mm.

2.2. Preliminary Design

Given the maximum braking force, a detailed analysis of the brake caliper design can be undertaken, beginning with the delineation of its geometric details. This study is limited to considering the case of a hydraulic brake system, particularly a monoblock fixed caliper with one or more pairs of opposing pistons acting symmetrically with respect to the brake disc plane. Such a configuration is known for its advantages, including more uniform wear of the inner and outer pads, and typically requiring fewer parts compared to a floating-caliper design [13]. A typical example of a fixed caliper design is illustrated in Figure 3, which considers the case of a six-piston caliper.

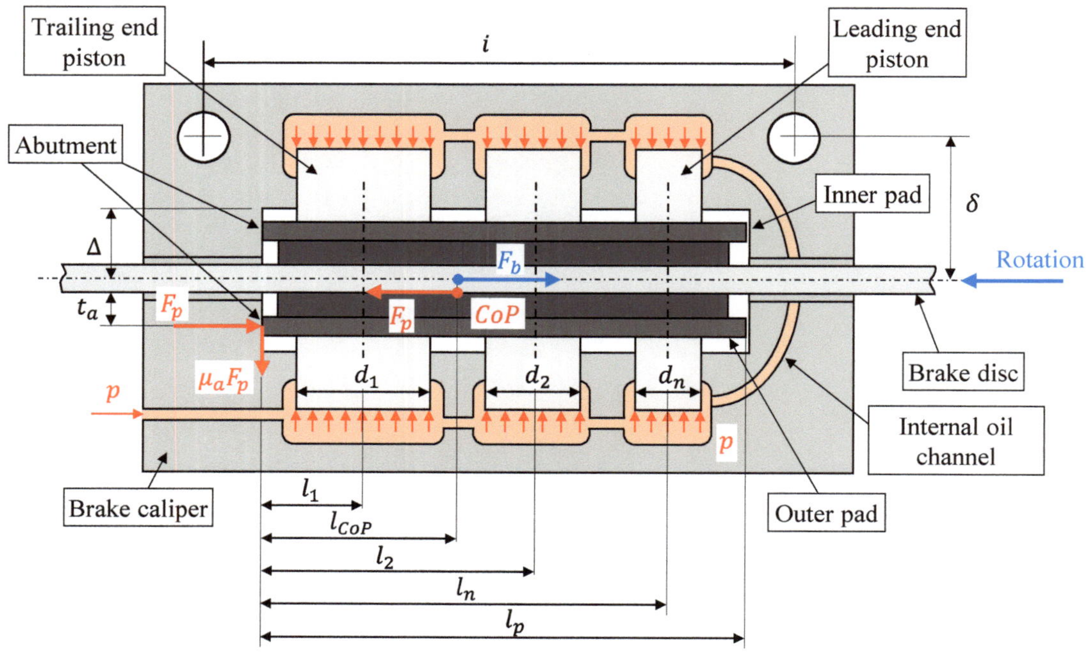

Figure 3. Schematic of a monoblock six-piston fixed caliper.

Temporarily introducing the assumption that the forces generated by friction in the abutment $\mu_a F_p$ and the forces generated by the seals of each piston are negligible, the friction force generated at the interface between the disc and one pad is given by (Figure 3):

$$F_p = \mu \, p \, \frac{1}{2} \sum_{i=1}^{n} \frac{\pi}{4} d_i^2 \tag{8}$$

where μ is the friction coefficient between the brake disc and the lining material of the brake pad, p is the oil pressure in the hydraulic brake line, n is the total number of pistons, and d_i their corresponding diameters. It can be observed that the sizing of the caliper must be performed to ensure a braking capacity equal to or greater than the design value. This requirement translates into the following relation, obtained by comparing Equation (8) with Equation (9):

$$2 \cdot \max\{F_p\} = \mu \, p_{max} \sum_{i=1}^{n} \frac{\pi}{4} d_i^2 \geq \frac{1}{2} \frac{r_{eff}}{r_b} \, \mu_x \, m_{tot} \, g \frac{b}{w} \left(1 - \frac{h}{b} \min\left\{\frac{\ddot{x}_G}{g}\right\}\right) \tag{9}$$

where the pressure p has been replaced with its maximum value p_{max}.

At this stage, the variables on the right-hand side of Equation (9) are completely defined, whereas those on the left-hand side need to be determined based on some design choices. The piston diameters d_i are somewhat constrained by the dimensions of the brake pad (or the backing plate) used, and larger diameters result in greater braking force, but also in an increase in the overall size and weight of the brake caliper. Typical values for FSAE applications range between 25 mm and 35 mm. Similar conclusions can be drawn for the number of pistons n, specifically, a higher number of pistons typically ensures greater force but also increases the overall size of the caliper. Additionally, the number of pistons depends on the type of pad used and can significantly affect the wear of the lining material [19].

Regarding the choice of brake pads, it is generally more practical and economical to use a commercial pad or lining material with known performance characteristics rather than to develop a new pad material tailored to a specific caliper design [13]. Subsequently, the selection among various commercial alternatives can be based on considerations related to cost, availability, and, most importantly, performance, usually in terms of friction coefficients (μ in Equation (9)), wear resistance, and dimensions/geometry. Finally, the choice of maximum pressure p_{max} must be made to ensure compliance with the inequality in Equation (9). Noteworthily, the oil pressure p_{max} typically represents the critical design load for the brake caliper since it affects the size and weight of the brake caliper more than F_b. Consequently, it is preferable to design the system to minimize the pressure if it remains sufficiently high to guarantee the desired braking force F_b.

With these characteristics defined, the overall dimensions of the caliper must be determined by selecting the size and position of the pistons. An efficient method for making this decision is proposed by Sergent [4]: he suggests using a simple model that assumes an in-plane 2D schematization of the brake caliper (as shown in Figure 3, where the curvature of the brake disc at its mean radius has been rectified) to evaluate the position of the center of pressure (*CoP*) of the contact forces between the disc and the pad and to optimize the position and size of the pistons to ensure uniform pad wear. Specifically, the caliper should be designed to ensure that the center of pressure is as close as possible to the centroid of the pad to avoid tapered wear and to ensure uniform wear across the entire surface of the pad [13].

According to this model, in the static case, where a braking force is applied with the vehicle stopped, the friction forces in the abutment $\mu_a F_p$ are negligible, and the position of the *CoP* on the pad coincides with the centroid of the pistons, calculated as the area-weighted average of the pistons as follows:

$$l_{CoP} = \frac{\sum_{i=1}^{n} d_i^2 l_i}{\sum_{i=1}^{n} d_i^2} \tag{10}$$

where l_i and l_{CoP} represent the positions of the pistons and the *CoP* with respect to the abutment, respectively. Uniform wear is guaranteed if the following relation is satisfied [12,13]:

$$l_{CoP} = \frac{l_p}{2} \tag{11}$$

It must be emphasized that this simplified model assumes, among other hypotheses, an infinitely stiff pad and caliper. Therefore, when the design of the caliper is finished and its final geometry known, the real location of the *CoP* is required to be checked through more accurate analyses, such as finite element analysis. Moreover, note that Equations (8) and (10) are valid only when the brake disc is stopped. Instead, when the pads are squeezed against the rotating brake disc, it drags them until they strike against the abutment, where a normal force F_p and a friction force $\mu_a F_p$ arise, the latter being no longer negligible due to the high value of the contact force F_p. Then, the friction force, located at a distance t_a from the pad–disc friction surface, opposes the motion of the pistons during the application of the load (Figure 3) and displaces the *CoP* compared to the static case (Equation (10)). Accordingly, referring to a single pad, the new friction force is less than that estimated by Equation (8) and can be calculated as follows:

$$F_p = \mu \left[p \, \frac{1}{2} \left(\sum_{i=1}^{n} \frac{\pi}{4} d_i^2 \right) - \mu_a F_p \right] \tag{12}$$

which, properly rearranged, becomes:

$$F_p = \frac{\mu}{1 + \mu \mu_a} p \left[\frac{1}{2} \left(\sum_{i=1}^{n} \frac{\pi}{4} d_i^2 \right) \right] \tag{13}$$

Instead, the l_{CoP} updated for the case of the rotating disc can be estimated by imposing the equilibrium of moments about the *CoP* itself:

$$\frac{1}{2}\left(\sum_{i=1}^{n}(l_i - l_{CoP})\cdot\frac{\pi}{4}d_i^2\right)p + t_a\cdot F_p + l_{CoP}\cdot\mu_a\cdot F_p = 0 \tag{14}$$

Finally, l_{CoP} is obtained by rearranging Equation (14), and by exploiting Equation (13):

$$l_{CoP} = t_a\mu + (1+\mu\mu_a)\frac{\sum_{i=1}^{n} l_i\cdot d_i^2}{\sum_{i=1}^{n} d_i^2} \tag{15}$$

Equation (15) allows for designing the pistons' diameters and locations since it points out explicitly the role of the main parameters affecting the position of the CoP. In particular, it must be noted that most of them (t_a, μ, μ_a, l_p, and, indirectly, d_i) depend on the shape and the material of the brake pad; so, the choice of the position and size of the pistons is specific for the considered brake pads. Moreover, other constraints must be considered, for example, technological feasibility and the need for a certain clearance between each piston to allow for the location of the seals. Eventually, a slightly larger diameter is typically chosen on the trail side compared to that on the leading side to compensate for the effects of the friction at the abutment ($l_{CoP}\cdot\mu_a\cdot F_p$ in Equation (14)), the lever arm of the normal force ($t_a\cdot F_p$ in Equation (14)), and the elastic deformation of the caliper [10,12]. A more detailed discussion on this topic can be found in [12,13].

In the present investigation, a fixed caliper with four opposite pistons ($n = 4$) was chosen, this configuration being a good compromise for minimum size of the caliper and optimal pressure distribution at the brake pad. The adopted commercial brake pads allow for a friction coefficient $\mu \approx 0.7$–0.8 at the interface with the brake disc and an estimated value of $\mu_a \approx 0.3$ at the abutment–backing plate interface. Furthermore, a new brake pad was considered and $t_a \approx 8$ mm was used in the calculations, this value being dependent on the geometries of the caliper and pad, as well as on pad's wear. Noteworthily, assuming that the value of l_{CoP} coincides with the midpoint of the pad, $l_{CoP} = \frac{l_p}{2} = 42$ mm allows us to observe that the terms in Equation (14) relevant to the friction at the abutment ($l_{CoP}\cdot\mu_a\cdot F_p$) and the lever arm of the normal force ($t_a\cdot F_p$) assume comparable values. Specifically, in this case, the product $l_{CoP}\cdot\mu_a = 12.6$ mm is greater than $t_a = 8$ mm.

At this point, it was possible to solve Equation (15) for the unknowns l_i and d_i by imposing that the *CoP* coincides with the midpoint of the pad (Equation (11)) to ensure uniform wear. Moreover, the diameters of the pistons d_i had to be chosen to guarantee the required maximum braking force F_b according to Equation (9). In doing so, it was also fundamental to consider only physically compatible solutions (e.g., solutions that do not locate a piston outside the pad or result in piston overlap) and to account for the necessary clearance between pistons to accommodate their seals.

In this work, commercial values for pistons' dimensions and their seals were used, resulting in diameters of $d_2 = 26$ mm and $d_1 = 30$ mm for the leading and trailing pistons, respectively. Concerning their position, the results are reported in Figure 4, which also shows the corresponding *CoP* for both the configurations with the static and rotating discs evaluated according to Equation (10) and Equation (15), respectively. The results show that the static *CoP* (empty red marker in Figure 4) is slightly to the left (toward the trail side) of the geometric center of the pad, while the *CoP* with a rotating brake disc (filled red marker in Figure 4) shifts to the right (toward the leading side), similarly to what was found in [12]. In this particular case, to align the *CoP* more closely to the center of the pad, it would have been necessary to reduce the friction coefficient at the abutment μ_a, the height of the pad's contact point on the abutment t_a, the clearance between the pistons, or to choose a smaller diameter on the leading side and a larger one on the trailing side. In the present investigation, the friction coefficient μ_a and the height t_a were assumed to be fixed, their values being mainly dependent on the chosen brake pad. On the other hand, the clearance between the pistons had already been minimized, and further reduction would have fallen below the required limit to accommodate the hydraulic seals of the pistons (details of which will be presented later). Finally, a larger diameter d_2 (on the trailing side) would have improved the *CoP* position but reduced the maximum braking force (Equation (9)). Conversely, a larger diameter d_1 (on the leading side) would have moved the *CoP* closer to the piston's axis, but it would have shifted the *CoP* on the leading side since the distance l_1 was constrained by the abutment geometry, the piston's clearance, and the piston's diameter d_1.

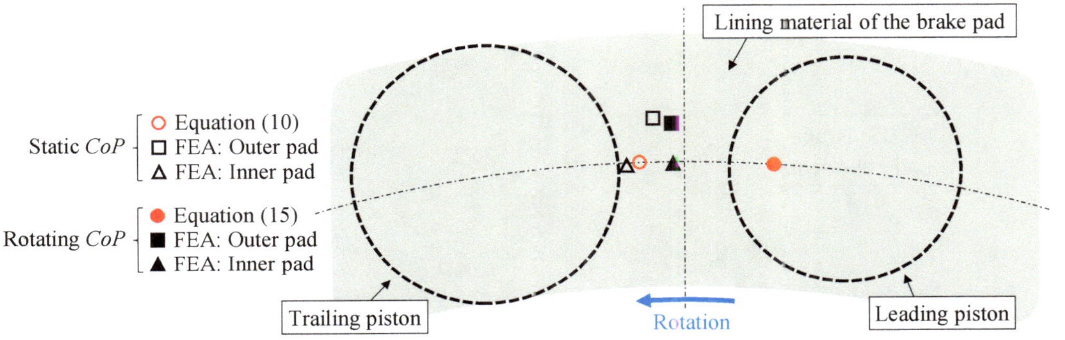

Figure 4. Location of CoP on the brake pad with static and rotating brake discs: comparison between values calculated using Equation (10), Equation (15), and finite element analyses.

Finally, it is also important to recall that Equation (15) is an approximate expression, highly useful in the preliminary design phase, but requiring subsequent verification through more detailed analyses, such as finite element analysis, which can account for the caliper's compliance and the contact between various components. Accordingly, Figure 4 also shows the positions of the *CoP* with static and rotating discs evaluated by the finite element analyses (black markers), which will be presented and discussed in the following paragraphs.

2.3. Oil Channels and Hydraulic Seals

In this study, the internal channels for oil distribution were designed to resemble many commercial solutions (Figure 5a). A diameter of 3.5 mm was chosen for the channels, which may appear relatively large. This size is typically selected for technological reasons related to drilling difficulties with very thin bits but ensures that the channels can be created using additive manufacturing without the need for supports, thereby eliminating the need for post-processing with machining tools. The position of the oil inlet was fixed near the smallest piston (leading piston, see Figure 5a), due to space constraints, and on the side of the mounting bolt because that region of the caliper is less stressed. Finally, a single bleeder was placed at the highest possible point, which depends on how the caliper is installed on the vehicle. In this specific case, it was positioned in the channel on the bridge, which connects the inner pistons to the outer pistons (Figure 5a).

The design of the pistons' seals (Figure 5b) is far more complex. Hydraulic sealing is usually achieved using a square-section O-ring protected by a dust seal, which helps to preserve and extend the life of the seal, especially in commercial solutions for standard road application. In FSAE applications, this dust seal is not essential, and its removal significantly reduces the space required for the seals as well as the size of the pistons and the caliper. The geometry of the seats for the hydraulic seals of the pistons is trapezoidal and is characterized, by design, by an outer diameter smaller than the outer diameter of the square-section O-ring it contains. This arrangement ensures that the O-ring is squeezed and compressed against the piston (Figure 5b), providing hydraulic sealing and, if properly designed, a small retraction force which ensures the pads do not remain in contact with the disc during brake release. In this study, the same seal seats' geometry (Figure 5c) was assumed for both diameters (26 mm and 30 mm) to ensure easier design and manufacturing. Since the focus of this work is not on the detailed sizing of the seals, these details are omitted. Further details can be found in the literature [20,21].

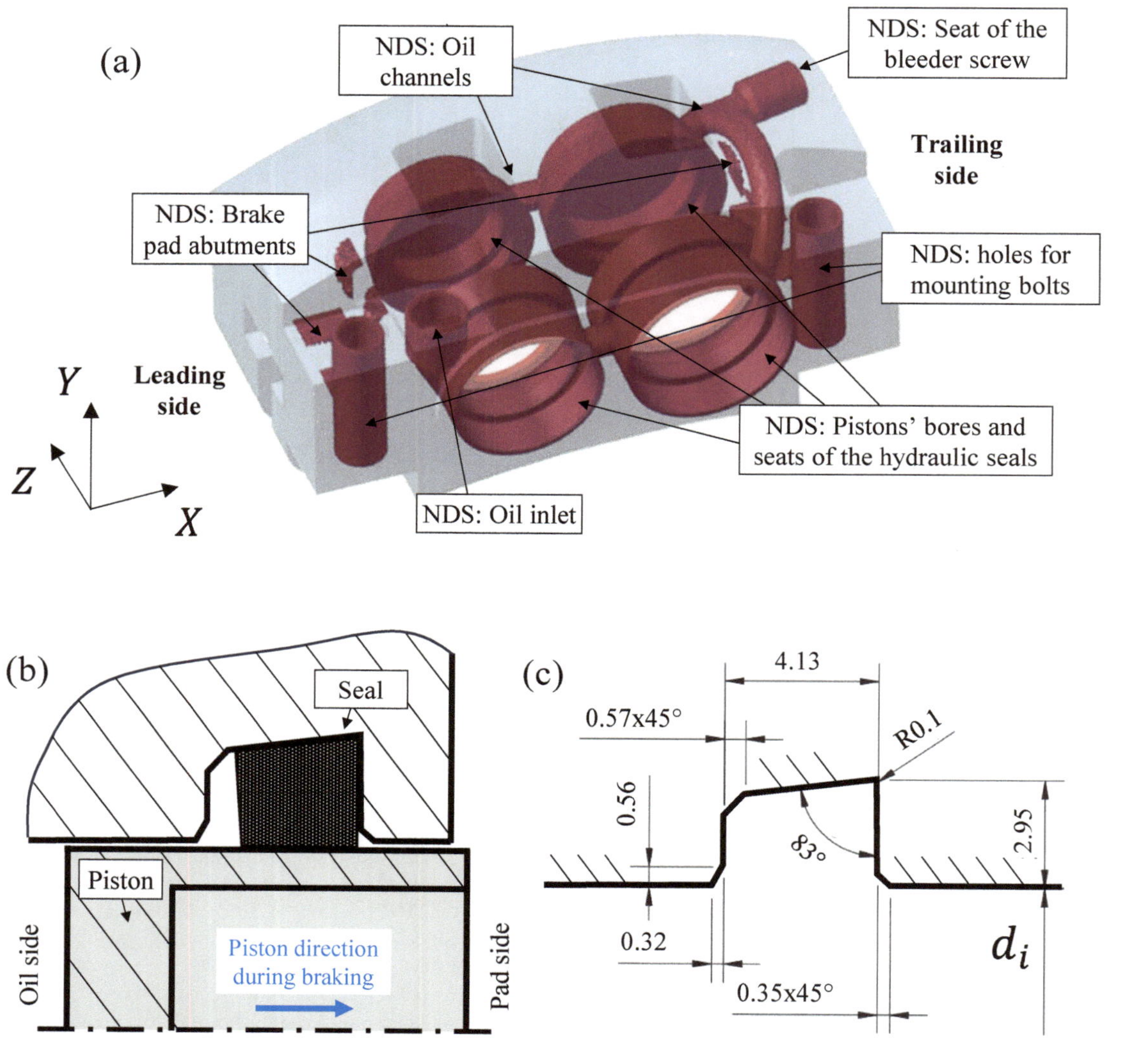

Figure 5. Geometry of the brake caliper: (**a**) main body with all internal features highlighted, (**b**) schematic of the working principle of the pistons' seal and (**c**) geometry and dimensions of their seats used in the present investigation (dimensions are in mm). Grey and red regions in a) represent the Design Space and Non-Design Space (NDS), respectively, adopted in the topology optimization.

2.4. Topology Optimization

The topology optimization required defining the Design Space and Non-Design Space, both depending on the maximum overall dimensions of the caliper summarized by the parameters i, Δ, and δ in Figure 3 and by the height h in Figure 1. In general, small values for i, Δ, and δ are preferred given that minimizing the volume of the caliper helps to reduce the overall mass and to reach a higher value of caliper's stiffness. In more detail, Δ is primarily dependent on the thicknesses of the brake pad and disc. These factors also influence the choice of offset δ (Figure 3), whose value must consider enough space for the pistons' stroke and for the oil chamber behind them. Similar considerations apply to the distance i (Figure 3). Lastly, the height of the caliper (h in Figure 1) must be minimized and chosen according to the space available between the wheel rim and the hub carrier.

The Non-Design Space includes regions of the caliper such as the pistons' bores, the seats for the seals, the brake pad abutments, the holes for the mounting bolts that secure the caliper to the hub carrier, and the oil inlet, as well as the internal oil channels and the seat of the bleeder screw (see Figure 5a). Attention should be given to ensuring tolerances and facilitating post-printing machining operations for the pistons' bores and their seals. To this end, it was decided to leave the bottoms of the pistons open on one side of the caliper and subsequently seal them with bore caps (see Figure 5a).

Concerning the material, the brake caliper was made of an aluminum alloy with elastic modulus $E = 71,000$ MPa, Poisson's ratio $\nu = 0.33$. Aluminum was preferred over titanium for this application due to its low density and high thermal conductivity that helps easily dissipating the heat generated during braking.

The model was generated within Altair®'s Hypermesh-OptiStruct and subsequently discretized with a mesh of four-node tetrahedral elements, with an average size of 1.5 mm. The applied loads are as shown in Figure 6a: a maximum braking force $F_b = 5200$ N applied at the theoretical *CoP* and distributed to the abutment using *RBE3* elements, a maximum pressure $p_{max} = 60$ bar on all internal walls and channels in contact with the oil, forces F_{c1} and F_{c2} to simulate the force exerted by the internal pressure on the bore caps, and *RBE2* rigid elements to constrain the displacements of the internal nodes to the holes for the mounting bolts. The optimization parameters were included to minimize compliance (maximize stiffness) as the objective function. Eventually, a constraint on the final volume of 100,000 mm^3, i.e., a final weight of roughly 300 g, and a second constraint on the maximum Von Mises stress to be less than or equal to 170 MPa were imposed. In particular, the Von Mises stress limit was set at approximately 35% of the yield strength of Scalmalloy®. This conservative value was chosen because the mesh used during the topology optimization phase was relatively coarse (to reduce computational time), and the peak stress values near the most critical notches were likely not fully converged. Selecting a higher target value at this stage could have resulted in geometries that were excessively stressed near the notches when analyzed with a refined mesh after the optimization process. Thereafter, the mesh generated by the optimization process (Figure 6b) was refined within Ansys® SpaceClaim and subsequently imported into Blender® and nTopology® for surface smoothing and final solid model generation (Figure 6c,d).

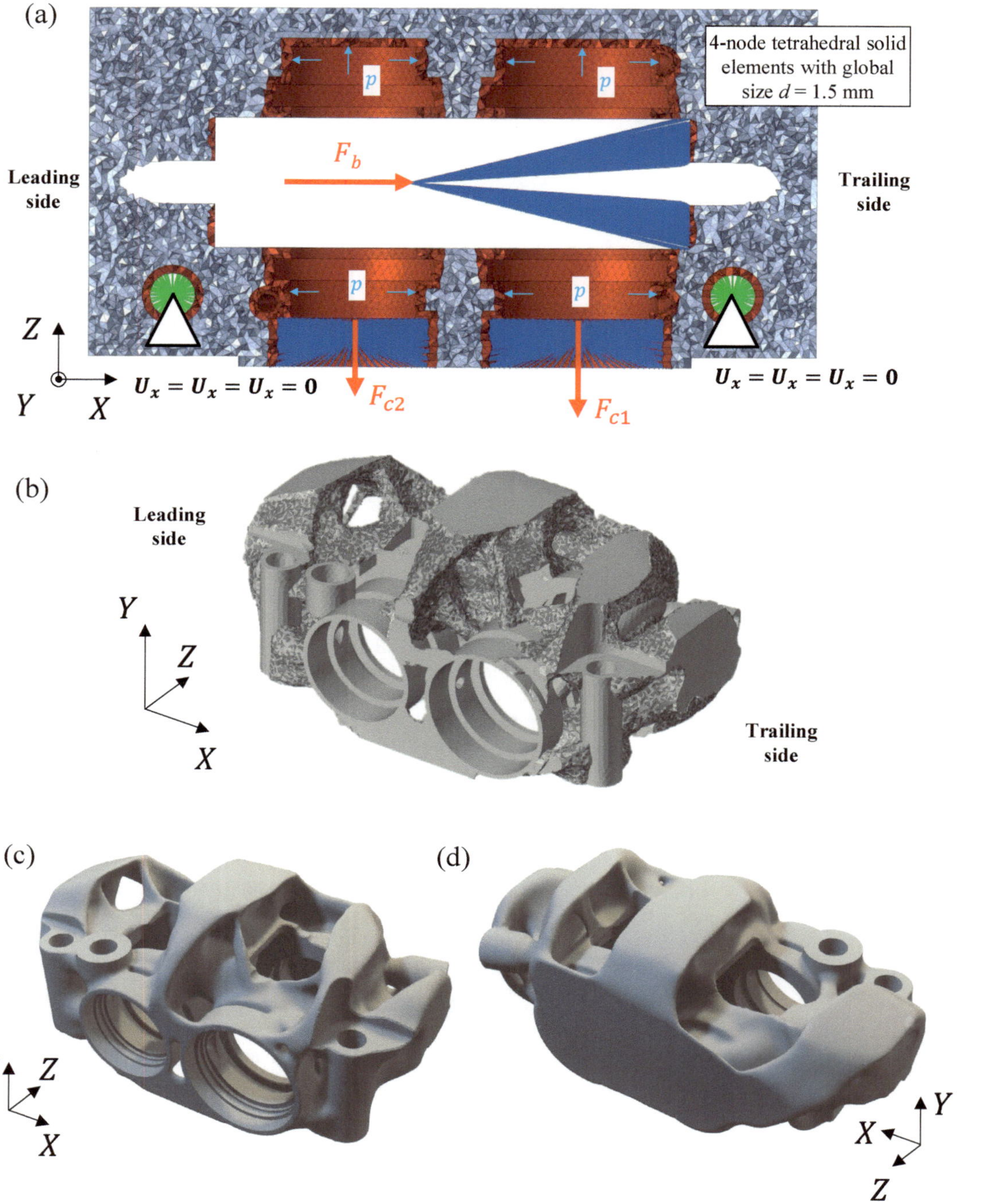

Figure 6. Topology optimization of the brake caliper: (**a**) section view of the FE model adopted for the topology optimization with details on the applied load and constraints, (**b**) output of the analysis, while (**c**,**d**) are two different views of the final CAD model.

2.5. Finite Element Analyses and Manufacturing

With the final CAD model (Figure 6c,d), a structural finite element model was carried out to verify the correctness of the real *CoP*'s location (Figure 7a,b). In particular, the aim of this FE analysis was to evaluate the actual position of the *CoP* with both static and rotating discs, considering contacts and stiffnesses of the various components comprising the caliper assembly (disc, pads, pistons, and the caliper body), and to check if it coincided with the geometric centroid of the pads. If discrepancies were found, the design choices would need to be reiterated, and the optimization process restarted.

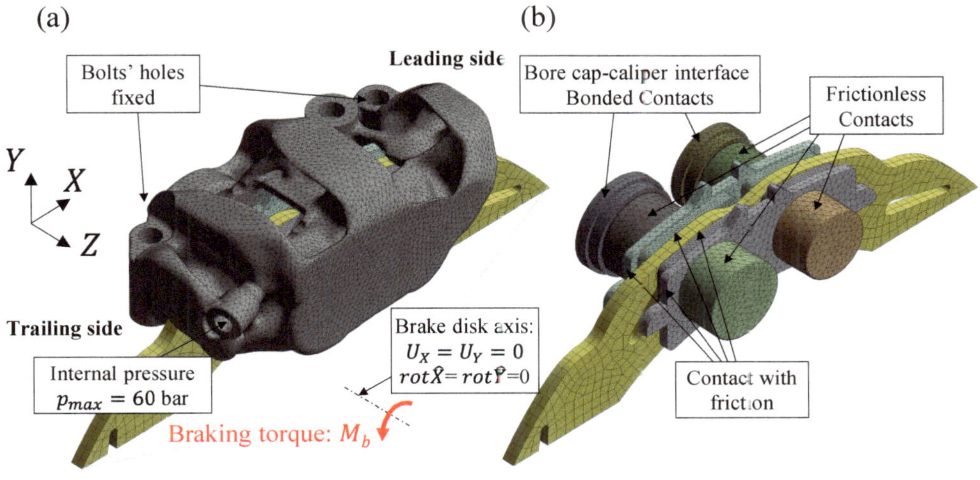

Figure 7. Finite element simulations: (**a**) numerical simulations of the entire brake caliper with (**b**) detailed view of the internal components.

To make the simulation as realistic as possible, the entire brake caliper assembly was modeled in Ansys® Workbench considering all the different components as separate bodies in contact with each other (Figure 7a,b). The hydraulic seals were omitted to avoid increasing the already significant complexity and computational time of the simulation, the latter being nonlinear due to the large number of contact regions. Consequently, the hydraulic seals were replaced with a unilateral holonomic frictionless constraint that precisely guides the axial movement of the pistons, which were placed in the cylinders without clearance. For simplicity, a unilateral holonomic frictionless constraint was also used at the piston–pad interface (Figure 7b). In contrast, a contact with friction was applied at both the pad–disc and the pad–caliper abutment interfaces (Figure 7b), with friction coefficients $\mu = 0.8$ and $\mu_a = 0.3$, respectively. Finally, structural continuity was assumed at all the bore cap–caliper interfaces ("Bonded contacts" in Figure 7b). The bore caps and caliper body were modeled in aluminum alloy, assuming an elastic modulus E = 71,000 MPa and a Poisson's ratio $\nu = 0.33$. The pistons were made of titanium alloy (E = 115,000 MPa, $\nu = 0.34$) due to its low thermal conductivity, which helps isolate the oil and the body of the caliper from the heat generated by the brake pad. The Young's modulus E = 210,000 MPa and Poisson's ratio $\nu = 0.3$ of steel were used for both the brake disc and the brake pads, the latter with a 4.5 mm thick backing plate made of steel behind the lining material.

After defining the materials and contacts, the model was discretized with a mesh of 10-node tetrahedral elements for the caliper body and 4-node tetrahedral elements for all other parts, with an average element size of 1.5 mm in both cases (Figure 7a,b). All nodal displacements at the nodes on the holes for the mounting bolts were constrained. The brake disc was restrained with a remote point at its center to block translations in the X and

Y directions and rotations about these axes. Conversely, translation along the Z-axis and rotation around the same axis were left free to correctly simulate the motion of the floating disc (Figure 7b). The loads, applied in two different load steps, allowed for the simulation of both the static and the rotating disc configurations in a single FE analysis. In the first load step, only the internal pressure $p_{max} = 60$bar was applied to all internal channels of the caliper, to the bore caps, and to the faces of the pistons in contact with the oil. In the second load step, an additional torque was applied to the disc to simulate the condition of the rotating disc.

At the end of each substep of the simulation, the nodal values of the pressure on the face of each pad in contact with the disc were extracted and used to calculate the *CoP*. The corresponding results are presented in Figure 4 for both the static disc case (open black markers) and the rotating disc case (filled black markers), and for both the outer pad (square markers) and the inner pad (triangle markers). The results indicate that the deformation of the caliper, particularly the 'opening up' due to the bending of the bridge [10], causes a slight upward shift of the *CoP* for the outer pad compared to the inner pad (compare square markers with triangle markers in Figure 4). Noteworthily, the *CoP* determined from numerical analyses is very close to the geometric centroid of the inner and outer pads in all simulated cases (see black markers in Figure 4), confirming the final caliper's geometry. As discussed above, if this had not been the case, meaning the *CoP* did not coincide with the geometric centroid of the pad, the position and size of the pistons would have needed to be iteratively adjusted, repeating the optimization and verification process until convergence was achieved.

The final geometry of the optimized caliper was additively manufactured in an aluminum alloy, specifically an Al-Mg-Sc alloy (Scalmalloy®), using laser powder bed fusion (LPBF). In particular, an EOS M400-1 3D printer was adopted, with no pre-heating, and the platform temperature at approximately 35 °C. The powder diameter ranged from 80 to 100 μm and the building direction was that given in Figure 8a. After the printing process, the caliper was shot-peened and sandblasted for support removal and then heat-treated for 240 min at 325 °C. The obtained component is shown in Figure 8a. More details and parameters are omitted for confidentiality reasons.

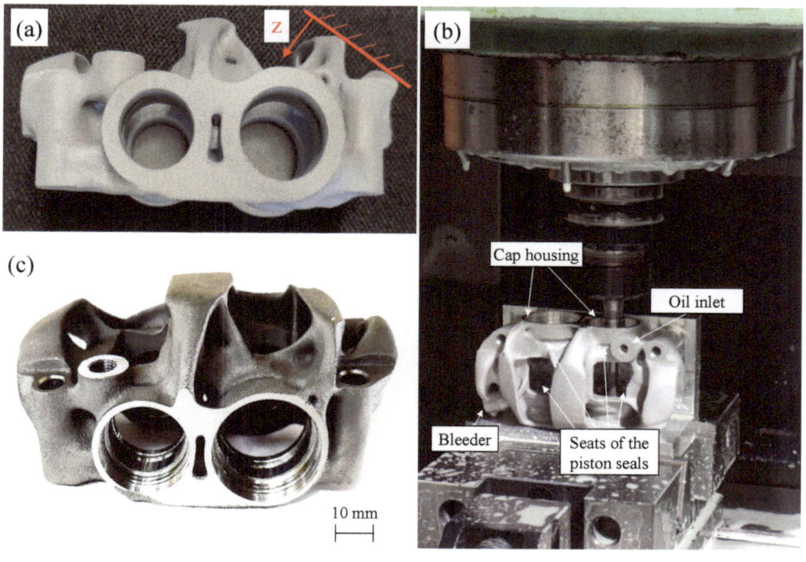

Figure 8. Manufacturing of the brake caliper: (**a**) as-built 3D-printed caliper with indication of the build direction z, (**b**) detail of the milling process for machining the pistons' seals on the inner pad side, and (**c**) the machined caliper.

Subsequently, the 3D-printed caliper was machined to refine regions requiring tight geometric tolerances. Specifically, the caliper was installed on a custom fixture designed for mounting on a three-axis CNC milling machine (Figure 8b). This setup enabled the machining of the seats for the seals of the pistons and the caps (which required the use of custom-designed tools), as well as the oil inlet and the bleeder. The latter, being inclined relative to the milling machine axes, was machined using a tilting vise. The final caliper is shown in Figure 8c.

Eventually, the entire brake caliper was assembled, and its overall dry weight was measured to be 429 g.

3. Experimental Tests

3.1. Strain Gauge Measurement with Static Disc

Tests were carried out on an inertial dynamometer, the details of which are described in [22], and aimed to validate the finite element analyses. Specifically, the test bench setup involved connecting the brake caliper to a hydraulic system that allows for continuously variable pressure application between 0 and 60 bar. Additionally, the bench is equipped with an electric motor and a flywheel mass, enabling the brake disc to rotate and replicate the kinetic energy of the vehicle, thereby simulating the actual braking conditions of the vehicle [22]. Strain and displacement measurements were planned by taking advantage of strain gauges and dial gauge, respectively. Concerning strain measurements, they were planned at different points on the caliper, the positions of which were chosen to easily place the strain gauges where the strain field was as elevated, uniform, and predominantly uniaxial (if possible). Based on the numerical analyses discussed previously, five points were identified, with a total of three axial strain gauges with a 3 mm base length (ε_1, ε_2, and ε_3 in Figure 9), one with 1 mm base length (ε_4 in Figure 9), and one 90° biaxial rosette strain gauge with a 2 mm base length (ε_5 and ε_6 in Figure 9). Additionally, five thermocouples were included to compensate for thermal expansion effects and ensure that the maximum operating temperature of the adhesives was never exceeded. Each of the six strain gauges was connected in a quarter-bridge configuration to an IMC-CRONOS PL2 Data AcQuisition system. Similarly, two thermocouples were connected to the same DAQ, while the remaining three were connected to a National Instruments® NI USB-9162 + NI 9211 module.

Before running the tests, the strain gauge readings were zeroed with the caliper fully unloaded and the mounting bolts loosened. Thereafter, the bolts were tightened and the static test, i.e., with the disc stopped, was conducted by applying by discretely increasing and decreasing the pressure in the brake line between 0 and 50 bar. The results, shown in Figure 10, present the experimental measurements of each strain gauge as markers and the corresponding numerical values as solid lines of the same colors. Notably, the measurements exhibited excellent agreement with the numerical estimates, with the error almost always being less than 5%. It should also be noted that the small (negligible) residual deformation at zero pressure is solely due to the bolt tightening (see Figure 10).

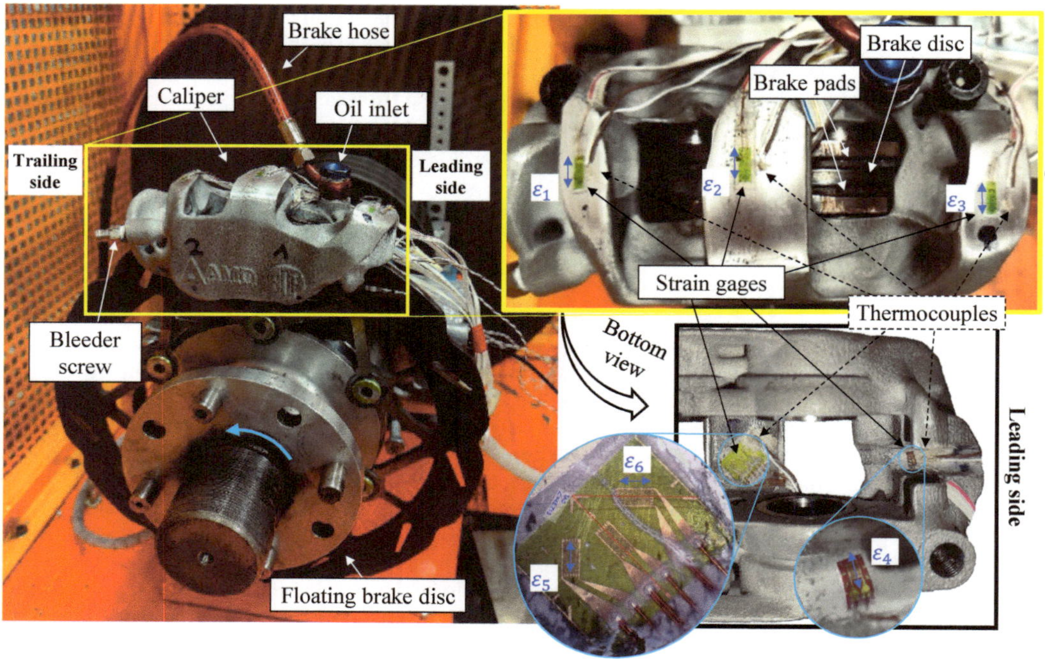

Figure 9. Brake caliper installed in the dynamometer described in [22]. Details of the strain gauges and thermocouples are given on the right side of the figure.

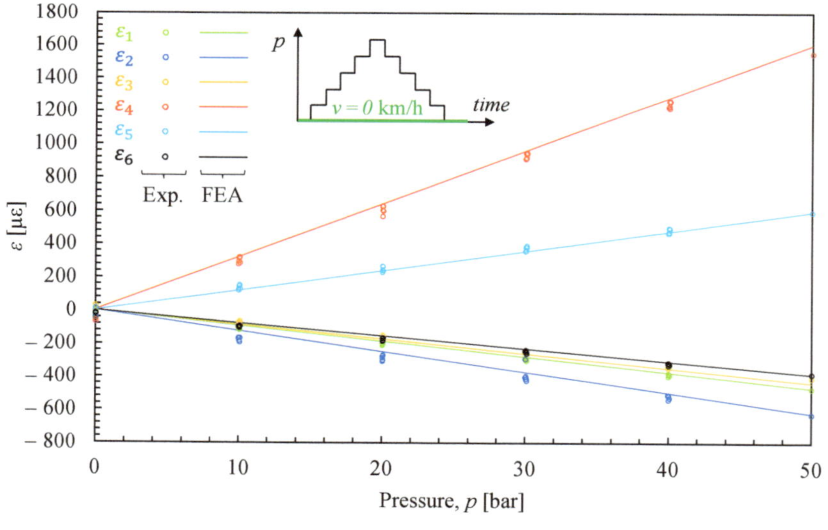

Figure 10. Comparison between experimental and numerical caliper's strains under static loading.

3.2. Dial Gauge Measurement with Static Disc

In addition to the strain gauge measurements, an experimental evaluation of the caliper's displacement field was conducted by placing analog dial gauges at three different points on the caliper (points A, B, and C in Figure 11). These points were chosen for the high values of the caliper's displacement when loaded and their easy access, which allows for the proper positioning of the dial gauges. It should be noted that the selected measurement points primarily capture the displacements resulting from the caliper's 'opening up' deformation component [10]. Furthermore, the specific design of the test bench does not permit absolute displacement measurements by mounting the dial gauges rigidly to the frame since the brake caliper is fixed to a floating hub carrier that allows it to move slightly relative to the frame [22]. For this reason, two dial gauges were placed on one side of the caliper (points A and B in Figure 11), while the third gauge was positioned in the opposite direction on the opposite face (point C in Figure 11). This arrangement allowed for the evaluation of the caliper's displacement field as the difference between the displacements measured at points A and B and that measured at point C, i.e., $f_A - f_C$ and $f_B - f_C$, so as to remove any rigid body motion of the caliper.

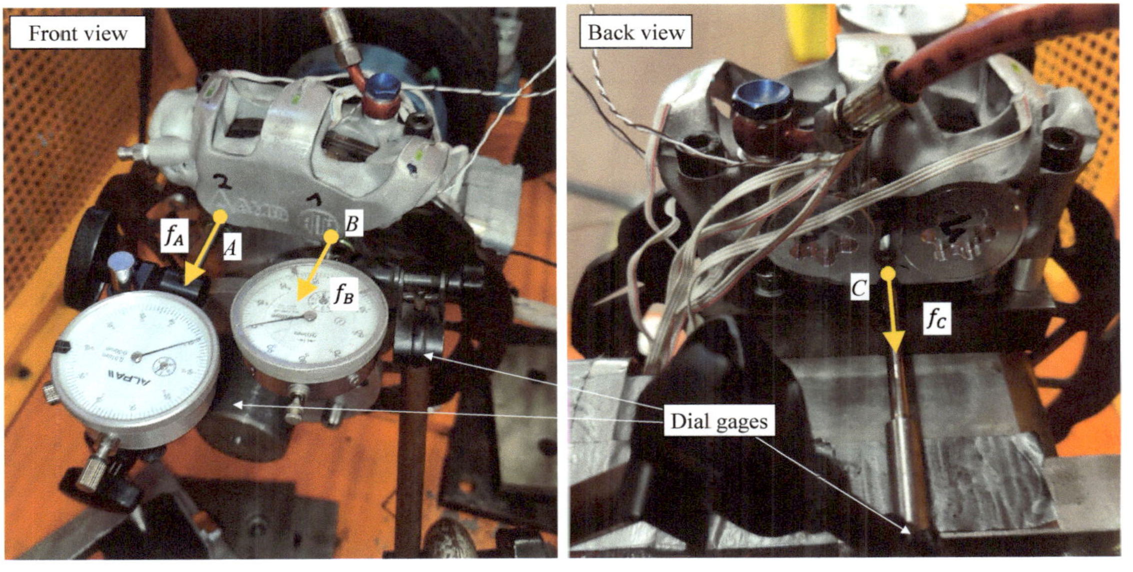

Figure 11. Position of the dial gauges for caliper displacement measurement with static disc.

Measurements were taken for some of the static pressure ramps between 0 and 50 bar previously discussed for the strain gauge measurements. The results, shown in Figure 12, exhibit excellent agreement between the experimental measurements (markers) and the corresponding finite element analysis estimations (solid lines of the same colors). Figure 12's experimental results also include error bars of ±0.02 mm, this value being considered representative of the measurement resolution. This accounts for the intrinsic resolution of the dial gauges, the precise positioning of the dial gauge, potential parallax errors, and the compliance of the dial gauge supports.

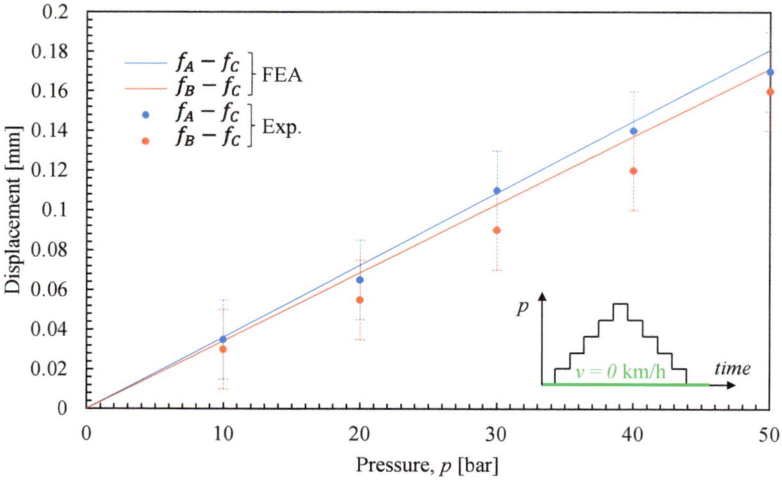

Figure 12. Comparison between experimental and numerical caliper's displacements with static disc.

3.3. Strain Gauge Measurement with Rotating Disc

Strain measurements were also conducted under real braking conditions, i.e., with rotating brake disc. The brake test bench facilitated the application of repeated braking ramps at a constant initial speed v_{max} and set pressure p_{max} [22]. In essence, the inertial dynamometer accelerated the brake disc (along with the flywheels simulating the vehicle's kinetic energy) from 0 km/h to a maximum speed v_{max} (defined by the operator up to a maximum of 90 km/h) and maintained this speed for a few seconds (see green line Figure 13). Subsequently, the motor turned off, and a constant pressure p_{max} (also set by the operator up to a maximum value of 60 bar) was applied and kept constant until the disc completely stopped (see black line Figure 13). Eventually, this sequence was repeated for a specified number of times. The advantage of these tests, as opposed to static tests with a static rotor, is that they allow for validating the numerical analyses under a non-zero braking force F_b. Specifically, the rotating disc induces a braking torque M_b, whose instantaneous value is measured by the dynamometer and used to calculate the friction force F_b acting on the pad as $F_b = M_b/r_b$ (see blue line Figure 13). To validate the models, tests were conducted with two different maximum pressures p_{max} (20 and 50 bar) and varying the initial speeds v_{max} between 30 and 70 km/h. A sampling frequency of 20 Hz was used. At the end of each test, the strain measurements (see red line Figure 13, which represents the measurement of strain gauge 4) were correlated with the corresponding p_{max} and friction force F_b values. Specifically, Figure 13 (in particular, the zoomed view at the bottom) shows that during the entire braking cycle, both the pressure p (black curve) and the braking force F_b (blue curve) vary continuously from zero to a maximum value and then return to zero at the end of the braking phase. Similarly, the strain measured by strain gauge 4 (analogous to the measurements from all other strain gauges) evolves continuously with p and F_b. It starts from a value slightly above zero, reaches a maximum at the highest pressure and braking force, and then stabilizes at a constant value for $p = p_{max}$ and $F_b = 0$, before finally returning to zero upon complete release of the pressure ($p = 0$). The initial value slightly above zero is attributed to the pad lightly touching the disc at zero pressure with the disc rotating, resulting in a slight braking force. This slight contact, while ideally avoidable (for instance, by providing a greater return force from the piston's hydraulic seals), makes the system more responsive by eliminating the latency at the beginning of the braking phase.

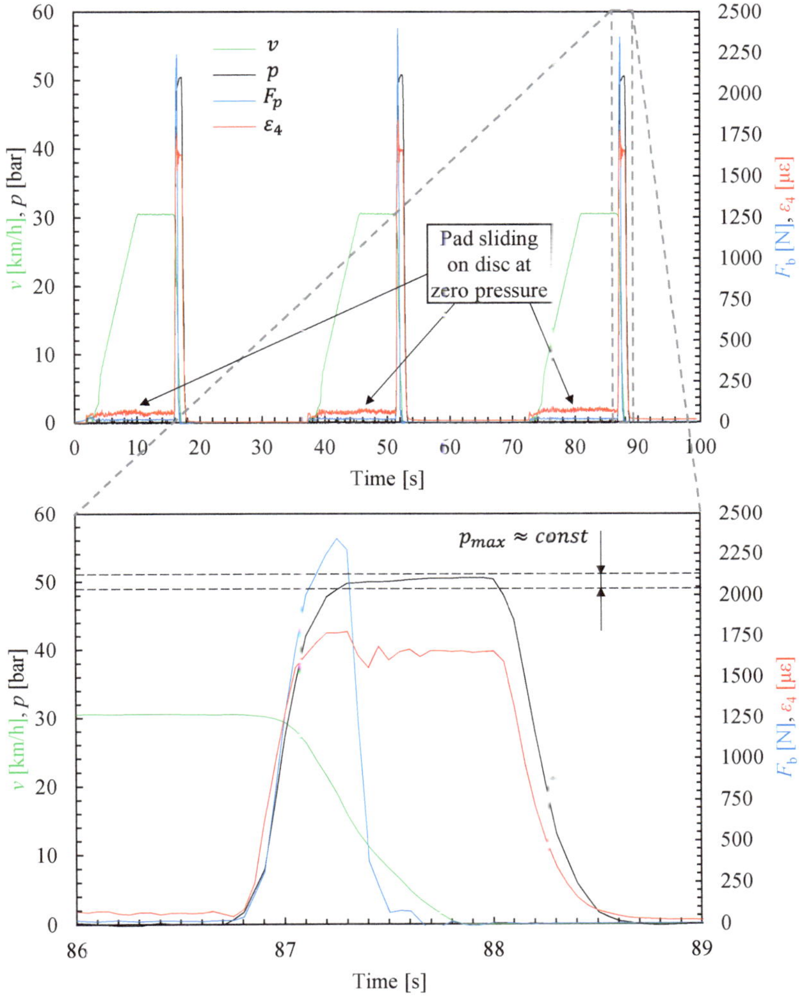

Figure 13. Example of data acquisition from a repetition of 3 ramps at a constant initial speed v_{max} = 30 km/h (green line, referred to the left axis) and set pressure p_{max} = 50 bar (black line, referred to the left axis). The same chart also gives the measured braking force F_b (blue line, referring to the right axis) and, for brevity, the strain gauge measurement of the strain gauge 4 only (red line, referring to the right axis). The chart below is a zoom of the third ramp.

Given that the caliper deformation is a function of two parameters, a two-dimensional comparison with numerical analyses required filtering the data. Only the portions of each braking cycle with approximately constant pressure ($p = p_{max} \pm 1$ bar) were extracted and analyzed (see dashed black lines in the lower section of Figure 13). Specifically, all deformation values within this window were used for comparison with the numerical analyses, the results of which are shown in Figure 14.

Figure 14 illustrates all the strain value as a function of the braking force F_b (with F_b = 0 being the case of static brake disc) at an almost constant pressure and equal to either 20 bar (blue markers in Figure 14) or 50 bar (red markers in Figure 14). Additionally, Figure 14 shows the strain values estimated from finite element analysis (as indicated in Figure 7) for six different pairs of (p_{max}, F_b) (see the six black markers in Figure 14). Intermediate values at constant p_{max} are estimated by assuming a linear relationship between the FEA results (see black dashed lines in Figure 14). Note that, for brevity and clarity, Figure 14 shows only the data from strain gauge 4, i.e., the most stressed gauge (see Figure 10), with the results from the other strain gauges being similar.

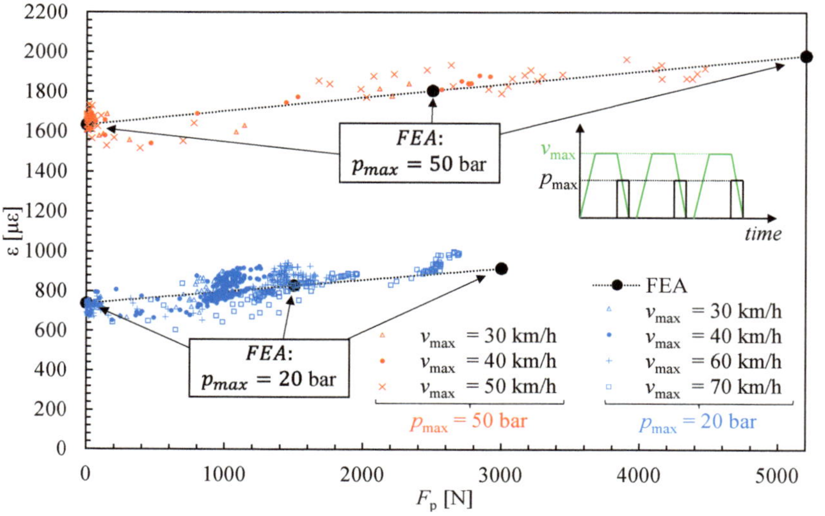

Figure 14. Comparison between experimental and numerical strains with rotating disc.

An analysis of the figure indicates that the experimental results (markers in Figure 14) are in good agreement with the numerical estimates (black dashed lines in Figure 14) for both 20-bar and 50-bar pressures. Despite the huge scatter of the data, the average trend is well captured by the numerical analyses, with maximum errors generally within ±15% for p_{max} = 20 bar and ±10% for p_{max} = 50 bar.

At the conclusion of the tests with the rotating disc, it was also possible to measure the thickness of the brake pad's lining material. As a result, this measurement confirmed that uniform wear occurred on both the inner and outer pads, further validating the efficacy of the analyses conducted and the correctness of the procedure used in the caliper's design.

4. Conclusions

This study presents a comprehensive and systematic approach to conceiving, designing for additive manufacturing, manufacturing, and testing a hydraulic brake caliper specifically tailored for Formula SAE race cars and made from an Al-Mg-Sc alloy (Scalmalloy®), an innovative Al-Mg-Sc alloy which was never adopted before to manufacture a brake caliper. The focus is on a topology-optimized monoblock fixed caliper with four opposing pistons, aiming to balance minimal mass with high braking force, which is crucial for competitiveness in Formula SAE. The design process began with a detailed theoretical analysis of the vehicle dynamic to define the design loads, including the required braking force and applied pressure. This was followed by a preliminary design phase that defined the caliper's main dimensions and key geometric features, e.g., the size and position of the pistons. Topology optimization was employed to minimize the caliper's mass while maintaining appropriate stiffness and structural integrity.

The final design was realized using laser powder bed fusion (LPBF) additive manufacturing by taking advantage of the high-strength, low-weight properties of Scalmalloy®. The caliper was subsequently subjected to comprehensive experimental validation. Tests conducted on an inertial dynamometer with both static and rotating brake discs confirmed the caliper's performance, demonstrating strong alignment with finite element analysis

(FEA) estimations. The strain and displacement measurements showed excellent correlation with numerical estimates, with errors generally within 5% and 15% for static and rotating brake discs, respectively. Furthermore, the design ensured uniform wear of the brake pads, as validated by post-test inspections, confirming the effectiveness of the design process. In conclusion, this study successfully demonstrates the conceptualization, design, manufacturing, and validation of a topology-optimized brake caliper for Formula SAE race cars made of Scalmalloy® by using LPBF technology.

Author Contributions: Conceptualization, F.C. and G.M.; Data curation, F.C. and L.V.; Formal analysis, F.C. and L.V.; Investigation, F.C. and L.V.; Methodology, F.C., L.V. and G.M.; Project administration, G.M.; Resources, F.C. and G.M.; Software, F.C. and L.V.; Supervision, G.M.; Validation, F.C. and L.V.; Visualization, L.V. and F.C.; Writing—original draft preparation, L.V.; Writing—review and editing, F.C., L.V. and G.M. All authors have read and agreed to the published version of the manuscript.

Funding: This research received no external funding.

Data Availability Statement: The original contributions presented in the study are included in the article, further inquiries can be directed to the corresponding author.

Acknowledgments: The authors would like to thank AMD Engineering S.r.l for its contribution to the caliper's development and the consultancy provided during the design phase, Alth S.n.c for the machining and realization of the auxiliary components, and Cinel S.r.l. for the metrology support. Finally, thanks are extended to APWORKS GmbH for the 3D printing.

Conflicts of Interest: The authors declare no conflicts of interest.

References

1. SAE International. *Formula SAE Rules 2024*; SAE International: Warrendale, PA, USA, 2023.
2. Carvalho, D.F.T.; Melo, C.A.P. *Simulation Model and Testing of a Formula SAE Brake System*; 2021-36-0430; SAE Technical Paper; SAE International: Warrendale, PA, USA, 2021. [CrossRef]
3. Chen, W.; Jia, D.; Huang, S. Optimized Design of Braking System in FSAE Racers. *J. Eng. Mech. Mach.* **2022**, *7*, 62–74.
4. Dalal, K.; Karnik, A. Monoblock Brake Caliper Design and Analysis. *Int. Res. J. Eng. Technol.* **2020**, *7*, 2419–2425.
5. da Silva, P.H.M.; Idehara, S.J. *Characterization of the Brake System of a Formula SAE Vehicle*; 2018-36-0148; SAE Technical Paper; SAE International: Warrendale, PA, USA, 2018. [CrossRef]
6. Farias, L.T.; Schommer, A.; Haselein, B.Z.; Soliman, P.; de Oliveira, L.C. *Design of a Brake Caliper Using Topology Optimization Integrated with Direct Metal Laser Sintering*; SAE Techincal Paper; SAE International: Warrendale, PA, USA, 2015. [CrossRef]
7. Tyflopoulos, E.; Lien, M.; Steinert, M. Optimization of Brake Calipers Using Topology Optimization for Additive Manufacturing. *Appl. Sci.* **2021**, *11*, 1437. [CrossRef]
8. Gupta, E.; Bora, D.K.S.; Rammohan, A. Design and analysis of brake system for FSAE race car. *Eng. Res. Express* **2022**, *4*, 025039. [CrossRef]
9. Phad, D.; Auti, T.; Joshi, R.; Jadhav, S.; Devasthali, S. Design and Analysis of a Brake Caliper. *Int. J. Mech. Eng. Res. Dev.* **2015**, *5*, 1–10.
10. Sergent, N.; Tirovic, M.; Voveris, J. Design optimization of an opposed piston brake caliper. *Eng. Optim.* **2014**, *46*, 1520–1537. [CrossRef]
11. Tirovic, M.; Sergent, N.; Campbell, J.; Roberts, P.; Vignjevic, R. Structural analysis of a commercial vehicle disc brake caliper. *Proc. Inst. Mech. Eng. Part D J. Automob. Eng.* **2012**, *226*, 613–622. [CrossRef]
12. Sergent, N. Analysis and Optimisation of Disc Brake Calipers. Ph.D. Thesis, Cranfield University, Bedford, UK, 2010.
13. Limpert, R. *Brake Design and Safety*; SAE International: Warrendale, PA, USA, 2011.
14. Ravi Kumar, L.; Prathiesh Lalan, R.A.; Shriram Naibal, B.; Chiranjeev Sanjay, P.; Gananathji Naveen Kishore, S.; Vasundharadevi, D. *Design, Modeling and Analysis of Customized Brake Caliper for SAE BAJA Vehicle*; SAE Technical Paper; SAE International: Warrendale, PA, USA, 2024. [CrossRef]
15. Ugemuge, M.; Das, S. *Design and Testing of Custom Brake Caliper of a Formula Student Race Car*; SAE Technical Paper; SAE International: Warrendale, PA, USA, 2019. [CrossRef]
16. Ugemuge, M.; Das, S. *Topology Optimisation of Brake Caliper*; SAE Technical Paper; SAE International: Warrendale, PA, USA, 2020. [CrossRef]
17. Meyer, J.; Barnes, J.E. Scalmalloy is too expensive and design optimisation only makes sense in aerospace. True or false? *Met. AM* **2019**, *5*, 127–135.
18. Limebeer, D.J.N.; Massaro, M. *Dynamics and Optimal Control of Road Vehicles*; Oxford University Press: Oxford, UK, 2018.
19. Antanaitis, D.; Sanford, J. *The Effect of Racetrack/High Energy Driving on Brake Caliper Performance*; SAE Technical Paper; SAE International: Warrendale, PA, USA, 2006. [CrossRef]
20. Chang, H.H. *On a Numerical Study for Rubber Seals*; SAE Technical Paper; SAE International: Warrendale, PA, USA, 1988. [CrossRef]

21. Anwana, O.D.; Cai, H.; Chang, H.T. *Analysis of Brake Caliper Seal-Groove Design*; SAE Technical Paper; SAE International: Warrendale, PA, USA, 2002. [CrossRef]
22. Vecchiato, L.; Negri, M.; Picci, G.; Viale, L.; Zaltron, G.; Giacometti, S.; Meneghetti, G. Design and Development of a Brake Test Bench for Formula SAE Race Cars. *Machines* **2024**, *12*, 135. [CrossRef]

vehicles

MDPI

Article

Prediction of Residual Wear Resources of Composite Brake Pads of a Modernized Brake System of Freight Wagons

Sergii Panchenko [1], Juraj Gerlici [2], Alyona Lovska [2], Vasyl Ravlyuk [3] and Ján Dižo [2,*]

[1] Department of Automation and Computer Telecontrol of Trains, Ukrainian State University of Railway Transport, Feuerbach Square 7, 61050 Kharkiv, Ukraine; panchenko074@ukr.net
[2] Department of Transport and Handling Machines, Faculty of Mechanical Engineering, University of Žilina, Univerzitná 8215/1, 010 26 Žilina, Slovakia; juraj.gerlici@fstroj.uniza.sk (J.G.); alyona.lovska@fstroj.uniza.sk (A.L.)
[3] Department of Wagon Engineering and Product Quality, Ukrainian State University of Railway Transport, Feuerbach Square 7, 61050 Kharkiv, Ukraine; ravvg@kart.edu.ua
* Correspondence: jan.dizo@fstroj.uniza.sk

Abstract: This research highlights the results of a comprehensive study of the efficiency of modernized brake systems operation of freight wagons. The inspection of the modernized elements of the lever brake system of bogies and the measurement of the wear parameters of composite brake pads during each cycle of the experimental wagons in the interval of mileage from 2.1 to 197.8 thousand km were carried out. A statistical approach was used to study the wear parameters of brake pads of modernized bogies brake systems determined during operational studies. This allowed appropriate dependencies of brake pad wear to be obtained. Based on the research results, a regression model was developed. This makes it possible to predict the residual wear resource of composite brake pads with modernized braking systems of bogies for the entire inter-repair period of operation of freight wagons guaranteed by the wagon repair company. The peculiarity of the model is that it considers the total and additional mileage of the freight wagon. This makes it possible to more accurately predict the residual lifetime of composite brake pads. It was established that, under the condition of uniform wear of brake pads, the average mileage of a freight wagon during the use of modernized brake systems of bogies can reach up to 284.57 thousand km, which increases the resource of composite pads' wear approximately by 2.59 times. The generated model was verified by the F-criterion. Approbation of experimental devices for uniform wear of composite pads in operation established that measures to modernize brake systems of freight wagons ensure the reliable and efficient operation of the brake lever system as a whole.

Keywords: railway transport; brake pad; pad lifetime prediction; regression wear model; brake system

Citation: Panchenko, S.; Gerlici, J.; Lovska, A.; Ravlyuk, V.; Dižo, J. Prediction of Residual Wear Resources of Composite Brake Pads of a Modernized Brake System of Freight Wagons. *Vehicles* **2024**, *6*, 1975–1994. https://doi.org/10.3390/vehicles6040097

Academic Editors: Ralf Stetter, Udo Pulm and Markus Till

Received: 23 October 2024
Revised: 22 November 2024
Accepted: 22 November 2024
Published: 25 November 2024

1. Introduction

The development of the railway industry in the modern conditions of the competitive environment is accompanied by an increase of the running speed of freight trains, axle load, the improvement of structures and materials for the manufacture of components of rail vehicles, an increase of the load factor of wagons, etc. [1–5]. However, special attention should be paid to the modernization of brake equipment elements of wagons as one of the most responsible factors from the ensuring traffic safety point of view [6–9].

The traffic safety analysis of the Wagon Management Department of the Ukrzaliznytsia Stock Company (Kyiv, Ukraine) has indicated in recent years that the braking equipment of freight wagons is too vulnerable in the current operating conditions. Currently, a lot of effort is being put to solve the problem of wedge-dual wear of brake pads [10–15], because this kind of wear leads to reduction of the working part of brake pads (Figure 1), and it is widespread in freight wagons bogies on the main transport network, including the tracks with a gauge of 1520 mm.

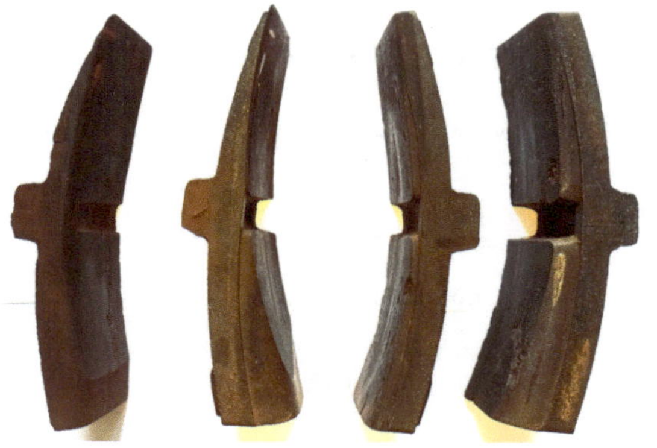

Figure 1. Unusable composite brake pads with wedge-dual wear and significant reduction of their working parts.

It is important to note that the uneven wear of brake pads can also affect the quality of the wheel surface of wheelsets if it is significant.

Triangles are the main element of the brake gears of freight wagons, and they work in such a way that, during the braking of a wagon, all brake pads of the bogie are pushed against the wheels. The bogie brake system (BBS) must be balanced with respect to the force load of each pad. Due to the dynamics of the interaction of the wheels with the unevenness of the rail track, this balance is disturbed, and the pads wear out most often in the upper part. As a result, this is the main reason for their wedge-dual wear. This problem causes a decrease of braking efficiency of freight wagons due to the reduction of the working part, i.e., the braking area of the contact between the brake pads and the bogie wheels. Furthermore, this leads to an increase in the frequency of repairs of rail vehicles running parts of rolling stock, additional costs of energy resources for train traction, and a total decrease in economic and environmental indicators during freight transportation.

In this regard, it is necessary to modernize the elements of the BBS to prevent the occurrence of wedge-dual wear of composite brake pads. This will make it possible to use the pads for the entire inter-repair period of operation of freight wagons.

The issues of train movement safety within railway transport are relevant and they depend on significant factors, particularly on the technical condition and the load of the brake system elements. Thus, for example, the study of the thermal load of the brake pad was carried out using the SolidWorks simulation software in the work [16]. Based on the obtained results, an alternative solution of using composite material in the form of a modified alkylbenzene resin is proposed, which helps to increase the friction coefficient of the pads.

The study in [17] established the cause of harmful wedge-dual wear of the pads, and identified possible ways to eliminate the specified deficiency, which leads to the premature replacement of brake pads that have not used their working mass resource in the inter-repair period. The work uses a statistical approach to planning experiments during field tests of freight wagons with a typical and a modernized BBS, which were included in one warehouse. However, the question of predicting the residual resource of brake pads remained ignored by the researchers.

Another approach was proposed by a group of authors in the work [18], where the issue of the uneven wear of brake pads was considered. To achieve the given task, a complex multifactorial model is proposed. However, cases were considered where the values of the studied parameters obey the normal law of distribution.

The author of the work in [19] developed a regression model of the friction of pads and wheels of industrial transport locomotives in conditions of structural uncertainty. This model is multifactorial, but it does not consider the wear of the pads by thickness during operation.

Similar studies were conducted on the processing of the statistical material of nodes of the transport and machine-building industry in the works in [20–22]. However, none of them reflected the features of prediction of the residual resource of elements to reduce planned and preventive repairs of brake systems of freight wagons.

In the production studies commissioned by the national Ukrzaliznytsia Stock Company (Kyiv, Ukraine), the development of design and technological documentation regarding the modernization of the freight wagons' BBS was carried out [23]. However, no statistical studies of the wear of brake pads of the fleet of freight wagon of private enterprises were conducted in this research work. Unlike the wagon of the Ukrzaliznytsia Stock Company (Kyiv, Ukraine), they have a permanent location, and they are operated in "softened" conditions.

One of the approaches to such statistical studies is proposed in [24]. There were estimated parameters such as the force of pressing the brake pads on the rolling surface of the wheel, the hardness of the material of the pads, etc., for the set value of the braking distance, which depends on the speed of movement, the slope of the rail track, and the radius of the curve during the braking of rail vehicles. Critical slopes of the braking distance in the case of the full-service braking of the rail vehicle were statistically established. However, the work did not consider the conditions when the braking area of the contact between the pads and the wheels decreases in the presence of their wedge-dual wear. Thus, it is not possible to reliably estimate the braking efficiency of the train under such conditions.

In the works in [25,26], the teams of authors presented the results of operational studies on the assessment of factors that cause the appearance of defects of thermal origin on the rolling surface of wheelsets due to the action of composite brake pads. To prevent the occurrence of such defects, it is suggested to use improved blocks, which make it possible to significantly reduce the number of wheel malfunctions. During the examination of the brake equipment of the freight train, various malfunctions of the mechanical and pneumatic parts of the brakes were discovered. An examination of the pads was carried out, during which their wedge-shaped wear was revealed because of the interaction of the upper end with the rolling surface of the wheels. However, the mentioned works did not consider the issue of analyzing the collected statistical material of non-normative pad wear and its processing to estimate the amount of wear, which affects the reduction of the braking efficiency of the freight train and traffic safety.

The work in [27] includes the results of tests of freight rail vehicles of industrial transport regarding the evaluation of braking efficiency, as well as a structural and dynamic analysis of the braking mechanism. The conducted research consisted of determining the type and parameters of the empirical dependence of the friction coefficient of the brake pad on the wheel rolling surface, depending on the speed and braking, as well as in determining the kinetic characteristics of the brake of the freight rail vehicles of industrial transport. However, it should be noted that the specialists did not take into account the possible wedge-dual wear (clinodual wear) of the pads of freight wagons in the research, which significantly affects the evaluation of the braking efficiency of industrial wagons.

The research in [28] highlights the analysis of performance indicators of the brake pads quality for various types of rail vehicle. Individual negative factors of composite brake pads are given, and their impact on the environment and processes causing damage to the rolling surfaces of rail vehicles' wheels are described. However, the authors did not pay attention to the issue of the effective use of the brake pad working partly under operational conditions.

As practice proves, the wedge-dual wear of brake pads occurs due to unprofessional actions of workers who perform the maintenance and repair of freight wagons in violation of regulatory technical documents [29]. That is, in some cases, brake pads previously removed from the wagons are again installed at their discretion. Therefore, their wear obviously cannot correspond to the normal distribution law, and similar studies [30–32] do not make sense.

The necessary cost reduction of the maintenance of rail vehicles in operation has ensured the consideration of brake pads as a commodity that is often purchased at the lowest price under the conditions of their satisfactory operation. However, this may not lead to a reduction of operation costs, and the choice of friction material may have a direct impact on the service lifetime of the wheel, which is usually much more expensive to replace than other parts of wagons. Similarly, if low-quality materials are used for the manufacture of brake pads, the resource of their working part will quickly decrease. It will lead to an increase in the braking distance of the train and a decrease in the level of traffic safety [33].

Publications [34–36] consider the peculiarities of the tribotechnical repairs operation. A solution of increasing the efficiency of their work is proposed. The implementation of the proposed solutions will help to increase the speed of movement, axle load, efficiency of the braking system of rail vehicles, etc. However, at the same time, there are several problems related to the wadge-dual wear of freight wagons' brake pads that need to be solved. Thus, the problems associated with the wear of brake pads and wheels of freight wagons actually exist [37,38], and in this direction work is being carried out related to the modernization of the elements of the brake lever system of freight wagons to ensure the safety of freight trains' operation by increasing the efficiency of their brakes.

As it is described in the overview above, the freight wagon transport is important, together with its safe operation under the various loads [39,40]. Modern designs of freight wagons should be supplemented by the reliable system of bogies with modern technical solutions of safety components [41,42].

Currently, typical freight wagons' brake pads for 1520 mm gauge are made of composite material. The main requirements for such pads are covered in the regulatory document [43]. Other regulatory documents apply to wagons of the European gauge, for example [44,45].

To improve the strength of brake pads, research is being carried out on their production from new materials.

Thus, the study [46] is aimed at analyzing the operation of tribotechnical units and justifying the introduction of promising materials in their construction. Due to such solutions, the speed of train movement increases significantly, the load on the axle of the wagon increases significantly, the resource of tribotechnical parts increases—composite brake pads and linings; bushings used in kinetostatic units, the operation of the braking system of the rolling stock is improved, etc. However, the problems of non-normative wear of elements of the tribotechnical unit—"brake pad-wheel" were not investigated in these works.

The publication in [47] examines the influence of low temperatures on the operation of brake pads. The requirements for the material of their manufacture are specified there. It is specified that it must have high strength and wear resistance to avoid cracks and fractures, as well as sufficient hardness to ensure minimal wear of the wheel during braking. However, the authors did not investigate the issue of wear of such pads in operation.

The work in [48] has the same drawback. It analyzes the use of the latest materials in pads, including foam–aluminum inserts. The author gives options for the execution of such blocks. However, the specifics of their operation and wear are not specified.

The study of literary sources in [16–48] makes it possible to conclude that the issues of analysis and processing of statistical material regarding the wear of brake pads for the purpose of prediction their residual resource in the inter-repair period of freight wagons operation are relevant and require further development.

The purpose of the study is to highlight the features of prediction of the residual resource of composite pads of modernized brake systems of freight wagons.

To achieve this goal, the following tasks are defined:

- To carry out an inspection of composite brake pads of freight wagons with the modernized BBS in operational conditions;
- To develop a regression model for prediction of the residual lifetime of composite brake pads with the modernized BBS;
- To verify the statistical hypotheses regarding the nature of the distribution of random wear values of composite brake pads and their relationship regarding the data belonging to one general whole.

2. Research Materials and Methods

To increase the reliable operation of the brakes of freight wagons, the modernization of the BBS was carried out. It is carried out with the aim of eliminating structural defects in the system of removing the brake pads from the rolling surfaces of the wheelsets. The location of the technological hinge B connecting the vertical lever with the triangle spacer was changed in the modernized BBS and was developed at the Ukrainian State University of Railway Transport (UkrSURT) (Figure 2). This hinge is located on the same line A-A with the hinges of the pendulum suspensions. Also, a curved rod is introduced. This rod keeps the BBS constantly in balance with its ends in the cylindrical sliders. This ensures strictly uniform gaps between the blocks and wheels, and it guarantees their normative wear.

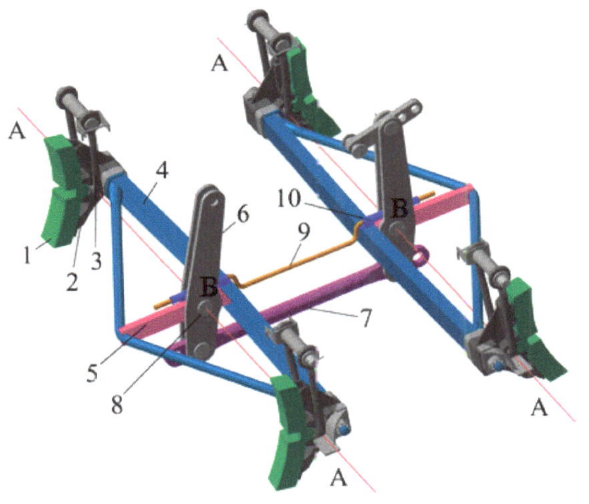

Figure 2. A computer model of the modernized BBS developed according to the technology of the UkrSURT with a device for uniform pad wear: 1—a brake pad; 2—a brake pad holder; 3—a pendulum suspension; 4—a triangle; 5—a spacer triangle; 6—a lever; 7—tightening of levers; 8—a hinged connection of vertical levers to struts of triangles, located on the line A-B-A; 9—a curved rod; 10—a cylindrical slider.

The modernized BBS developed according to the UkrSURT technology works as follows. When the brakes are released, triangles with holders and brake pads move under the influence of gravitational forces on pendulum suspensions, so that they move the pads away from the wheels. Thanks to the balance of the brake system relative to the hinges of the pendulum suspensions, the pads move away from the wheels evenly. In case of accidental forces due to oscillations and tilts of the wagon during movement, the curved rod works. Due to the placement of its ends in cylindrical sliders, it does not allow the triangle to tilt, and therefore the pads to rest with the ends of the upper or lower parts on the wheels. At this time, reaction forces are created in the sliders, which are balanced on those parts of the curved rod that are contained in the sliders due to the symmetrical location of the sliders relative to the hinge hole. The parts of the rods which are bent vertically downward and are located near the ends of the cylindrical sliders keep the rod from longitudinal displacement and falling out from the action of longitudinal forces on it during braking and the action of random forces from oscillations and tilts of the wagon.

Thanks to the guide device, the horizontal movement of adjacent triangles and brake pads in the BBS occurs. Thus, the uniformity of the clearances between the blocks and wheels is ensured when the brakes of freight wagons are released.

The approbation tests were performed under operating conditions in order to verify the reliable operation of the modernized BBS and they were developed at the Department of Wagon Engineering and Product Quality of the Ukrainian State University of Railway Transport, Kharkiv, Ukraine. By the order of the Department of Wagon Management of the Ukrzaliznytsia Stock Company (Kyiv, Ukraine), modernized BBSs were installed on ten open freight wagons built for the Donetsk Railway by the Kryukiv Wagon Building Plant (Kremenchuk, Ukraine) of a specified mileage and they were unhooked on special tracks for inspections [23,49].

A commission inspection was carried out with an examination of the condition and performance evaluation of the experimental devices of the modernized BBS with the measurement of the gaps between the blocks and wheels and the thickness of the blocks to determine their wear parameters. The statistical material obtained in this way is used both to determine the reliability indicators of experimental devices as well as to evaluate their parameters that affect the intensity of wear of brake pads [50,51].

Measurements were carried out in accordance with the developed "Program and methodology for conducting scientific production research of brake systems and wheels of freight wagons".

Examinations showed that no damage or wear was found on any of the experimental devices of the modernized BBS. The experimental devices of the modernized BBS ensure the normative remoteness of brake pads from the wheels in freight wagons. In this manner, the inclination of the brake pads upper ends and harmful friction of the upper ends of the pads on the wheels during movement in trains without braking is eliminated. Thanks to the experimental devices, the working mass of the pads wears out only during braking. This leads to significantly less overall wear of the composite brake pads (Figure 3). The obtained wear values of the composite brake pads of the examined test wagons proved the probability of their lifespan until the wagons have traveled at least 210,000 km. This means that it is before the first depot repair of new wagons without replacing the brake pads. It should be noted that the periodical adjustment of the lever transmission mechanism by rearranging the rollers in the hinged unit of the BBS is needed.

The values of the pads' wear of their upper and lower parts as the most worn during the milage of 121.3 thousand km of freight wagons were processed by means of the Statistica 12.5 software (Informer Technologies, Altamor Drive, Los Angeles, CA, USA). The resulting histograms are created based on X-quadrate criterion [6] depicted in Figure 4.

It can be seen in the histograms (Figure 4) that there is a rather large range of wear values of composite brake pads—from the minimum value of 14 mm to the maximum value of 32 mm. This indicates that the nature and intensity of brake pad wear is influenced by a large number of various factors requiring the special research.

The information obtained during the research of the brake pads' wear depending on the freight wagons' mileage in operating conditions was subject to careful processing. The general scheme of the freight wagons' technical condition of freight wagons is possible to visualize with the help of statistical processing methods. In this way, it is possible to create the favorable conditions for the further serial introduction of updated freight wagons designs into a production at wagon-building plants or their modernization at wagon repair enterprises of the Ukrzaliznytsia Stock Company (Kyiv, Ukraine).

Calculations were made for the wear's average value of all examined wagons with the modernized BBS [23].

(a) (b)

Figure 3. A sample a composite brake pad with uniform wear with regard to the thickness: (**a**) the first brake pad in the running direction (marked as Л1); (**b**) the second brake pad in the running direction (marked as П2).

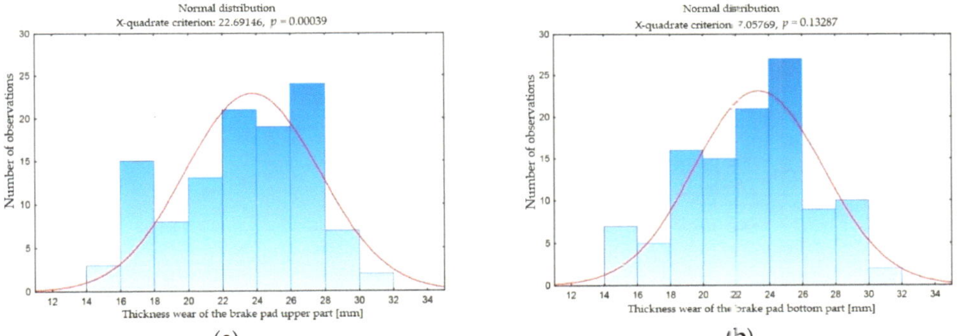

(a) (b)

Figure 4. The results of the statistical processing regarding the wear of composite brake pads during the mileage of 121.3 thousand km of freight wagons: (**a**) the brake pad upper part; (**b**) the brake pad lower part.

One bogie includes eight brake pads, and the total number of shoes under study was of 80 units. Tables 1 and 2 show the average parameter values for the pads of each wagon. The reliability of the sample was checked using the Student criterion.

$$n = \frac{t^2 \cdot \sigma^2}{\delta^2}, \quad (1)$$

where t is determined from the ratio $\Phi(t) = \gamma/2$; $\Phi(t)$ is the Laplace function; i is a tabled value; σ is the root mean square deviation of the random variable under study, which must be known in advance, even before experimental measurements; δ^2 is an absolute error of the measurement result.

Table 1. Average values of measured wear at the upper part of composite brake pads.

Inventory Number of an Examined Wagon	The Sequence of Thickness Reduction of the Upper Part of the Brake Pads y_i [mm] with Increasing Mileage S [Thousand km]						
	4.60	**16.20**	**24.10**	**74.10**	**121.30**	**164.60**	**197.80**
61138707	0.75	3.63	5.75	17.38	25.50	33.25	40.63
61139481	0.50	3.13	5.88	16.00	23.63	32.75	40.63
61138970	0.50	4.13	6.75	13.63	22.00	29.75	37.13
61139168	0.13	2.88	6.25	16.00	24.25	33.13	39.50
61139317	0.88	3.25	6.13	16.13	23.88	31.63	38.25
61139176	1.13	3.63	6.88	16.00	24.88	32.38	38.38
61139556	0.25	3.25	5.38	15.13	22.75	30.63	37.25
61139531	0.25	3.00	6.25	16.00	22.75	31.25	39.25
61140083	1.13	3.88	6.25	17.50	25.00	33.25	39.63
61140307	0.38	3.00	5.63	16.00	23.38	30.63	37.38
Average value for all wagons	0.59	3.38	6.12	15.98	23.80	31.87	38.80

Table 2. Average values of measured wear at the bottom part of composite brake pads.

Inventory Number of an Experimental Wagon	The Sequence of Thickness Reduction of the Bottom Part of the Brake Pads y_i [mm] with Increasing Mileage S [Thousand km]						
	4.60	**16.20**	**24.10**	**74.10**	**121.30**	**164.60**	**197.80**
61138707	0.50	3.25	5.25	13.25	20.63	29.50	36.50
61139481	0.63	3.25	6.25	15.88	23.38	31.75	39.38
61138970	0.63	4.00	7.00	14.00	22.38	30.25	37.00
61139168	0.38	3.38	6.13	15.38	23.63	31.38	38.25
61139317	0.63	3.25	5.88	14.5	22.25	29.75	36.38
61139176	0.38	2.63	6.00	14.75	22.25	29.75	36.50
61139556	0.63	3.13	5.88	15.38	23.13	30.88	38.00
61139531	0.63	2.88	5.50	18.00	25.88	33.25	40.50
61140083	0.38	2.50	5.13	14.88	22.75	30.63	37.50
61140307	0.88	3.13	5.88	16.13	24.13	30.75	37.25
Average value for all wagons	0.57	3.14	5.89	15.22	23.04	30.79	37.73

As is known from the statistical theory [52–54], there are two sources of information when considering any task using the laws of mathematical statistics. The first source is the results of observations (examinations). In addition, the observation process can be adjusted based on the previous results (the so-called sequential analysis). The second source is a priori information about the properties of the object under the study, i.e., it is the wear and tear accumulated at this moment for a certain mileage of a wagon. This information is displayed in the model, which is selected for the task consideration.

3. The Research Results

The wear value was determined considering the theory of statistical inferences [55] based on the data given in Tables 1 and 2, where n_i is the average value of brake pad wear for all wagons. Its value equals 60.27 mm for the upper part and it equals 58.19 mm for the bottom part of the pads. The median interval is from 24.1 to 74.1 thousand km, and the frequency equals 15.98 mm for the upper part and 15.22 mm for the bottom part of the pads.

The median is obtained by the formula:

$$M_e = x_{M_e} + \frac{\frac{\sum n_i}{2} - sn_{M_{e-1}}}{n_{M_e}} \cdot h \tag{2}$$

where x_{M_e} is the beginning of the median interval, n_{M_e} is the median interval frequency, and $sn_{M_{e-1}}$ is the accumulated frequency of the interval preceding the median.

The histograms of relative frequencies of the variation series were created (Figure 5) based on the results of the performed calculation. A horizontal axis of these histograms represents the partial intervals of the length h, and their heights equal to the relative frequency densities f_i. The area of the i-th partial rectangle is determined by the formula:

$$hf_i = h\left(\frac{\omega_i}{h}\right) = \omega_i, \tag{3}$$

where ω_i is the relative frequency of the options belonging to the i-th interval.

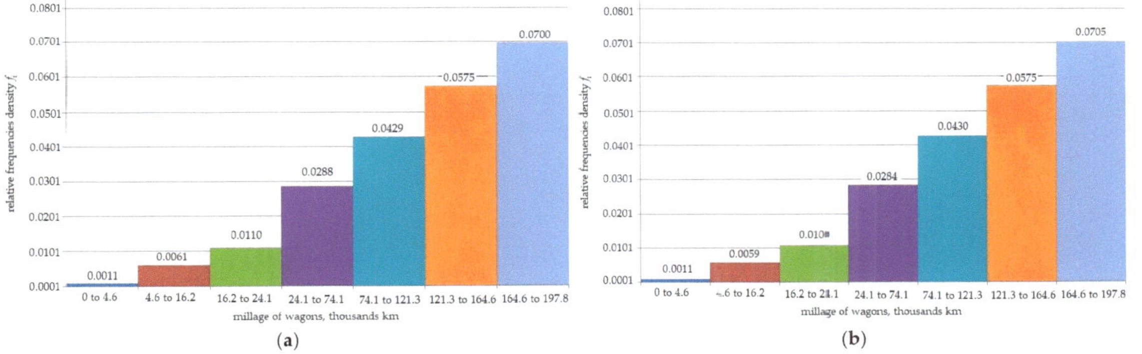

(a) (b)

Figure 5. Histograms of the density distribution of the average wear of the composite brake pads depending on the mileage of the examined wagons: (**a**) the upper part of the pads; (**b**) the bottom part of the pads.

The area of the histogram of relative frequencies equals the sum of all relative frequencies, i.e., 1.0.

All calculations were performed using the Excel software (Microsoft Corporation, Redmond, Washington, DC, USA) and the results are summarized in Tables 3 and 4.

To characterize the properties of a statistical distribution in the mathematical statistics, the concept of an empirical distribution function is introduced as follows:

$$F_*(x) = \frac{n_x}{n} = \sum_{i:x_i<x} \frac{n_i}{n}, \tag{4}$$

where n is the volume of a sample; n_x is the frequency of the variant of the value χ_i, which is smaller than x.

The empirical function $F_*(x)$ was used to estimate the theoretical distribution function of the general population. Their difference consists of the fact that the theoretical function $F(x)$ determines the probability of the event $x_i < x$, and the empirical function $F_*(x)$ represents the relative frequency of this event [52,54].

Table 3. The results of the calculated average wear of the upper part of the composite brake pads of the modernized BBS.

Mileage Interval (Thousand km)	a_{i-1}	a_i	x_i	n_i	$x_i n_i$	$x_i^2 n_i$	ω_i	f_i	sn_i	$s\omega_i$
0 to 4.6	0	4.6	2.3	0.59	1.36	3.12	0.0049	0.0011	0.59	0.0049
4.6 to 16.2	4.6	16.2	10.4	3.38	35.15	365.58	0.028	0.0061	3.97	0.0329
16.2 to 24.1	16.2	24.1	20.2	6.12	123.62	2497.2	0.0508	0.011	10.09	0.0837
24.1 to 74.1	24.1	74.1	49.1	15.98	784.62	38,524.74	0.1326	0.0288	26.07	0.2163
74.1 to 121.3	74.1	121.3	97.7	23.8	2325.3	227,177.9	0.1974	0.0429	49.87	0.4137
121.3 to 164.6	121.3	164.6	143	31.87	4557.4	651,709.6	0.2644	0.0575	81.74	0.6781
164.6 to 197.8	164.6	197.8	181.2	38.8	7030.6	1,273,937	0.3219	0.07	120.54	1
Total				120.54	14,858	2,194,216	1			

Note: a_{i-1}—the left limits of the bogie run; a_i—the right limits of the bogie run; x_i—the average mileage of wagons; n_i—the average value of pad wear for all wagons; $x_i n_i$—weights of statistical distribution; $x_i^2 n_i$—weights of the square of the statistical distribution; ω_i—relative frequency; f_i—relative frequency density; $s_i n_i$—accumulated frequencies; $s_i \omega_i$—accumulated relative frequencies. The width of the interval $h' = 4.6$; sample mean arithmetic value of the sample $x'_{cp} = 123.26$; the arithmetic mean of the squares of the sample values $x'_{cp}{}^2 = 18,203$; sample variance $D'_B = 3010.2$; sample mean square deviation $\sigma'_B = 54.865$; the modus $M'_o = 6.283$; the median $M'_e = 1.1332$.

Table 4. The results of calculating the average wear of the bottom part of the composite brake pads of the modernized BBS.

Mileage Interval (Thousand km)	a_{i-1}	a_i	x_i	n_i	$x_i n_i$	$x_i^2 n_i$	ω_i	f_i	sn_i	$s\omega_i$
0 to 4.6	0	4.6	2.3	0.57	1.31	3.02	0.0049	0.0011	0.57	0.0049
4.6 to 16.2	4.6	16.2	10.4	3.14	32.66	339.62	0.027	0.0059	3.71	0.0319
16.2 to 24.1	16.2	24.1	20.2	5.89	118.98	2403.36	0.0506	0.011	9.6	0.0825
24.1 to 74.1	24.1	74.1	49.1	15.22	747.3	36,692.53	0.1308	0.0284	24.82	0.2133
74.1 to 121.3	74.1	121.3	97.7	23.04	2251	219,923.5	0.198	0.043	47.86	0.4113
121.3 to 164.6	121.3	164.6	143	30.79	4403	629,624.7	0.2646	0.0575	78.65	0.6759
164.6 to 197.8	164.6	197.8	181.2	37.73	6836.7	1,238,806	0.3242	0.0705	116.38	1.0
Total				116.38	14,391	2,127,792	1			

Note: the interval width $h'' = 4.6$; the sample mean arithmetic value of the sample $x''_{cp} = 123.26$; the arithmetic mean of the squares of the sample values $x''_{cp}{}^2 = 18,203$; the sample variance $D''_B = 3010.2$; the sample mean square deviation $\sigma''_B = 54.865$; the modus $M''_o = 6.283$; the median $M''_e = 1.1332$.

The empirical distribution function tends to the theoretical one in probability (converges in probability to $F_*(x)$ in the case of a large sample volume:

$$\lim_{n \to 0} P\left(|F_*(x) - F(x)| < \varepsilon\right) = 1 \quad \forall x, \quad \forall \varepsilon > 0. \tag{5}$$

at which, a sample of size n with the values $x_1, x_2, \ldots, x_p$ of the random variable x is considered. The average indicators called as the sample numerical characteristics are used to characterize the most important properties of the statistical distribution.

Furthermore, empirical distribution functions (Figure 6) were created for the upper and bottom parts of the composite brake pads with the modernized BBS of examined wagons based on the calculation results listed in Tables 3 and 4.

Today, there are several different software products, which implement the method of least squares (LSM). Their application is very diverse: statistics, econometrics, assessment of measurement errors, etc. [52,53,56,57].

In order to understand the practical implementation of one of them, the task of determining and predicting the composite brake pads' wear of the modernized BBS of freight wagons using the least squares method (LSM) was considered.

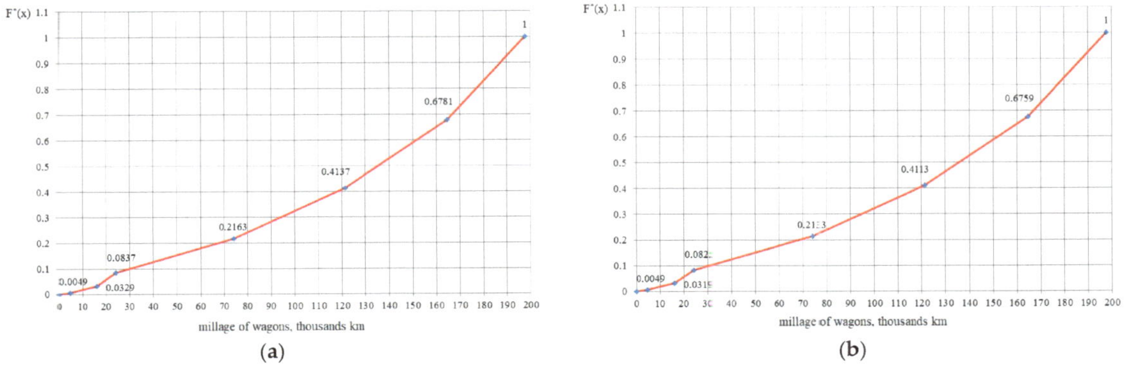

Figure 6. The graph of the empirical distribution function of the composite brake pads' average wear depending on the mileage of the examined wagons: (**a**) the upper part of the pads; (**b**) the bottom part of the pads.

The obtained results of the experiments were analyzed to compare the analytical functions in the intervals with the well-known Weierstrass theorem. The point of this procedure is to assess whether the approximate dependence of the composite brake pads' wear on the mileage of the wagons is reliably selected. At the same time, the number of members of the polynomial series is determined from the real possibilities of carrying out scheduled and preventive repairs of wagons, although this may be the reason for the disagreement according to the criteria, such as the Fisher's criterion, the Student's criterion, etc. Generally, many statisticians believe that several of known criteria do not always give a positive conclusion from the calculation results' consistency point of view. Despite that, research is confirmed in practice [58,59]. Therefore, a form of the approximate function of the regression analysis can be chosen considering the convenience of its further use. On the contrary, other experts believe that the appearance of this function should be justified.

In the performed study, a combined approach was chosen to solve this problem. The preliminary analysis of the process was investigated based on the statistical data. Then, the mathematical model has the following form:

$$\hat{y} = \beta_0 + \beta_1 \cdot x + \beta_2 \cdot x^2. \quad (6)$$

This is a nonlinear model of the second order. In a significant number of cases of choosing the type of approximation models of the braking processes of wagons, a reliable dependence of pad wear on mileage is obtained. It corresponds to the linear model of regression analysis [48,49], which satisfactorily describes the process of the studied pad wear during braking. Therefore, the formulation (6) can be reduced to linearity. To do this, it is considered that $x_1 = x$ and $x_2 = x^2$. This results in a linear model (for linear models, there are powerful algebraic tools for their research), which looks as follows:

$$\hat{y} = \beta_0 + \beta_1 \cdot x_1 + \beta_2 \cdot x_2 \quad (7)$$

There will be deviations between the values $\hat{y}$ calculated by the model and the experimental calculations $\hat{y}_i$. This deviation will be denoted as:

$$\hat{u}_i = \hat{y} - \hat{y}_i, \; i = 1, 2, \ldots, n \, . \quad (8)$$

The formulation (8) is called residues. They include the influence of unaccounted factors, namely variables, random disturbances, observation errors, etc. Their values may vary from one observation to another.

LSM makes it possible to find the following values (estimates), b_0, b_1, b_2, of the model's initial parameters β_0, β_1, β_2. This is because of the necessary criterion for selecting the coefficients of the model, which must consider the circumstance of the obtained regression function (if it is presented on a graph). It will pass as closely as possible between the experimentally obtained variables:

$$U = \sum_{i=1}^{n} U_i^2 \to \min. \tag{9}$$

Furthermore, this model will be written in the following form:

$$\hat{y} = b_0 + b_1 \cdot x_1 + b_2 \cdot x_2. \tag{10}$$

when, the partial derivatives by values b_0, b_1, b_2 are considered and they equal zero, a system of three equations with three unknowns is obtained. Its solution leads to the determination of values b_0, b_1, b_2. As, in our case:

$$U = \sum_{i=1}^{n} U_i^2 = \sum_{i=1}^{n} (\hat{y}_i - b_0 - b_i \cdot x_{i1} - b_2 \cdot x_{i2})^2, \tag{11}$$

the following is obtained:

$$\begin{cases} \dfrac{\partial U}{\partial b_0} = -2 \cdot \sum\limits_{i=1}^{n} (\hat{y}_i - b_0 - b_i \cdot x_{i1} - b_2 \cdot x_{i2}) = 0, \\ \dfrac{\partial U}{\partial b_1} = -2 \cdot \sum\limits_{i=1}^{n} (\hat{y}_i - b_0 - b_i \cdot x_{i1} - b_2 \cdot x_{i2}) \cdot x_{i1} = 0, \\ \dfrac{\partial U}{\partial b_2} = -2 \cdot \sum\limits_{i=1}^{n} (\hat{y}_i - b_0 - b_i \cdot x_{i1} - b_2 \cdot x_{i2}) \cdot x_{i2} = 0. \end{cases} \tag{12}$$

The system of Equation (12) results in the following system of algebraic equations:

$$\begin{cases} b_0 n + b_1 \cdot \sum\limits_{i=1}^{n} x_{i1} + b_2 \cdot \sum\limits_{i=1}^{n} x_{i2} = \sum\limits_{i=1}^{n} \hat{y}_i, \\ b_0 \cdot \sum\limits_{i=1}^{n} x_{i1} + b_1 \cdot \sum\limits_{i=1}^{n} x_{i1}^2 + b_2 \cdot \sum\limits_{i=1}^{n} x_{i1} \cdot x_{i2} = \sum\limits_{i=1}^{n} x_{i1} \cdot \hat{y}_i, \\ b_0 \cdot \sum\limits_{i=1}^{n} x_{i2} + b_1 \cdot \sum\limits_{i=1}^{n} x_{i1} \cdot x_{i2} + b_2 \cdot \sum\limits_{i=1}^{n} x_{i2}^2 = \sum\limits_{i=1}^{n} x_{i2} \cdot \hat{y}_i. \end{cases} \tag{13}$$

The solution of this system (13) leads to the unknown coefficients b_0, b_1, b_2:

$$b_0 = \hat{y} - b_1 \cdot x_1 - b_2 \cdot x_2. \tag{14}$$

$$b_1 = \frac{\sum\limits_{i=1}^{n} (x_{i1} - \overline{x}_1) \cdot (\hat{y}_i - \overline{y}) \cdot \sum\limits_{i=1}^{n} (x_{i2} - \overline{x}_2)^2}{\sum\limits_{i=1}^{n} (x_{i1} - \overline{x}_1)^2 \cdot \sum\limits_{i=1}^{n} (x_{i2} - \overline{x}_2)^2 - \left(\sum\limits_{i=1}^{n} (x_{i1} - \overline{x}_1) \cdot (x_{i2} - \overline{x}_2)\right)^2} - \frac{\sum\limits_{i=1}^{n} (x_{i2} - \overline{x}_2) \cdot (\hat{y}_i - \overline{y}) \cdot \sum\limits_{i=1}^{n} (x_{i1} - \overline{x}_1) \cdot (x_{i2} - \overline{x}_2)}{\sum\limits_{i=1}^{n} (x_{i1} - \overline{x}_1)^2 \cdot \sum\limits_{i=1}^{n} (x_{i2} - \overline{x}_2)^2 - \left(\sum\limits_{i=1}^{n} (x_{i1} - \overline{x}_1) \cdot (x_{i2} - \overline{x}_2)\right)^2}. \tag{15}$$

$$b_2 = \frac{\sum\limits_{i=1}^{n} (x_{i2} - \overline{x}_2) \cdot (\hat{y}_i - \overline{y}) \cdot \sum\limits_{i=1}^{n} (x_{i1} - \overline{x}_1)^2}{\sum\limits_{i=1}^{n} (x_{i1} - \overline{x}_1)^2 \cdot \sum\limits_{i=1}^{n} (x_{i2} - \overline{x}_2)^2 - \left(\sum\limits_{i=1}^{n} (x_{i1} - \overline{x}_1) \cdot (x_{i2} - \overline{x}_2)\right)^2} - \frac{\sum\limits_{i=1}^{n} (x_{i1} - \overline{x}_1) \cdot (\hat{y}_i - \overline{y}) \cdot \sum\limits_{i=1}^{n} (x_{i1} - \overline{x}_1) \cdot (x_{i2} - \overline{x}_2)}{\sum\limits_{i=1}^{n} (x_{i1} - \overline{x}_1)^2 \cdot \sum\limits_{i=1}^{n} (x_{i2} - \overline{x}_2)^2 - \left(\sum\limits_{i=1}^{n} (x_{i1} - \overline{x}_1) \cdot (x_{i2} - \overline{x}_2)\right)^2}. \tag{16}$$

Substituting their values into the general form of regression results in the so-called regression line with coefficients b_1 and b_2, which are called regression coefficients $\hat{y}$ for x_1 and $\overline{y}$ for x_2, respectively.

The found point (b_0, b_1, b_2) is the point of satisfaction of the mentioned condition. However, regarding the mathematical analysis, there is a theorem that makes it possible to determine sufficient conditions for the extremum of a function. In our case, the minimum of a function $U(b_0, b_1, b_2)$.

Thus, a regression model was developed for predicting the residual life of composite brake pads with the modernized BBS, which considers the mileage of the freight wagons—general and additional. The additional mileage is the distance covered by the wagon during maintenance with uncoupling or repair, i.e., shunting work at a station, shunting operations associated with the supply of wagons (loaded or empty) along the tracks to industrial enterprises, or from them to the main railway tracks.

Figure 7 shows the graphs of the linear regression created based on the results of experimental data processing (Tables 5 and 6). This makes it possible to predict the residual resource of composite brake pads for their maximum thickness of 10 mm [29] for the modernized BBS in the case of their use for the entire inter-repair operation period of freight wagons. Under the condition of installing the modernized BBS, the average mileage of a freight wagon in case of the uniform wear of brake pads can reach 284.57 thousand km.

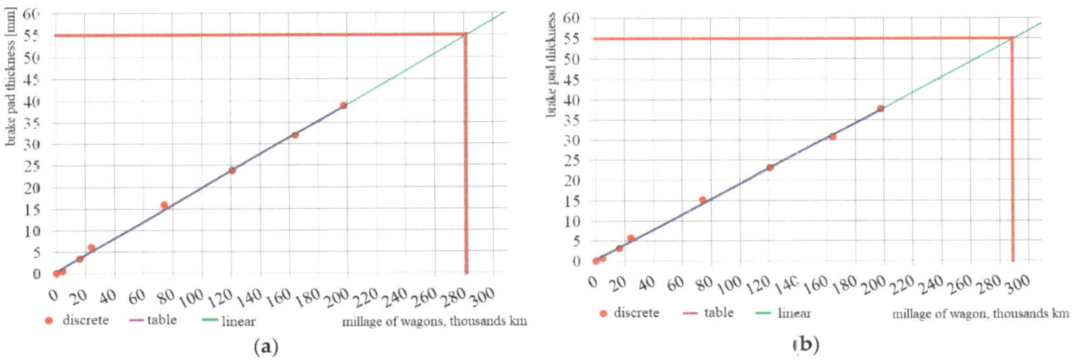

Figure 7. Graphs of the linear regression with a prediction of the composite brake pads' life extension depending on their average wear, taking into account the mileage of the examined wagons: (**a**) the upper part of the brake pads; (**b**) the bottom part of the brake pads.

Table 5. The results of calculating the average residual life of the upper part of composite brake pads with the modernized BBS.

x	$\hat{y}$	x^2	$x\hat{y}$	y_{lin}	d_{bid}	d^2
0	0	0	0	0.46936	−0.4694	0.22029
4.6	0.59	21.16	2.714	1.3607	−0.7707	0.59397
16.2	3.38	262.44	54.756	3.60843	−0.2284	0.05218
24.1	6.12	580.81	147.492	5.13921	0.98079	0.96195
74.1	15.98	5490.81	1184.12	14.8277	1.15229	1.32777
121.3	23.8	14,713.7	2886.94	23.9737	−0.1737	0.03016
164.6	31.87	27,093.2	5245.8	32.3639	−0.4939	0.24393
197.8	38.8	39,124.8	7674.64	38.7971	0.00295	8.7×10^{-6}
602.7	120.54	87,287	17,196	-	-	3.4303

It can be seen in the linear regression graphs (Figure 7) that it is not necessary to replace the composite brake pads by using modernized BBSs with devices for uniform removal of the pads during the inter-repair period of the wagon operation. The combined criterion applied in the studies confirms that the mileage of freight wagons repaired according to the standard frequency should be of 110,000 km after the last depot repair, and of 160,000 km for the last major repair [49].

A similar technique is used in the case of finding m regression coefficients $b_0, b_1, b_2, \ldots, b_m$.

To estimate the density of the connection between x and y, the correlation coefficient and the coefficient of determination is used. It shows how much the variation of the variable x explains the variation of y. The standard error of the residuals is applied to evaluate how well the regression line explains the relationship between x and y. It shows the deviation of the empirical values from the regression line.

Table 6. The results of calculating the average residual life of the bottom part of composite brake pads with the modernized BBS.

x	$\hat{y}$	x^2	$x\hat{y}$	y_{lin}	d_{bid}	d^2
0	0	0	0	0.37035	−0.3704	0.13716
1	2	3	4	5	6	7
4.6	0.57	21.16	2.622	1.23599	−0.666	0.44354
16.2	3.14	262.44	50.868	3.4189	−0.2789	0.07778
24.1	5.89	580.81	141.949	4.90553	0.98447	0.96917
74.1	15.22	5490.81	1127.8	14.3146	0.90538	0.8197
121.3	23.04	14,713.7	2794.75	23.1968	−0.1568	0.02459
164.6	30.79	27,093.2	5068.03	31.3451	−0.5551	0.30811
197.8	37.73	39,124.8	7462.99	37.5927	0.13728	0.01885

Testing the hypothesis about its adequacy is an important subsequent step in the creation of a regression model. To do this, the F-criterion [51–53] is considered. Furthermore, the t-test was applied to test the hypothesis, whether the estimated regression parameters are statistically significant or different from zero. The F-criterion of the model adequacy has the form:

$$F_{1,n-2} = \frac{(n-2)\cdot \sum\limits_{i=1}^{n}(\hat{y}_i-\overline{y})^2}{\sum\limits_{i=1}^{n}(\hat{y}_i-\overline{y})^2}. \quad (17)$$

The observed value of the t-test to test the significance of the correlation coefficient is determined by the expression:

$$t_{n-2} = \frac{r\cdot\sqrt{n-2}}{\sqrt{1-r^2}}. \quad (18)$$

The observed value of the t-test for testing the hypothesis:

$$H_0: \ \ b_1 = \beta_i \ (H_1 : b_i = \beta_i) : \ t_{n-2} = \frac{b_i - \beta_i}{\hat{\sigma}_{b_i}}, \ \ i = 0,1.$$

$$\hat{\sigma}_{b_0} = \hat{\sigma}_{\varepsilon}\cdot\sqrt{\frac{\sum\limits_{i=1}^{n}x_i^2}{n\cdot\sum\limits_{i=1}^{n}(x_i-\overline{x})^2}}, \hat{\sigma}_{b_1} = \frac{\hat{\sigma}_{\varepsilon}}{\sqrt{\sum\limits_{i=1}^{n}(x_i-\overline{x})^2}}, \hat{\sigma}_{\varepsilon}^2 = \frac{\sum\limits_{i=1}^{n}(y_i-\hat{y}_i)^2}{n-2}. \quad (19)$$

And, the observed value of the t-test for testing the hypothesis:

$$H_0: \ \ b_i = 0 \ (H_1 : b_i \neq 0) \ \ t_{n-2} = \frac{b_i}{\hat{\sigma}_{b_i}}.$$

If the straight line is placed so that the points are approximately equally placed on both sides of it, then the parameters of the linear equation can be determined quite simply.

LSM is most often used to determine the parameters of the equation of form $\overline{y} = k\cdot x + b$. At the same time, the condition of the sum of the squared deviations (distances) of all investigated points from the ordinates calculated according to the equation of the straight line ε_i, is minimal. In other words, the straight line should pass as close as possible to the peaks of the empirical regression line. This means that the parameters k and b of the regression equation must be determined from the expression:

$$\varepsilon_i = \sum_{i=1}^{n} (y_i - \widetilde{y}_i) = \min \tag{20}$$

where y_i are the ordinates of the studied points; $\widetilde{y}_i$—the ordinates of the calculation points determined by the regression equation $\overline{y} = kx + b$ as follows:

$$\varepsilon_i = \sum_{i=1}^{n} [y_i - (k \cdot x + b)]^2 = F(k, b)\min. \tag{21}$$

A necessary condition for the extremum of this function is the equality of partial derivatives taken by the parameters k and b:

$$\frac{\partial F}{\partial k} = 0 \text{ and } \frac{\partial F}{\partial b} = 0, \quad [F(u)]' = F_u(u) * u'. \tag{22}$$

From here:

$$\begin{aligned} \frac{\partial F}{\partial k} &= -2 \cdot \sum_{i=1}^{n} [y_i - (k \cdot x_i + b)] \cdot x_i = 0, \\ \frac{\partial F}{\partial k} &= -2 \cdot \sum_{i=1}^{n} [y_i - (k \cdot x_i + b)] = 0. \end{aligned} \tag{23}$$

Simplifying these expressions, the system of linear equations is obtained as follows:

$$\begin{aligned} \sum_{i=1}^{n} y_i \cdot x_i &= k \cdot \sum_{i=i}^{n} x_i^2 + b \cdot \sum_{i=1}^{n} x_i, \\ \sum_{i=1}^{n} y_i \cdot x_i &= k \cdot \sum_{i=i}^{n} x_i + b \quad n. \end{aligned} \tag{24}$$

The parameters k and b are found by substituting the numerical values of the corresponding quantities into the system.

Similar calculations are performed using the Cochrane quality criterion. However, due to this condition, the number of measurement points is increased to 53. The results of the calculations are described below.

The validity of the null hypothesis was checked according to the Cochrane quality criterion. The ratio of the maximum corrected variance to the sum of all corrected variances was considered:

$$G_{ChC} = \frac{S_{max}^2}{S_1^2 + S_2^2} = \frac{116.65}{116.65 + 109.66} = 0.5154. \tag{25}$$

The distribution of this random variable depends on the number of degrees of freedom $k = n - 1 = 52$, where n is the sample size, and it equals the value of 53.

The critical point G_{cr} (0.05;52;2) = 0.6543 was found according to the table in the appendix of the documents [51,57,58] at the significance level of 0.05, the number of degrees of freedom being 52, and the sample value $l = 2$ (wear from above and below).

As $G_{cr} > G_{ChC}$ and their difference is insignificant, there is no reason to reject the null hypothesis. Therefore, the final general variance under the condition of homogeneity of variances is established as follows:

$$D(x) = \frac{1}{2} \cdot \left(S_1^2 + S_2^2\right). \tag{26}$$

Thus, the corrected variances are practically equal. i.e., $S_1^2 \approx S_2^2$.

Whereas the data in the task of sampling the wear of the upper and bottom parts of the composite brake pads have the same volume, the comparison of two average normal general populations is considered [54]. When the notation $d_i = (X_1)_i - (X_2)_i$ is introduced, the corrected mean square deviation equals:

$$s_d = \sqrt{\frac{\sum d_i^2 - \frac{[\sum d_i]^2}{n}}{n-1}} = 0.342598\,. \quad (27)$$

In order to test the null hypothesis for a given significance level of α = 0.05, it is necessary to determine the value of the observed criterion:

$$T_{ChC} = \frac{\overline{d} \cdot \sqrt{n}}{s_d} = 13.90331\,, \quad (28)$$

where $\overline{d} = \sum d_i / n$ is the average difference.

Based on the critical points of the distribution, the given level of significance α = 0.05, and the number of degrees of freedom $k = n - 1 = 52$, the value of the critical point $t_{BV,cr}$ = 2.01 is found. Hence, it can be concluded based on the findings $|T_{ChC}| > t_{BV,cr}$ that the obtained results are a confirmation of the fact that the wear of the upper and bottom parts of the composite brake pads has different values on average.

4. Discussion

To eliminate the wedge-dual wear of the brake pads of freight wagons with a gauge of 1520 mm, the modernization of the BBS is proposed. This modernization consists of changing the location of the technological hinge B (Figure 2), which connects the vertical lever with the strut of the triangle. The specified hinge is located on the same line A-A with the hinges of the pendulum suspensions. In addition, a curved rod is introduced, which, with its ends in the sliders, keeps the BBS constantly in balance, which ensures strictly uniform gaps between the blocks and wheels and guarantees their normative wear.

To justify the proposed modernization, complex studies of the technical condition of the BBS during the maintenance of freight wagons in the fleets of the sorting station were carried out.

Based on the collected statistical material (Tables 1 and 2), a regression model was built. It allows the residual resource of composite brake pads to be predicted with a modernized BBS. The peculiarity of this model is that it considers the total and additional mileage of freight wagons.

As part of the study, graphic dependencies of the composite brake pads' wear with the modernized BBS were formed in terms of thickness, which depends on the mileage of freight wagons. It was established that there is no need to replace the composite brake pads, considering the use of modernized BBSs during the inter-repair period of the wagon's operation.

To check the adequacy of the created model, its verification was carried out using the F-criterion. The results of the calculation established that the hypothesis of adequacy is not rejected.

This study has certain advantages in comparison with known ones. For example, in contrast to works [16,17,32,33], the team of authors predicted the residual resource of brake pads, and did not determine their load in operation. In contrast to works [18,28], within the framework of this study, solutions aimed at increasing the resource of brake pads in operation are proposed. The model proposed by the authors in this research considers the brake pads' wear by thickness during operation, which was not studied in works [19,30,31]. In comparison with the research results highlighted in works [20–22], this presented research proposes the improvement of the BBS to increase the resource of the brake pads. In contrast to works [23,24,27], the most unfavorable type of brake pads' wear was considered and a solution was proposed for its elimination. In comparison with studies [25,26,29], the authors evaluated the brake pads' wear considering the proposed modernization of the BBS and not typical designs. Unlike the works [46–48], the authors examined the most unfavorable operating conditions of brake pads. This will allow solutions aimed at increasing their service life in the future to be proposed.

The described methodology can be also applied to other means of railway transport, such as trams, passenger wagons, etc. [59–62].

Despite the fact described above, this presented study has a certain limitation. At this stage, the authors did not consider the influence of over-normalized modes on the brake pads' wear of the modernized BBS, or the influence of the inherent degree of freedom of the cargo loaded in a wagon on the braking efficiency, or the malfunction of the harness device of the automatic coupler, etc.

One of the main future challenges of this study consists of the fact that the efficiency of the modernized design of the BBS has been proven only on an open wagon type so far. The effectiveness of such a BBS on other wagon types has not been considered. However, it should be noted that this wagon type makes up more than 50% of the inventory of the Ukrzaliznytsia Stock Company (Kyiv, Ukraine). These questions will be considered as a further development of the research.

The results of this study will contribute to increasing the efficiency of the operation of rail vehicles and the profitability of railway transport as a whole. The conducted research will also contribute to improving the environmental friendliness of rail transport [63,64].

5. Conclusions

1. Comprehensive experimental studies of the technical condition of modernized brake systems in operational conditions during the maintenance of freight wagons in the fleets of the sorting station were carried out. During each run of experimental freight wagons with the modernized BBS, the parameters of composite brake pads were measured during maintenance at control points in the range of 2.1 to 197.8 thousand km mileage.

2. A regression model was developed for predicting the residual life of composite brake pads with the modernized BBS, which considers the total and additional mileage of the freight wagon.

It was established that, under the condition of using the modernized BBS on freight wagons, the average mileage of a wagon with uniform wear of composite brake pads can reach up to 284.57 thousand km. It was established according to the results of mathematical calculations that the use of the modernized BBS helps to increase the resource of composite brake pads by approximately 2.59 times.

3. The verification of statistical hypotheses was carried out according to the nature of the distribution of random values of the wear of composite brake pads with the modernized BBS and their relationship regarding the data belonging to one general inventory.

It was established that the average wear of composite brake pads of the upper and bottom part of the brake pads has different values. It is proved according to the results of the calculations that the heterogeneity of the dispersion does not depend on the quantitative indicator of the sample of this experiment.

Author Contributions: Conceptualization, A.L. and V.R.; methodology, A.L., V.R. and J.D.; software, S.P., J.G. and V.R.; validation, A.L. and J.D.; formal analysis, A.L. and J.D.; investigation, J.G. and A.L.; resources, A.L., V.R. and J.D.; data curation, A.L. and V.R.; writing—original draft preparation, A.L., V.R. and J.D.; writing—review and editing, S.P. and J.G.; visualization, V.R. and J.D.; supervision, S.P. and J.G.; project administration, J.G., A.L. and J.D.; funding acquisition, J.G., A.L. and J.D. All authors have read and agreed to the published version of the manuscript.

Funding: This publication was supported by the Cultural and Educational Grant Agency of the Ministry of Education of the Slovak Republic under the project KEGA 024ZU-4/2024: Deepening the knowledge of university students in the field of construction of means of transport by carrying out professional and scientific research activities in the field. It was also supported by the Slovak Research and Development Agency of the Ministry of Education, Science, Research, and Sport under the project VEGA 1/0513/22: Investigation of the properties of railway brake components in simulated operating conditions on a flywheel brake stand. Funded by the EU NextGenerationEU under the Recovery and Resilience Plan for Slovakia under the project No. 09I03-03-V01-00131.

Data Availability Statement: Data are contained within the article.

Conflicts of Interest: The authors declare no conflicts of interest.

References

1. Kardas-Cinal, E. Statistical analysis of dynamical quantities related to running safety and ride comfort of a railway vehicle. *Sci. J. Silesian Univ. Technol. Ser. Transp.* **2020**, *106*, 63–72. [CrossRef]
2. Gerlici, J.; Lovska, A.; Vatulia, G.; Pavliuchenkov, M.; Kravchenko, O.; Solcansky, S. Situational adaptation of the open wagon body to container transportation. *Appl. Sci.* **2023**, *13*, 8605. [CrossRef]
3. Fischer, S.; Szürke, S.K. Detection process of energy loss in electric railway vehicles. *Facta Univ. Ser. Mech. Eng.* **2023**, *21*, 81–99. [CrossRef]
4. Kazemian, M.; Rad, M.M.; Shadfar, M.; Doost, A.M.; Raisi, E.H.; Fischer, S. Optimum train weighing in motion using inertial sensors. *Acta Polytech. Hung.* **2024**, *21*, 221–240. Available online: https://acta.uni-obuda.hu/Kazemian_MovahediRad_Shadfar_Doost_HadizadehRaisi_Fischer_141.pdf (accessed on 10 July 2024). [CrossRef]
5. Panchenko, S.; Gerlici, J.; Vatulia, G.; Lovska, A.; Rybin, A.; Kravchenko, O. Strength assessment of an improved design of a tank container under operating conditions. *Commun. Sci. Lett. Univ. Zilina* **2023**, *25*, B.186–B.193. [CrossRef]
6. Koch, S.; Köppen, E.; Gräbner, N.; von Wagner, U. On the influence of multiple equilibrium positions on brake noise. *Facta Univ. Ser. Mech. Eng.* **2021**, *19*, 613–632. [CrossRef]
7. Panchenko, S.; Gerlici, J.; Vatulia, G.; Lovska, A.; Ravlyuk, V.; Harusinec, J. Studying the load of composite brake pads under high-temperature impact from the rolling surface of wheels. *EUREKA Phys. Eng.* **2023**, 155–167. [CrossRef]
8. Leitner, B. Autoregressive models in modelling and simulation of transport means working conditions. In Proceedings of the 14th International Conference Transport Means, Kaunas, Lithuania, 20–22 October 2010; pp. 21–24.
9. Sventekova, E.; Leitner, B.; Dvorak, Z. Transport critical infrastructure in Slovak republic. In Proceedings of the 8th International Multi-Conference on Complexity, Informatics and Cybernetics, IMCIC 2017, Orlando, FL, USA, 21–24 March 2017; pp. 212–215.
10. Adeli, H.; Mirabadi, A.; Yousefi, S.; Fazel, S. Novel Approach for Reducing Train Wheel and Brake Shoe Wear Using Speed Profile Optimization. *Transp. Res. Rec.* **2024**, *2678*, 241–250. [CrossRef]
11. Casagrande, R.B.; De Souza, A.R.A.; Finimundi, A.V.; Pereira, C.H.S.; Masotti, D.; Rombaldi, R.J.; Gotardo, T. *Development of Brake Shoe with Carbon Footprint Reduction*; SAE Technical Papers; SAE International: Warrendale, PA, USA, 2024. [CrossRef]
12. Wang, B.; Wu, Q.; Luo, S.; Spiryagin, M.; Ma, W.; Ding, H.; Yang, Z. Wheel tread temperature assessment and its impact on rolling contact fatigue under long-term braking conditions. *Tribol. Int.* **2024**, *200*, 110162. [CrossRef]
13. Vakulenko, I.; Plitchenko, S.; Bolotova, D.; Asgarov, K. Influence hot plastic deformation on the structure and properties of carbon steel of the railway wheel. *Sci. J. Silesian Univ. Technol. Ser. Transp.* **2023**, *121*, 257–266. [CrossRef]
14. Kou, L.; Sysyn, M.; Liu, J. Influence of crossing wear on rolling contact fatigue damage of frog rail. *Facta Univ. Ser. Mech. Eng.* **2024**, *22*, 25–44. [CrossRef]
15. Szalai, S.; Herold, B.; Kurhan, D.; Németh, A.; Sysyn, M.; Fischer, S. Optimization of 3D printed rapid prototype deep drawing tools for automotive and railway sheet material testing. *Infrastructure* **2023**, *8*, 43. [CrossRef]
16. Ambikaprasad, O.C.; Abhijeet, A.R. Failure Analysis of Brake Shoe in Indian Railway Wagon. *IPASJ Int. J. Mech. Eng.* **2015**, *3*, 37–41. Available online: https://www.scribd.com/document/294429738/Failure-Analysis-of-Brake-Shoe-in-Indian-Railway-Wagon (accessed on 5 June 2024).
17. Ravlyuk, V.G. Investigation of features of dual wear of pads in brake system of freight cars. *Sci. Transp. Prog.* **2019**, *80*, 111–126. (In Ukrainian) [CrossRef]
18. Koropets, P.A.; Chernikov, V.D.; Kostiukevych, A.I. On uneven wear of moving friction pairs. *Visn. Volodymyr Dahl East Ukr. Natl. Univ.* **2010**, *147*, 41–45. (In Ukrainian)
19. Koptovec, A.N. Identification of mine locomotive brake state in terms of structural uncertainty. *Geo-Tech. Mech.* **2014**, *119*, 241–246. (In Russian)
20. Bakardzhyiev, R.O.; Komarov, A.O. Features of evaluation of results regression analysis. *Collect. Work. Kirovohrad Natl. Tech. Univ. Mach. Agric. Prod. Ind. Mach. Build. Autom.* **2015**, *28*, 255–260. Available online: https://dspace.kntu.kr.ua/server/api/core/bitstreams/e05008bf-abe4-497f-97dd-5c66717b7ee0/content (accessed on 4 June 2024). (In Ukrainian).
21. Gorobchenko, O.; Tkachenko, V. Statistical analysis of locomotives traction motors performance. In Proceedings of the 6th International BAPT Conference "Power Transmissions 2019", Varna, Bulgaria, 19–22 June 2019. [CrossRef]
22. Leng, L.; Zhu, W. Compound Regression and Constrained Regression: Nonparametric Regression Frameworks for EIV Models. *Am. Stat.* **2019**, *74*, 226–232. [CrossRef]
23. *Development of Design and Technological Documentation for the Modernization of Brake Lever Gears of Freight Wagon Bogies: Report on R&D (Zvit pro NDR) (Zaklyuch.)*; Ukrainian State University of Railway Transport, Ukrajinskij Derzhavnij Universytet Zaliznychnoho Transportu: Kharkiv, Ukraine, 2012; 53p, No. R0111U008972.
24. Zharov, A.; Makas, A. Methods of statistical processing results of brake distance measurement when evaluating efficiency of rolling stock brake systems. *VNIIZHT Bull.* **2009**, *5*, 29–33. (In Russian)

25. Muradian, L.A.; Shaposhnik, V.Y.; Shikunov, O.A. Malfunctions of braking equipment and defects of wheel-sets of freight wagons. *Bull. Railw. Transp. Certif.* **2021**, *67*, 5–15. Available online: https://crust.ust.edu.ua/items/c573639c-fae1-484e-a432-763ab223c03c (accessed on 2 September 2024). (In Ukrainian).
26. Muradian, L.A.; Shaposhnik, V.Y.; Vinstrot, B.U.; Mukovoz, S.P. Testing of promising brake pads on the railways of Ukraine. *Locomot. Inf.* **2015**, *7/8*, 20–22. Available online: https://crust.ust.edu.ua/items/bf014671-d296-4017-93d0-eca4560f1faf (accessed on 28 May 2024). (In Ukrainian).
27. Koptovets, O.; Haddad, J.S.; Brovko, D.; Posunko, L.; Tykhonenko, V. Identification of the conditions of a mine locomotive brake system as well as its functional and morphological model with the stressed closed kinematic circuit. In Proceedings of the XIV International Research and Practice Conference "Ukrainian School of Mining Engineering". Berdiansk, Ukraine, 7–11 September 2020; Available online: https://doi.org/10.1051/e3sconf/202020101033 (accessed on 21 November 2024). [CrossRef]
28. Mazur, V.L.; Naidek, V.L.; Popov, Y.S. Comparison of cast iron and compositional with iron insert brake pad inserts for railway rolling stock. *Met. Cast. Ukr.* **2021**, *29*, 30–39. (In Ukrainian) [CrossRef]
29. Instructions for Operating Rolling Stock Brakes on the Railways of Ukraine. *CT-CV-CL-0015. [Effective from 10-28-1997 No. 264-C.*; Polygraph Service LLC.: Kyiv, Ukraine, 2004; 146p.
30. Vakkalagadda, M.R.K.; Srivastava, D.K.; Mishra, A.; Racherla, V. Performance analyses of brake blocks used by Indian Railways. *Wear* **2015**, *328–329*, 64–76. [CrossRef]
31. Vineesh, K.P.; Vakkalagadda, M.R.K.; Tripathi, A.K.; Mishra, A.; Racherla, V. Non-uniformity in braking in coaching and freight stock in Indian Railways and associated causes. *Eng. Fail. Anal.* **2016**, *59*, 493–508. [CrossRef]
32. Zhang, Y.; Zhang, M. The application status of unit brakes on metro vehicles in China. *IOSR J. Mech. Civ. Eng.* **2018**, *3*, 17–23. [CrossRef]
33. Hodges, T. A life-cycle approach to braking costs. *Int. Railw. J.* **2012**, *52*, 47–48. Available online: https://www.railjournal.com/in_depth/a-life-cycle-approach-to-braking-costs/ (accessed on 27 August 2024).
34. Cruceanu, C. *Brakes for Railway Vehicles*, 1st ed.; Matrix Room: Bucharest, Romania, 2007. (In Romanian)
35. Cruceanu, C. Train braking. In *Reliability and Safety in Railway*; Perpinya, X., Ed.; IntechOpen: London, UK, 2012.
36. Bucur, F.; Socalici, A.; Berghian, A.B.; Baneasa, C.B.; Pascu, L. The tribology of composite materials used for manufacturing brake shoes. *Materiale Plastice* **2022**, *59*, 13–20. [CrossRef]
37. Kiss, I.; Cioata, V.; Alexa, V.; Ratiu, S. Investigations on the selection of friction materials destined to railway vehicles applications. *Ann. Fac. Eng. Hunedoara-Int. J. Eng.* **2016**, *14*, 231.
38. Kiss, I. The chemical composition of phosphorous cast irons behavior in the manufacturing of brake shoes meant for the rolling stock. *Acta Tech. Corviniensis-Bull. Eng.* **2016**, *9*, 77–84. Available online: https://www.proquest.com/scholarly-journals/chemical-composition-phosphorous-cast-irons/docview/1806389126/se-2 (accessed on 30 August 2024).
39. Vatulia, G.; Lovska, A.; Myamlin, S.; Stanovska, I.; Holofieieva, M.; Horobets, V.; Nerubatskyi, V.; Krasnokutskyi, Y. Revealing the effect of structural components made of sandwich panels on loading the container transported by railroad. *East. Eur. J. Enterp. Technol.* **2023**, *121*, 48–56. [CrossRef]
40. Fischer, S.; Liegner, N.; Bocz, P.; Vinkó, Á.; Terdik, G. Investigation of track gauge and alignment parameters of ballasted railway tracks based on real measurements using signal processing techniques. *Infrastructure* **2023**, *8*, 26. [CrossRef]
41. Leitner, B.; Figuli, L. Fatigue life prediction of mechanical structures under stochastic loading. In Proceedings of the 22nd Slovak-Polish Scientific Conference on Machine Modelling and Simulations, MMS 2017, Sklene Teplice, Slovakia, 5–8 September 2017.
42. Szczepański, E.; Gołębiowski, P.; Kondracka, B. Evaluation of the technological process of wagon processing at shunting stations using the simulation model. *Sci. J. Silesian Univ. Technology. Ser. Transp.* **2023**, *120*, 249–267. [CrossRef]
43. *TU U 6-05495978.017-2001*; Composite Brake Pads with a Mesh-Wire Frame for Railway Freight Cars. OJSC "Tribo": Lviv, Ukraine, 2001; 27p. (In Ukrainian)
44. *UIC Code 541-3*; Brakes—Disc Brakes and Their Applications-General Conditions for the Certification of Brake Parts (2017). International Union of Railways: Paris, France, 2017.
45. *UIC 541-4*; Brakes. Brakes with Composite Brake Blocks—General Conditions for the Certification of Composite Brake Blocks (2010). International Union of Railways: Paris, France, 2010.
46. Pascu, L.V. Research on Improving the Quality of Brake Shoes Intended for Rolling Stock. Ph.D. Thesis, University Politehnica Timisoara: Timișoara, Romania, 2015. (In Romanian)
47. Zhang, Y.; Chen, Y.; He, R.; Shen, B. Investigation of tribological properties of brake shoe materials—phosphorous cast irons with different graphite morphologies. *Wear* **1993**, *166*, 179–186. (In Ukrainian) [CrossRef]
48. Jamshinsky, M.M. Analysis of brake pads and ways of evaluating their promising designs. *Bull. East. Ukr. Natl. Univ.* **2015**, *1*, 204–209. (In Ukrainian)
49. *STP 04-010:2018*; Freight Cars. Maintenance and Repair System According to Technical Condition. Ukrzaliznytsia Stock Company: Kyiv, Ukraine, 2019; 25p. (In Ukrainian)
50. Kellrich, M.B. *The Basics of Wagon Reliability: A Study Guide*; UkrSURT: Kharkiv, Ukraine, 2013; 106p. (In Ukrainian)
51. Bidjuk, P.I.; Tkach, B.P.; Kharrinhton, T. *Mathematical Statistics*, 1st ed.; Personal: Kyiv, Ukraine, 2018. (In Ukrainian)
52. Javala, L.L.; Slyusarchuk, Y.M.; Khromyak, Y.Y.; Tsymbal, V.M. *Probability Theory, Mathematical Statistics and Probabilistic Processes: Textbook*; Lviv Polytechnic: Lviv, Ukraine, 2015; 364p. (In Ukrainian)

53. Kamkina, L.V.; Nadtochiy, A.A.; Grishin, A.M.; Stogniy Yu, D. *Fundamentals of Scientific Research: A Textbook*; NmetAU: Dnipropetrovsk, Ukraine, 2013; 89p. (In Ukrainian)
54. Thrane, C. *Applied Regression Analysis*, 1st ed.; Taylor and Francis Group: London, UK, 2019.
55. Sarip, S. Design development of lightweight disc brake for regenerative braking finite element analysis. *Int. J. Appl. Phys. Math.* **2013**, *3*, 52–58. Available online: https://www.academia.edu/26538825/Design_development_of_lightweight_disc_brake_for_regenerative_braking_finite_element_analysis (accessed on 3 September 2024). [CrossRef]
56. Sarma, K.V.S.; Vishnu Vardhan, R. *Multivariate Statistics Made Simple*, 1st ed.; Taylor and Francis Group: New York, NY, USA, 2018.
57. Donchenko, V.S.; Sidorov, M.V.-S. *Probability Theory and Mathematical Statistics: A Textbook*; Kyiv University: Kyiv, Ukraine, 2015; 400p. (In Ukrainian)
58. Lebedev, E.O.; Livinska, G.V.; Rozora, I.V.; Sharapov, M.M. *Mathematical Statistics: A Study Guide*; Kyiv University: Kyiv, Ukraine, 2016; 159p. (In Ukrainian)
59. Lan, Q.; Dhanasekar, M.; Handoko, Y.A. Wear damage of out-of-round wheels in rail wagons under braking. *Eng. Fail. Anal.* **2019**, *102*, 170–186. [CrossRef]
60. Jóvér, V.; Gáspár, L.; Fischer, S. Investigation of tramway line No. 1, in Budapest, based on dynamic measurements. *Acta Polytech. Hung.* **2022**, *19*, 65–76. Available online: https://acta.uni-obuda.hu/Jover_Gaspar_Fischer_121.pdf (accessed on 24 June 2024). [CrossRef]
61. Lu, B.; Chen, X.; Qu, B.; Zhang, H. Research on wheel-shoe wear for high friction composite brake shoes based foundation brake rigging in railway wagon. *Key Eng. Mater.* **2016**, *667*, 530–535. [CrossRef]
62. Jover, V.; Fischer, S. Statistical Analysis of Track Geometry Parameters on Tramway Line No. 1 in Budapest. *Balt. J. Road Bridge Eng.* **2022**, *17*, 76–106. [CrossRef]
63. Presciani, P.; Rinchi, M.; Pugi, L. Banchi per la certificazione del componenti frenanti. *Ing. Ferrov.* **2003**, *58*, 285–294.
64. Panchenko, S.; Gerlici, H.; Lovska, A.; Ravlyuk, V. Substantiation of the environmental efficiency of the device for parallel retraction of brake shoes for freight wagons. In Proceedings of the Current problems in rail vehicles—PRORAIL 2023, Žilina, Slovakia, 20–23 September 2023; pp. 109–118. [CrossRef]

vehicles

MDPI

Review

Revolutionizing Automotive Design: The Impact of Additive Manufacturing

Anis Hamza *, Kamel Bousnina, Issam Dridi and Noureddine Ben Yahia

Mechanical, Production and Energy Laboratory (LMPE), Higher National School of Engineering of Tunis (ENSIT), University of Tunis, Avenue Taha Hussein, Montfleury, Tunis 1008, Tunisia; kamel.bousnina@ensit.u-tunis.tn (K.B.); issam.dridi@ensit.u-tunis.tn (I.D.); noureddine.benyahia@ensit.u-tunis.tn (N.B.Y.)
* Correspondence: anis7amza@gmail.com

Abstract: Design for Additive Manufacturing (DfAM) encompasses two primary strategies: adapting traditional designs for 3D printing and developing designs specifically optimized for additive manufacturing. The latter emphasizes consolidating assemblies and reducing weight, leveraging complex geometries and negative space through advanced techniques such as generative design and topology optimization. Critical considerations in the design phase include printing methods, material selection, support structures, and post-processing requirements. DfAM offers significant advantages over conventional subtractive manufacturing, including enhanced complexity, customization, and optimization, with transformative applications in aerospace, medical devices, and automotive industries. This review focuses on the automotive sector, systematically examining DfAM's potential to redefine vehicle design, production processes, and industry standards. By conducting a comprehensive analysis of the existing literature and case studies, this research identifies gaps in the integration of additive manufacturing into broader manufacturing frameworks. The study contributes to the literature by providing insights into how 3D printing is currently reshaping automotive production by offering a forward-looking perspective on its future implications for the industry.

Keywords: additive manufacturing (AM); design for additive manufacturing (DfAM); lightweight parts; automobile

Academic Editors: Ralf Stetter, Udo Pulm and Markus Till

Received: 9 January 2025
Revised: 24 February 2025
Accepted: 28 February 2025
Published: 3 March 2025

Citation: Hamza, A.; Bousnina, K.; Dridi, I.; Ben Yahia, N. Revolutionizing Automotive Design: The Impact of Additive Manufacturing. *Vehicles* **2025**, *7*, 24. https://doi.org/10.3390/vehicles7010024

1. Introduction

The automotive industry is at the forefront of a technological revolution, driven by the rapid adoption of additive manufacturing (AM), commonly referred to as 3D printing. AM is fundamentally reshaping vehicle design and production by introducing new levels of flexibility, efficiency, and sustainability. Unlike traditional subtractive manufacturing processes that rely on costly molds, dyes, and tooling, AM builds components layer by layer, allowing for greater design freedom and material efficiency. This shift is significantly reducing barriers to entry in automotive production, as high initial costs and rigid manufacturing constraints have long hindered innovation and market competitiveness [1]. One of the most profound advantages of AM in the automotive sector is its ability to streamline prototyping and product development. Engineers can now rapidly produce and test multiple design iterations, accelerating innovation cycles and reducing time-to-market. The flexibility of AM also enables the production of highly customized and lightweight components, which is crucial for improving vehicle performance and fuelling efficiency [2]. For instance, polymer-based composite materials manufactured using AM techniques have been widely explored for their potential in lightweight structural applications, reducing overall vehicle weight without compromising durability or safety [3].

Beyond prototyping and lightweighting, AM is transforming the supply chain by facilitating on-demand production and reducing dependency on large-scale warehousing. This is particularly beneficial for automotive spare parts, where traditional inventory models often lead to excessive costs and logistical inefficiencies [4]. By enabling localized and decentralized manufacturing, AM minimizes lead times and ensures that critical components are readily available when needed. Moreover, the use of AM in producing complex tooling solutions, such as custom jigs and fixtures, enhances assembly line efficiency and precision while lowering production expenses [1]. The rise in metal additive manufacturing (MAM) has further expanded the scope of AM in the automotive industry, particularly in the fabrication of high-performance engine components and structural elements. Unlike conventional metal casting and machining processes, MAM allows for the creation of intricate geometries and optimized internal structures that improve component strength, thermal efficiency, and material utilization. This is particularly advantageous for internal combustion engine (ICE) components, where performance optimization and weight reduction are critical factors [5,6]. Additionally, AM facilitates the integration of novel materials and multi-material printing techniques, enabling the development of components with enhanced mechanical and thermal properties.

Despite its numerous advantages, the widespread adoption of AM in the automotive industry is not without challenges. Issues such as scalability, consistency in material properties, and regulatory compliance remain key concerns [7]. The need for standardization in AM processes and materials is essential to ensure reliability and safety in mass production. Additionally, cost considerations, particularly for high-performance metal AM systems, continue to be a limiting factor for broader industrial implementation [8]. However, ongoing research, advancements in printing technologies, and increased investment from automotive manufacturers are addressing these challenges, paving the way for a more sustainable and efficient future in automotive manufacturing.

This article delves into the transformative impact of additive manufacturing on automotive design, exploring its benefits, challenges, and future directions. By analyzing the latest developments and real-world applications, we aim to highlight how AM is shaping the next generation of vehicle production and driving innovation in the automotive sector.

2. Understanding Design for Additive Manufacturing (DfAM)

2.1. Definition and Principles (DfAM)

2.1.1. What Is DfAM?

Additive manufacturing (AM), or 3D printing, is a transformative technology that creates three-dimensional objects from digital files, gaining traction across various industries due to declining costs and increased capabilities. Its applications range from rapid prototyping to on-demand production, but the full potential of AM is realized only when designs are specifically tailored for its processes, adhering to the principles of Design for Additive Manufacturing (DfAM). This approach allows for significant cost reductions, particularly in low- to medium-volume production, by eliminating the need for expensive tooling and enabling the creation of complex geometries. Industries such as aerospace, consumer products, and medical devices benefit greatly from AM's capabilities for mass customization and efficient spare part manufacturing. As AM technologies continue to evolve, they are reshaping manufacturing practices and paving the way for more efficient and customizable production methods.

2.1.2. Key Principles of DfAM

Design for Additive Manufacturing (DfAM) involves customizing product designs specifically for 3D printing to maximize the advantages of additive manufacturing technologies [9]. The design phase is crucial; as it requires careful consideration of how parts will be manufactured, leading to stronger, lighter components with improved performance. One of DfAM's key benefits is the ability to consolidate complex assemblies into single parts, simplifying production and reducing costs. DfAM also optimizes support structures and build directions, minimizing material usage and print times. Intelligent design features, such as conformal cooling ducts, enhance functionality and process reliability. To succeed in AM, designers should avoid features like thin walls and intricate channels that complicate printing. Instead, using thicker cross-sections and managing internal cavities can prevent issues during production. Overall, DfAM represents a strategic approach that integrates thoughtful design considerations from the outset to improve the efficiency and quality of additive manufacturing processes.

2.1.3. Benefits of DfAM

DfAM involves optimizing design parameters to fully leverage the capabilities of specific AM technologies, each of which offers unique advantages. For instance, in Laser Powder Bed Fusion (LPBF), design parameters such as lattice structures and topology optimization enable lightweighting and high strength-to-weight ratios, making it ideal for aerospace and automotive applications. Fused deposition modeling (FDM) benefits from design considerations like minimal support structures and layer orientation, which reduce material waste and improve production efficiency, particularly for prototyping and functional parts. Stereolithography (SLA) excels in achieving high surface finish and intricate geometries, making it suitable for detailed components in medical devices and consumer products. Electron beam melting (EBM) allows for high-temperature-resistant designs with complex internal features, often used in the production of turbine blades and orthopedic implants. This scope is based on widely adopted AM technologies—LPBF, FDM, SLA, and EBM—due to their diverse applications and ability to demonstrate the transformative potential of DfAM. By detailing how specific design parameters align with the strengths of each technology, this section highlights the tailored advantages of DfAM across industries. DfAM has emerged as a critical area of research, focusing on optimizing the design process to fully leverage the capabilities of AM technologies. Tang and Zhao [10] provide a comprehensive review of design methods tailored for AM, emphasizing strategies such as topology optimization, lattice structures, and multi-material designs to enhance functional performance. These methods enable designers to create components with superior mechanical, thermal, and aerodynamic properties, which are often unachievable through traditional manufacturing techniques. For instance, topology optimization allows for the redistribution of material within a design to achieve optimal strength-to-weight ratios, while lattice structures can reduce weight without compromising structural integrity.

Building on this, Pradel et al. [11] introduce a framework for mapping DfAM knowledge, addressing the integration of AM-specific principles like part consolidation and lightweighting into traditional design workflows. Their framework highlights the importance of understanding AM processes and materials early in the design phase to maximize the benefits of AM, such as reduced assembly requirements and improved product performance. Wiberg et al. [12] further explore the available design methods and software tools for DfAM, highlighting the role of simulation, generative design, and topology optimization in enabling efficient AM-compliant designs. They note that while software tools have advanced significantly, there is still a need for more user-friendly interfaces and integrated workflows to make DfAM accessible to a broader range of designers and engineers. Despite the advancements, challenges remain in the widespread adoption of DfAM. Durakovic [13] identifies key barriers, including high initial costs, material limitations, and the need for specialized expertise, while also underscoring the benefits of reduced material waste and increased design freedom. For example, the limited range of materials suitable for AM can restrict the application of DfAM in certain industries, such as aerospace and automotive, where high-performance materials are often required. Additionally, the lack of standardized design guidelines and training programs can hinder the adoption of DfAM in traditional manufacturing environments. Alfaify et al. [14] contribute to this discussion with a systematic review of DfAM, emphasizing its role in promoting sustainability through material efficiency and energy savings. They argue that DfAM can significantly reduce the environmental impact of manufacturing by minimizing material waste and enabling the production of lightweight components that consume less energy during their lifecycle.

To enhance AM efficiency, it is important to minimize overhanging geometries that require support structures, as these complicate post-processing. Optimizing designs by using rounded edges instead of sharp corners can improve print quality (See Figure 1). Following minimum feature size guidelines ensures components are within the AM system's capabilities, contributing to better overall quality. Additionally, simplifying complex surface finishes is essential since AM has limitations in replicating intricate details compared to traditional methods. By adhering to these design principles, the effectiveness and quality of the additive manufacturing process can be significantly improved, leading to superior final products.

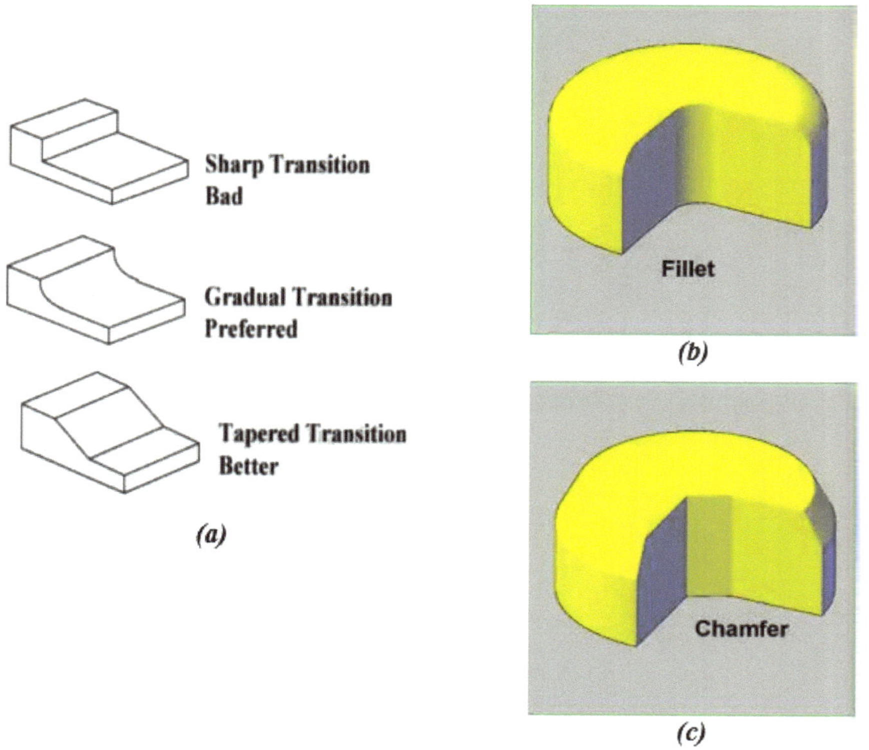

Figure 1. Illustrations of design simplicity (**a**) Using DfAM transitions should be tapered rather than sharp; (**b**) filets should have rounded corners for load bearing; and (**c**) chamfers should have slanted edges facilitating assembly.

Topology optimization is a design approach that uses simulation software to enhance material distribution within a specified space, aiming to create lighter structures while maintaining optimal mechanical performance. This process systematically removes unnecessary material, preserving essential load paths to minimize mass and improve strength (See Figure 2). It is particularly beneficial for AM, as it allows for the creation of complex geometries that are difficult to achieve with traditional methods. By leveraging the flexibility of AM, designers can produce efficient load-bearing shapes, maximizing performance while minimizing material usage.

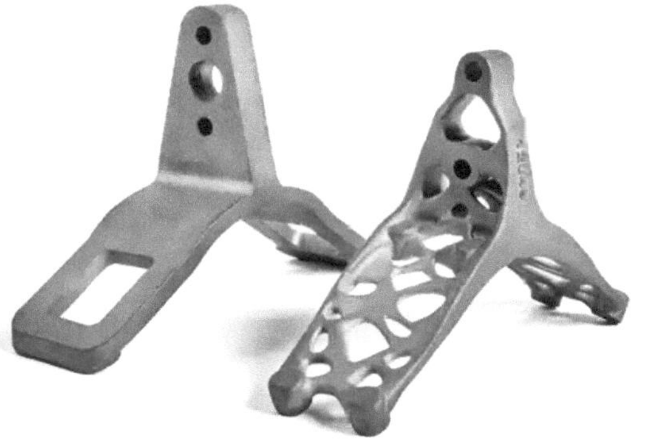

Figure 2. Examples of design topology optimization: The final ENDY additive bracket part (**left**) weighs about 40% less than the original (**right**), weighing in at 152 g instead of 250 g [15].

Accomplishing portion union through a solid plan includes engineers recognizing components that would customarily exist as partitioned pieces and consistently joining them inside the show. This all-encompassing approach permits the consolidation of complex inner structures and pathways that would be unreasonable or inconceivable to make utilizing routine strategies (See Figure 3). Within the domain of moving congregations, the required person-pieces to be joined is forestalled, as the complete component can be coordinated into a single printed substance. Interfacing that was once dependent on welds or clasp, presently consistently changes into morphological moves inside the bounded-together body geometry (See Figure 4). A striking advantage of this approach lies within the exceptional decrease in back structures, accomplished by killing overhanging geometries through portion solidification.

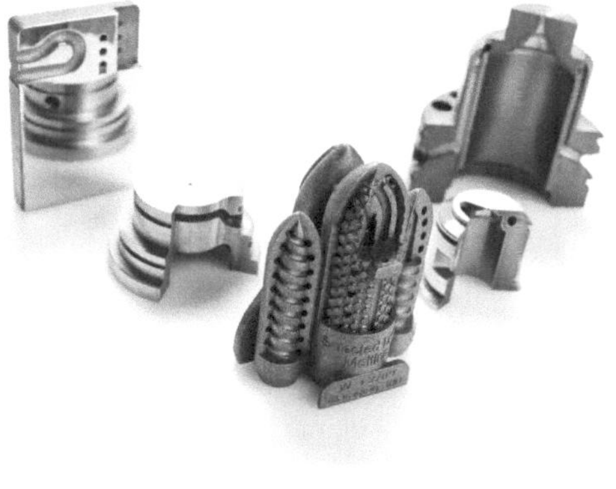

Figure 3. Examples of conformal cooling channel [16].

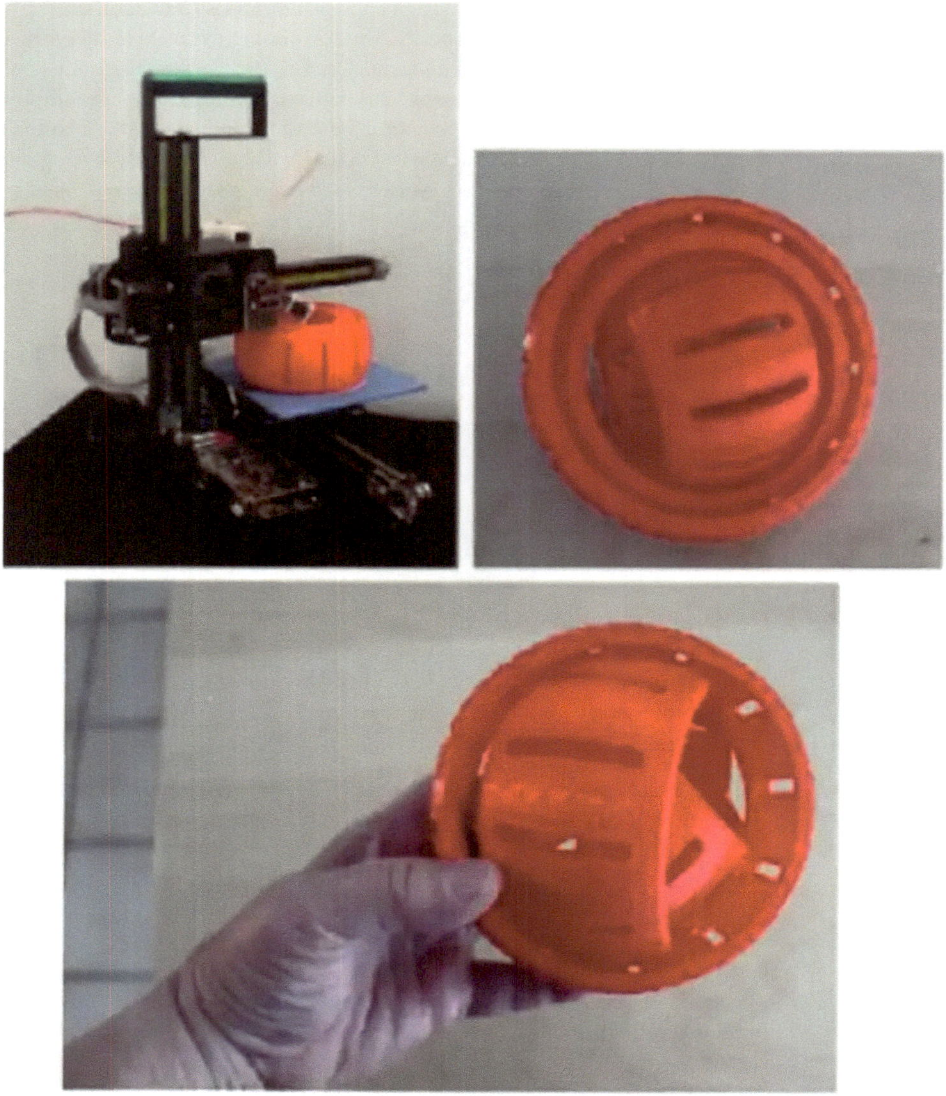

Figure 4. Examples of print moveable parts as a single piece [17].

Post-processing efforts are minimized with monolithic designs, as they eliminate the need for assembly and additional finishing work, simplifying inspection and repair. This approach enhances manufacturing efficiency and the robustness of the final product. Functional integration in 3D printing (See Figure 5) includes strategies like embedding conduits for cooling or fluid management, which optimizes performance. Lattice structures also contribute to multi-functionality in applications such as heat exchangers and biomedical implants. Overall, these design strategies improve the utility of printed parts, allowing for customization and tailored geometries that traditional manufacturing cannot achieve. The integration of hardware inside 3D printed parts has become conceivable with progressed printing procedures. This incorporates inserting circuitry and wiring inside the printed structures, opening up conceivable outcomes for applications such as Internet of Things (IoT) sensors and keen prosthetics. Moreover, plans made with post-processing in intellect contribute to useful integration. Contemplations for afterward forms, such as machining inner strings or connection focuses, are joined into the beginning plan to include mechanical capacities consistently. In outline, accomplishing useful integration through 3D printing includes an all-encompassing approach that envelops assorted plan components to optimize the utility and execution of printed parts over different applications.

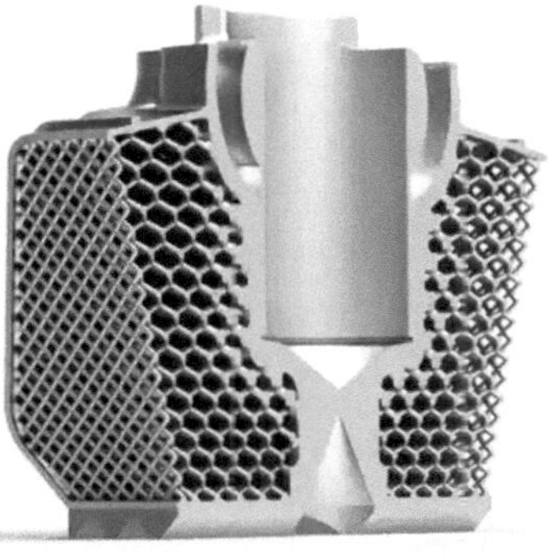

Figure 5. Examples of lattice structured heat exchanger [16].

Table 1 outlines key concepts and principles of Design for Additive Manufacturing (DfAM) as applied in the automotive industry. It defines each concept (e.g., lightweighting, part consolidation, topology optimization) and provides specific examples of their implementation in automotive production, such as lattice structures for lightweight chassis or 3D-printed engine brackets. The table also links these principles to relevant AM technologies (e.g., LPBF, FDM, SLA), illustrating how DfAM drives innovation and efficiency in automotive manufacturing.

Table 1. Key concepts and principles of DfAM in automotive production.

Concept/Principle	Definition	Example in Automotive Industry	Relevant AM Technology
Lightweighting	Reducing component weight while maintaining strength and functionality.	Use of lattice structures in car chassis to reduce weight without compromising safety.	LPBF, EBM
Part Consolidation	Combining multiple parts into a single component to simplify assembly.	Integration of 20+ parts into a single 3D-printed engine bracket, reducing assembly time and cost.	LPBF, FDM
Topology Optimization	Optimizing material distribution to achieve the best performance-to-weight ratio.	Redesign of suspension components to improve load-bearing capacity and reduce material usage.	LPBF, SLA
Customization	Tailoring designs to meet specific user or application requirements.	Production of custom-fit ergonomic seats using 3D scanning and AM.	FDM, SLA
Complex Geometries	Designing intricate shapes that are difficult or impossible to produce traditionally.	Creation of aerodynamic ducts with internal channels for improved airflow in electric vehicles.	LPBF, SLA
Material Efficiency	Minimizing waste by using only the necessary material for production.	Use of generative design to produce lightweight, material-efficient brake calipers.	LPBF, EBM

2.2. Tools and Software for Designing Additive Manufacturing (DfAM)

The integration of advanced tools and software into DfAM has been instrumental in unlocking the full potential of AM technologies. These tools enable designers to optimize geometries, reduce material usage, and improve functional performance, while addressing the unique constraints and opportunities of AM processes. Saadlaoui et al. [18] conducted a comparative study of topology optimization methods using industrial codes, highlighting the importance of selecting the right software to achieve optimal design outcomes. Their work demonstrates how different tools can influence the efficiency and effectiveness of topology optimization, particularly in the context of AM, where complex geometries and material distributions are often required.

Reddy et al. [19] further explore the capabilities of topology optimization software for AM, providing a comprehensive review of current tools and their applications. They emphasize the role of these tools in enabling designers to create lightweight, high-performance components by iteratively refining designs based on structural and functional requirements. The authors also present a real-world example to illustrate how topology optimization software can be applied to solve complex engineering challenges, showcasing its potential to revolutionize product design and manufacturing.

Dalpadulo et al. [20] assess the effectiveness of computer-aided design (CAD) tools for topology optimization in the context of automotive components. Their study identifies key features and limitations of existing software, providing valuable insights for designers and engineers working in the automotive industry. The authors highlight the importance of user-friendly interfaces, robust simulation capabilities, and seamless integration with AM processes to ensure the successful adoption of DfAM tools. They also stress the need for continuous improvement in software development to address the evolving demands of AM applications.

Tyflopoulos and Steinert [21] contribute to this discussion with a comparative study of commercial software for topology optimization. Their research evaluates the performance of various tools in terms of usability, computational efficiency, and design outcomes. The findings reveal significant differences between software packages, underscoring the importance of selecting the right tool for specific applications. The authors also discuss the challenges of balancing computational complexity with design accuracy, particularly when working with large-scale or highly complex models.

Fuchs et al. [22] address the necessary advances in CAD tools to fully leverage the design freedom offered by AM. They argue that current software often falls short in supporting the unique requirements of AM, such as the ability to handle complex geometries, graded materials, and multi-functional designs. The authors propose a roadmap for future developments in CAD tools, emphasizing the need for enhanced simulation capabilities, better integration with AM processes, and more intuitive user interfaces. They also highlight the importance of collaboration between software developers, researchers, and industry professionals to drive innovation in DfAM tools.

In summary, the development and adoption of advanced tools and software are critical to the success of DfAM. These tools enable designers to overcome the limitations of traditional manufacturing methods and fully exploit the design freedom offered by AM. However, challenges remain in terms of usability, computational efficiency, and integration with AM processes. Continued advancements in software development, coupled with collaboration between stakeholders, will be essential to address these challenges and unlock the full potential of DfAM. As the field continues to evolve, the role of tools and software in enabling innovative, efficient, and sustainable designs will become increasingly important, shaping the future of manufacturing across industries.

2.2.1. Generative Design Tools

Generative plan instruments play an essential part within the domain of plan-for-added substance-fabricating DfAM, with numerous devices leveraging progressed generative plan calculations. These apparatuses engage architects to input particular parameters and limitations, inciting the era of various plan cycles independently. For illustration, Autodesk Generative Plan utilizes cloud-based calculations to investigate a large number of plan conceivable outcomes consistently coordination with Auto-desk Combination 360. This collaborative approach improves the plan workflow, giving creators inventive arrangements. Another essential device is Altair Rouse, which employs topology optimization to make lightweight and fundamentally productive plans. With its user-friendly interface and real-time input on plan changes, Altair Rouse coordinating consistently with different CAD computer programs, encouraging proficient collaboration all through the plan handle. These generative plan tools exemplify the cutting-edge capabilities accessible for originators locked in within the added substance-fabricating scene.

2.2.2. Topology Optimization Tools

Topology optimization devices play a vital part within the added substance-fabricating scene, utilizing modern programs to deliberately modify strong fabric into productive geometric formats based on indicated stack and boundary conditions. These programs analyze components and calculate optimized cellular structure plans to upgrade execution and effectiveness. One such device is SIMULIA Tosca, a measured framework eminent for its capabilities in topology, measuring, shape, and dot optimization for mechanical structures. Outstandingly, it consistently coordinates with noticeable limited component examination (FEA) and strength solvers like Abaqus, Ansys, and MSC Nastran through Tosca, advertising a comprehensive arrangement for basic optimization. Another vital topology optimization device is OptiStruct by Altair, recognized for its capable solver devoted to basic optimization. This apparatus goes past topology optimization, expanding its capabilities to shape and estimate optimization. With strong execution in dealing with complex optimization challenges, OptiStruct is broadly utilized in businesses such as aviation and cars, where proficiency and auxiliary judgment are fundamental. These topology optimization devices speak to cutting-edge arrangements for engineers and architects looking to optimize and upgrade the execution of components in different businesses.

2.2.3. Support Structure Tools

Bolster structure apparatuses are fundamental within the domain of added sub-stance fabricating, advertising capabilities to optimize 3D printing models successfully. One striking suite of devices is Polydevs, giving a comprehensive set of functionalities for optimization. This suite envelops seeing, altering, repairing, arranging, and investigation, advertising an all-encompassing approach to improve 3D printing preparation. Polydevs stands out for its capacity to auto-generate and fine-tune bolstering structures custom-fitted to different geometries. Moreover, it encourages a smoother wrapping-up handle by making contact focuses that effortlessly break off, streamlining post-printing methods. Another essential back structure instrument is Ultimaker Cura, famous for its user-friendly interface that disentangles the setup of back structures. This instrument permits clients to compare components such as fabric utilization, print time for distinctive introductions, and bolster settings, empowering educated decision-making amid the pre-printing stage. Ultimaker Cura presents tree bolster structures as an eminent highlight, optimizing fabric utilization effectively. Generally, these back structure apparatuses contribute altogether to the added substance-fabricating workflow, guaranteeing the fruitful creation of complicated and complex 3D printed models with ideal bolstering.

Table 2 summarizes academic studies on the application of various software tools in Design for Additive Manufacturing (DfAM). It highlights the software used, the focus of each study, key findings or outcomes, and corresponding references. The table provides a concise overview of how tools like Altair 2021, ANSYS 2020 R1, SolidWorks 2019, nTopology

3.20, Autodesk Fusion 360, and PTC Creo 7.0 are utilized to optimize designs, improve performance, and streamline production in additive manufacturing, particularly in the automotive industry.

Table 2. Summary of academic studies on DfAM software applications.

Software	Study Focus	Key Findings/Outcomes	Reference
Altair	Topology optimization for light weighting	Achieved 40% weight reduction in automotive components while maintaining structural integrity.	Liu et al. (2018) [23]
ANSYS	Thermal analysis of 3D-printed parts	Identified optimal cooling strategies for engine components, improving thermal efficiency.	Leary et al. (2014) [24]
SolidWorks	Design automation for AM	Developed a workflow to automate part design, reducing prototyping time by 30%.	Rosen (2007) [25]
nTopology	Lattice structure optimization	Enhanced energy absorption in crash-resistant automotive parts using generative design.	Maskery et al. (2017) [26]
Autodesk Fusion	Multi-material AM design	Enabled complex multi-material designs for customized automotive interiors.	Gibson et al. (2015) [27]
PTC Creo	Integration of AM in traditional manufacturing	Demonstrated hybrid-manufacturing workflows for cost-effective production.	Thompson et al. (2016) [28]

2.3. Challenges and Limitations in Designing for Additive Manufacturing (DfAM)

There are wrangles about proposing that rather than advertising plan flexibility, plan-for-added substance-fabricating DfAM presents design limitations, compelling engineers to function inside the limitations of customary AM frameworks. Whereas it is genuine that AM has certain challenges and impediments, counting fabric properties, print determination, post-processing prerequisites, and high costs, endeavors are underway to address and overcome these issues. The current fabric choices for 3D printing are somewhat limited compared to conventional fabricating strategies, especially within the setting of high-strength, engineering-grade applications. There is continuous inquiry into the points to extend usable amalgam materials by creating progressed metal powders and optimizing printing parameters. Furthermore, the layer thickness in 3D printing can lead to a stair-stepping impact on surfaces, constraining geometric complexity. Whereas better resolutions are conceivable, they frequently come with slower construct speeds. Post-processing remains a common method for accomplishing craved wrap-ups, including exercises like sanding, penetrating, machining, or coating. Investigating endeavors are coordinated towards minimizing these manual steps to upgrade effectiveness and diminish costs. In addition, the high costs related to added substance-fabricating hardware and materials posture challenges for cost-effective small-scale generation. Activities are being embraced to make machines and materials more reasonable, with a center on standardizing forms and advancing parts commonality. Companies like Unionfab are effectively contributing to tending to these challenges by creating reasonable, high-quality AM arrangements, subsequently making 3D printing administrations available to small businesses and autonomous creators working on a budget.

3. Design for Additive Manufacturing in the Automobile Industry

3.1. Lightweighting and Part Consolidation

The concept of additive manufacturing design includes two distinct approaches: one involves modifying an existing conventional design to adapt to the requirements of 3D printing, while the other involves creating the design from scratch with a dedicated Design for Additive Manufacturing (DfAM) mindset. Plans made for added substance fabricating are regularly sought after two essential goals: gathering combination and light-weighting (See Figure 6). These goals are instrumental in tackling the benefits of 3D printing innovation to their fullest degree. Additively fabricated plans display an affinity for biomimicry, drawing motivation from characteristic shapes and structures, and grasp complex geometries that optimize negative space. Accomplishing these complex shapes is made conceivable through progressed plan techniques such as generative planning or topology optimization. These approaches improve the auxiliary productivity of components but also capitalize on the special capabilities of added substance fabricating.

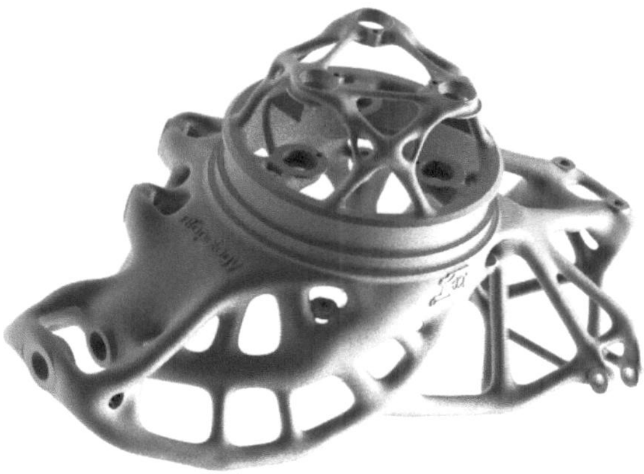

Figure 6. Lightweight and organic-looking topology optimized automotive upright bracket designed by Lions Racing Team eV [15].

3.2. *Customization and Complex Geometries*

3.2.1. Customized Automotive Parts

Assembly consolidation is a key facet of additive design, aiming to streamline the number of components in a given assembly. This not only simplifies the manufacturing process but also contributes to overall efficiency and durability. Light-weighting, another crucial aspect, involves the strategic use of materials to reduce the weight of components without compromising their structural integrity. This is particularly significant in industries such as aerospace and automotive industries, where minimizing weight contributes to enhanced fuel efficiency and performance.

3.2.2. Complex Geometries in Automotive Design

Biomimicry, inspired by nature's design principles, plays a pivotal role in additive design. Mimicking natural structures and forms often leads to more efficient and sustainable solutions. Additionally, the incorporation of complex geometries, made possible through generative design and topology optimization, allows designers to optimize negative space, ensuring that materials are used judiciously without sacrificing functionality.

The initial stages of the design process for additive manufacturing necessitate careful consideration of various factors. Printing processes, including the specific 3D printing technology to be employed, materials compatibility, and the need for support structures, are crucial elements. The design phase also extends its focus to down-stream steps, encompassing procedures like depowering and post-processing, which are essential for refining and preparing the final printed product. In summary, additive design embraces a dual approach, either adapting existing designs for 3D printing or originating from a DfAM perspective. The overarching goals involve assembly consolidation and light-weighting, with a strong emphasis on biomimicry and intricate geometries facilitated by advanced design techniques. Throughout the design process, careful consideration of printing processes, materials, and post-processing steps ensures the successful translation of concepts into high-quality, additively manufactured products.

In response to the intensifying demands of highly competitive markets, where considerations of manufacturing cost, reliability, and development cycles are paramount, innovative design techniques have emerged. Among these, the Design for X methodology (DfX) has evolved as a comprehensive strategy denoting a particular perspective to enhance product design and development. The X in DfX signifies a specific aspect targeted for improvement throughout the design process [29]. This strategic approach analyzes the entire product lifecycle from the early stages of design to achieve defined objectives, aiming to encapsulate essential knowledge about the process, materials, and product during the design phase [29].

Within the realm of design strategies, DfAM represents the application of DfX principles to the AM process, aspiring to fully exploit the capabilities of AM technology for design purposes [30]. DfAM encompasses a suite of tools oriented toward AM design methodologies, with three prevalent methodological approaches: the opportunistic approach, the restricted approach, and the dual approach [31–33]. The opportunistic approach places no limits on the complexity of AM, while the restricted approach considers boundaries related to materials, performance, machine characteristics, and manufacturability. The dual approach combines both opportunistic and restrictive techniques. In the automotive industry, DfAM approaches are examined, and reintroduced through two primary methods: component DfAM and assembly DfAM.

3.2.3. Case Studies of Customization and Complex Geometries

In essence, DfAM serves as a generalized design strategy or tool that enables the optimization of functional performance and critical product lifecycle aspects, such as manufacturability, cost, and reliability, within the constraints of additive manufacturing technology, particularly tailored for the automobile industry [34]. This approach persists due to the substantial design freedom offered by AM technology, and DfAM methods are essential to fully harness the specific characteristics of AM processes [35].

Typical DfAM methods or tools include topology optimization (determining the best layout of structural components in a vehicle), design for multiscale structures (incorporating lattice or cellular structures), multi-material design (enabling optimal material choices in each structural component for enhanced performance and cost-effectiveness), mass customization, part consolidation, and other design processes that benefit from AM-enabled features [36]. It is crucial to note that DfAM appliies to various 3D printing technologies in diverse ways, as different technologies require distinct optimization strategies. DfAM stands as a well-deployed additive solution viewed as a new tool to enhance manufacturing functionality and cost.

For DfAM to effectively function in the automotive sector, it necessitates a two-way street. An example of this is when the industry establishes a budget for production to obtain a component or an entire vehicle and begins the printing process. However, this approach often falls short as it lacks feedback or optimization. A well-deployed additive solution should embody a two-way street, extending from management to operations and vice versa. This workflow is adopted by automakers to ensure that DfAM is optimally utilized. The fundamental principles of DfAM in alignment with the automotive sector are elucidated in Table 3.

Table 3. Basic principles of DfAM in line with the automotive sector.

Principles of DfAM	Resource Reduction			Quality Enhancement		
	Weight and Cost	Printing Time	Touch Time	Strength	Aesthetic	Functionality
Additive thinking	■	■		■	■	
Design for orientation		■	■	■	■	
Contour design	■			■		
Segment parts	■	■				
Adding hardware			■	■	■	■
Minimizing complication	■	■	■		■	
Critical surface treatment						■

The extensive production durations, post-processing requirements, and surface quality constraints associated with AM have traditionally limited its integration into mainstream automotive manufacturing, often confining it to the prototype stage. Addressing these limitations requires substantial advancements, and recent breakthroughs in DfAM have displayed significant potential [14]. Combining topology optimization with an offset and a build time limit emerges as a promising approach to maximize the benefits of AM while mitigating concerns related to support material wastage and extended print times.

A case study illustrates the practical application of this approach in the design of a high-performance vehicle upright, highlighting the efficacy of topology optimization [14]. Topology optimization and the development of lattice structures represent DfAM approaches aimed at reducing material usage in automotive components with high stiffness requirements. To further optimize printing time and material volume, DfAM methods such as design by direct assembly, build orientation, and support reduction are employed. These DfAM technologies, particularly well-suited for extrusion-based additive manufacturing techniques, empower design engineers to create lightweight components for diverse applications, spanning automotive, aerospace, and electronics [37].

3.3. Tooling and Spare Parts Production

While AM (additive manufacturing) techniques in traditional automobile manufacturing struggle to match the production rates achieved by conventional methods like injection molding, die-casting, or stamping, there is significant potential to leverage AM for producing the necessary tools for these processes. Traditional tooling involves significant investments due to the complexity and time required for machining durable materials. However, certain AM processes present a promising avenue for efficiently developing various types of molds and dyes at a reduced cost and within a shorter timeframe [38].

Topology optimization has garnered considerable interest for its inherent weight-saving potential during the design phase. Numerous instances, including vehicle frames/chassis, suspensions, engine support, transmission, cross-car beams, doors, engine cradles, brake calipers, etc., have been explored for their optimization using topology optimization approaches [39–48]. The Solid Isotropic Material Penalization scheme (SIMP) is one of the widely utilized topology optimization methods, demonstrated in optimizing the body-in-white (BIW) of an electric vehicle [49]. This approach, like others, aims to reduce compliance (increase stiffness) while adhering to constraints such as mass or volume. Another study displayed the optimization of a vehicle's spring lower seat using the SIMP method, effectively reducing mass while maintaining strength through a combination of topology and topography optimization approaches [50]. The hardware in 3D printers has evolved to the point where it is no longer the bottleneck of performance; rather, the part's shape often becomes the constraining element in its overall performance. Consequently, a significant research debate on DfAM has emerged [51]. While the design requirements for various 3D printing technologies serve as a crucial starting point for DfAM, they do not encapsulate the entire picture, a fact that is not widely reported. DfAM comprises three major concepts: assembly and sub-assembly part consolidation, geometric freedom and biomimicry, with considerable overlap between them. These concepts collectively contribute to unlocking the full potential enabled by 3D printing technology in the realm of additive manufacturing for the automotive sector.

4. Discussion

If one desires to unlock the tremendous potential of additive manufacturing, it is imperative to consider DfAM right from the inception of the design phase. In previous times, 3D printing was solely perceived as a means for rapid prototyping or as an alternative technique for fabricating spare parts without causing substantial alterations to the geometry. However, in today's era, additive manufacturing has significantly evolved to the extent that it is now employed in the production of critical components that not only meet but also surpass the most rigorous engineering requirements. This remarkable level of advancement can only be attained by fully capitalizing on the fundamental advantages offered by this technology. There exist various approaches that a skilled engineer specialized in DfAM can employ to exploit the distinctive capabilities of additive manufacturing production. On certain occasions, these approaches are utilized to enhance the ergonomics or aesthetics of a product, which is particularly prevalent in the realm of consumer goods. Conversely, other approaches have the potential to elevate the performance and functionality of a design within more advanced industries, such as the aerospace sector.

Based on our research in the application of artificial intelligence in the field of mechanical engineering, the integration of artificial intelligence (AI) into DfAM presents exciting possibilities for the automotive industry. AI can enhance the design process by enabling advanced generative design algorithms that analyze vast datasets to produce optimized geometries tailored for additive manufacturing. This capability allows designers to explore a wider range of design alternatives that might not be immediately apparent through traditional methods. Additionally, AI can facilitate real-time decision-making during the design phase by predicting performance outcomes based on simulated conditions, thus, reducing the need for extensive physical prototyping. Machine learning techniques can also be employed to analyze historical manufacturing data, identifying patterns that lead to improved material selection and process parameters, ultimately resulting in higher-quality components with reduced waste. Furthermore, AI-driven tools can automate the optimization of support structures and build orientations, streamlining the production process and minimizing post-processing efforts. As this research field evolves, the collaboration between AI and DfAM is expected to redefine automotive manufacturing, fostering innovation in lightweight structures and complex assemblies while enhancing overall efficiency and sustainability in production practices.

Despite its numerous advantages, the widespread adoption of AM in the automotive industry faces several challenges, including scalability, consistency in material properties, and regulatory compliance. ASTM's Additive Manufacturing Technology standards, such as ASTM F2924-14 for titanium alloys [52] and ASTM F3001-14 for nickel alloys [53], provide critical guidelines for material selection, process optimization, and quality control, ensuring consistency and reliability across industries. These standards are essential for addressing challenges related to material properties, scalability, and regulatory compliance in automotive applications. Meanwhile, the Wohlers Report 2024 (Analysis. Trends. Forecasts. Three-dimensional Printing and Additive Manufacturing State of the Industry) [54] offers a comprehensive analysis of the AM landscape, highlighting key trends such as the growing adoption of AM in automotive manufacturing, advancements in multi-material printing, and the increasing role of AI and automation in optimizing AM processes. Together, these resources underscore the importance of standardization and provide valuable insights into the current state and future direction of additive manufacturing. By adhering to these standards and leveraging industry insights, automotive manufacturers can overcome barriers to adoption and fully realize the potential of AM in their production processes.

5. Conclusions

In conclusion, the field of DfAM offers a wide range of capabilities that surpass those of subtractive techniques in terms of complexity, customization, and optimization. These advantages have already been proven and demonstrated through successful applications in various industries such as aerospace, medical devices, and automotive. Although certain challenges and obstacles currently exist, it is important to note that these are not inherent or fundamental limitations. Rather, they are temporary roadblocks that can be overcome through continued ingenuity, extensive research efforts, and ongoing technological advancements. By consistently pushing the boundaries and exploring new possibilities, the potential of additive manufacturing will continue to expand, ultimately revolutionizing the way products are designed and manufactured. Moreover, the utilization of the suggested DfAM technique is incredibly groundbreaking within the confines of the automotive sector. This revolutionary approach implies a significant transformation in the overall production paradigm, introducing AM as an alternative method within the manufacturing process rather than solely as a tool for rapid prototyping purposes.

Funding: This research received no external funding.

Conflicts of Interest: The authors declare no conflict of interest.

References

1. Bubna, P.; Humbert, M.P.; Wiseman, M.; Manes, E. Barriers to Entry in Automotive Production and Opportunities with Emerging Additive Manufacturing Techniques. In Proceedings of the SAE 2016 World Congress and Exhibition, Detroit, MI, USA, 12–14 April 2016. [CrossRef]
2. Leal, R.; Barreiros, F.M.; Alves, L.; Romeiro, F.; Vasco, J.C.; Santos, M.; Marto, C. Additive manufacturing tooling for the automotive industry. *Int. J. Adv. Manuf. Technol.* **2017**, *92*, 1671–1676. [CrossRef]
3. Salifu, S.; Desai, D.; Ogunbiyi, O.; Mwale, K. Recent development in the additive manufacturing of polymer-based composites for automotive structures—A review. *Int. J. Adv. Manuf. Technol.* **2022**, *119*, 6877–6891. [CrossRef]
4. Dalpadulo, E.; Petruccioli, A.; Gherardini, F.; Leali, F. A Review of Automotive Spare-Part Reconstruction Based on Additive Manufacturing. *J. Manuf. Mater. Process.* **2022**, *6*, 133. [CrossRef]
5. Gray, J.; Depcik, C. Review of Additive Manufacturing for Internal Combustion Engine Components. *SAE Int. J. Engines* **2020**, *13*, 617–632. [CrossRef]
6. Sarzynski, B.; Sniezek, L.; Grzelak, K. Metal Additive Manufacturing (MAM) Applications in Production of Vehicle Parts and Components—A Review. *Metals* **2024**, *14*, 195. [CrossRef]
7. Wiese, M.; Thiede, S.; Herrmann, C. Rapid manufacturing of automotive polymer series parts: A systematic review of processes, materials, and challenges. *Addit. Manuf.* **2020**, *36*, 101582. [CrossRef]
8. Tuazon, B.J.; Custodio, N.A.V.; Basuel, R.B.; Delos Reyes, L.A.; Dizon, J.R.C. 3D Printing Technology and Materials for Automotive Application: A Mini-Review. *Key Eng. Mater.* **2022**, *913*, 3–16. [CrossRef]
9. Bousnina, K.; Hamza, A. Reducing the energy consumed and increasing energy efficiency in the turning process. *Int. J. Mod. Manuf. Technol.* **2020**, *12*, 23–28.
10. Tang, Y.; Zhao, Y.F. A survey of the design methods for additive manufacturing to improve functional performance. *Rapid Prototyp. J.* **2016**, *22*, 569–590. [CrossRef]
11. Pradel, P.; Zhu, Z.; Bibb, R.; Moultrie, J. A framework for mapping design for additive manufacturing knowledge for industrial and product design. *J. Eng. Des.* **2018**, *29*, 291–326. [CrossRef]
12. Wiberg, A.; Persson, J.; Ölvander, J. Design for additive manufacturing—A review of available design methods and software. *Rapid Prototyp. J.* **2019**, *25*, 1080–1094. [CrossRef]

13. Durakovic, B. Design for additive manufacturing: Benefits, trends, and challenges. *Period. Eng. Nat. Sci.* **2018**, *6*, 179–191 [CrossRef]
14. Alfaify, A.; Saleh, M.; Abdullah, F.M.; Al-Ahmari, A.M. Design for additive manufacturing: A systematic review. *Sustainability* **2020**, *12*, 7936. [CrossRef]
15. Advancing eVTOL with Additive Features. Available online: https://www.aero-mag.com/advancing-evtol-with-additive (accessed on 18 November 2022).
16. Culp, E. A Closer Look at the Advantages of Conformal Cooling Technology. Available online: https://www.etmm-online.com/a-closer-look-at-the-advantages-of-conformal-cooling-a-404359 (accessed on 10 May 2022).
17. Riad, K. Photocuring Epoxy with Quantum Dots for 3D Printing. Master's Thesis, Concordia University Montreal, QB, Canada, February 2016. [CrossRef]
18. Saadlaoui, Y.; Milan, J.L.; Rossi, J.M.; Chabrand, P. Topology Optimization and Additive Manufacturing: Comparison of Conception Methods Using Industrial Codes. *J. Manuf. Syst.* **2017**, *43*, 178–186. [CrossRef]
19. Reddy, K.; Ferguson, S.N.; Frecker, M.; Simpson, T.W.; Dickman, C.J.I. Topology Optimization Software for Additive Manufacturing: A Review of Current Capabilities and a Real-World Example. In Proceedings of the ASME 2016 International Design Engineering Technical Conferences, Charlotte, NC, USA, 21–24 August 2016.
20. Dalpadulo, E.; Pini, F.; Leali, F. Assessment of Computer-Aided Design Tools for Topology Optimization of Additively Manufactured Automotive Components. *Appl. Sci.* **2021**, *11*, 10980. [CrossRef]
21. Tyflopoulos, E.; Steinert, M. A Comparative Study of the Application of Different Commercial Software for Topology Optimization. *Appl. Sci.* **2022**, *12*, 611. [CrossRef]
22. Fuchs, D.; Bartz, R.; Kuschmitz, S.; Vietor, T. Necessary advances in computer-aided design to leverage additive manufacturing design freedom. *Int. J. Interact. Manuf.* **2022**, *16*, 1633–1651. [CrossRef]
23. Liu, J.; Gaynor, A.T.; Chen, S.; Kang, Z.; Suresh, K. Current and Future Trends in Topology Optimization for Additive Manufacturing. *Struct. Multidiscip. Optim.* **2018**, *57*, 2457–2483. [CrossRef]
24. Leary, M.; Merli, L.; Torti, F.; Mazur, M.; Brandt, M. Optimal Topology for Additive Manufacture: A Method for Enabling Additive Manufacture of Support-Free Optimal Structures. *Mater. Des.* **2014**, *63*, 678–690. [CrossRef]
25. Rosen, D.W. Computer-Aided Design for Additive Manufacturing of Cellular Structures. *Comput.-Aided Des. Appl.* **2007**, *4*, 585–594. [CrossRef]
26. Maskery, I.; Aboulkhair, N.T.; Aremu, A.O.; Tuck, C.J.; Ashcroft, I.A. Compressive Failure Modes and Energy Absorption in Additively Manufactured Double Gyroid Lattices. *Addit. Manuf.* **2017**, *16*, 24–29. [CrossRef]
27. Gibson, I.; Rosen, D.; Stucker, B. *Additive Manufacturing Technologies: 3D Printing, Rapid Prototyping, and Direct Digital Manufacturing*, 2nd ed.; Springer: New York, NY, USA, 2015. [CrossRef]
28. Thompson, M.K.; Moroni, G.; Vaneker, T.; Fadel, G.; Campbell, R.I. Design for Additive Manufacturing: Trends, Opportunities, Considerations, and Constraints. *CIRP Ann.* **2016**, *65*, 737–760. [CrossRef]
29. Briard, T.; Segonds, F.; Zamariola, N. G-DfAM: A methodological proposal of generative design for additive manufacturing in the automotive industry. *Int. J. Interact. Manuf.* **2020**, *14*, 875–886. [CrossRef]
30. Vaneker, T.; Bernard, A.; Moroni, G.; Gibson, I.; Zhang, Y. Design for additive manufacturing: Framework and methodology. *CIRP Ann.* **2020**, *69*, 578–599. [CrossRef]
31. Markus, D.; Lorin, A.; Thomas, N.; Sven, M. Identifying an opportunistic method in design for manufacturing: An experimental study on successful a on the manufacturability and manufacturing effort of design concepts. *Procedia CIRP* **2021**, *100*, 720–725. [CrossRef]
32. Prabhu, R.; Simpson, T.W.; Miller, S.R.; Meisel, N.A. Fresh in My Mind! Investigating the effects of the order of presenting opportunistic and restrictive design for additive manufacturing content on students' creativity. *J. Eng. Des.* **2021**, *32*, 187–212. [CrossRef]
33. Prabhu, R.; Leguarda, R.L.; Miller, S.R.; Simpson, T.W.; Meisel, N.A. Favoring Complexity: A Mixed Methods Exploration of Factors That Influence Concept Selection When Designing for Additive Manufacturing. *J. Mech. Des.* **2021**, *143*, 102001. [CrossRef]
34. Orquéra, M.; Campocasso, S.; Millet, D. Design for Additive Manufacturing Method for a Mechanical System Downsizing. *Procedia CIRP* **2017**, *60*, 223–228. [CrossRef]
35. Kumke, M.; Watschke, H.; Vietor, T. A new methodological framework for design for additive manufacturing. *Virtual Phys. Prototyp.* **2016**, *11*, 3–19. [CrossRef]
36. Jung, Y.; Lim, S.; Kim, J.; Min, S. Lightweight design of electric bus roof structure using multi-material topology optimisation. *Struct. Multidisc. Optim.* **2020**, *61*, 1273–1285. [CrossRef]
37. Sakthivel Murugan, R.; Vinodh, S. Application of Design for Additive Manufacturing to an Automotive Component. In Proceedings of the 1st International Conference on Industry 4.0 and Advanced Manufacturing, Bangalore, India, 11–12 January 2024; pp. 169–183.

38. Mantovani, S.; Barbieri, S.; Giacopini, M.; Croce, A.; Sola, A.; Bassoli, E. Synergy between topology optimization and additive manufacturing in the automotive field. *J. Eng. Manuf.* **2021**, *235*, 555–567. [CrossRef]
39. Chuang, C.-H.; Chen, S.; Yang, R.-J.; Vogiatzis, P. Topology optimization with additive manufacturing consideration for vehicle load path development. *Int. J. Numer. Methods Eng.* **2018**, *113*, 1434–1445. [CrossRef]
40. Bousnina, K.; Hamza, A.; Ben Yahia, N. An Integration of PSO-ANN and ANFIS Hybrid Models to Predict Surface Quality, Cost, and Energy (QCE) During Milling of Alloy 2017A. *J. Eng. Res.* **2023**. [CrossRef]
41. Gibson, I.; Rosen, D.; Stucker, B. *Additive Manufacturing Technologies: 3D Printing, Rapid Prototyping, and Direct Digital Manufacturing*; Springer: New York, NY, USA, 2021.
42. Wang, Y.; Rosen, D.W. Machine Learning for Additive Manufacturing. *J. Manuf. Syst.* **2020**, *56*, 241–251. [CrossRef]
43. Marchesi, T.R.; Lahuerta, R.D.; Silva, E.C.N.; Tsuzuki, M.S.G.; Martins, T.C.; Barari, A.; Wood, I. Topologically Optimized Diesel Engine Support Manufactured with Additive Manufacturing. *IFAC Pap.* **2015**, *48*, 2333–2338. [CrossRef]
44. Liang, M.; Hu, J.; Li, S.; Chen, Z. Topology optimization of transmission gearbox under multiple working loads. *Adv. Mech. Eng.* **2018**, *10*, 1687814018813454. [CrossRef]
45. Li, C.; Kim, I.Y. Topology, size and shape optimization of an automotive cross car beam. *J. Automob. Eng.* **2015**, *229*, 1361–1378. [CrossRef]
46. Li, G.; Xu, F.; Huang, X.; Sun, G. Topology Optimization of an Automotive Tailor-Welded Blank Door. *J. Mech. Des.* **2015**, *137*, 055001. [CrossRef]
47. Li, C.; Kim, I.Y.; Jeswiet, J. Conceptual and detailed design of an automotive engine cradle by using topology, shape, and size optimization. *Struct. Multidisc. Optim.* **2015**, *51*, 547–564. [CrossRef]
48. Tyflopoulos, E.; Lien, M.; Steinert, M. Optimization of Brake Calipers Using Topology Optimization for Additive Manufacturing. *Appl. Sci.* **2021**, *11*, 1437. [CrossRef]
49. Yang, S.; Qi, C.; Hu, P.; Wei, Z.Y.; Wang, Y.L. Topology Optimization of Electric Vehicle Body in White Based on SIMP Method. *Adv. Mater. Res.* **2011**, *308–310*, 606–609. [CrossRef]
50. Kong, Y.; Abdullah, S.; Omar, M.; Haris, S.M. Topological and Topographical Optimization of Automotive Spring Lower Seat. *Lat. Am. J. Solids Struct.* **2016**, *13*, 1388–1405. [CrossRef]
51. Clare Scott. 3D Printing an Improved DMLS Automotive Component Using Topology Optimization and DfAM. 3DPrint.com | The Voice of 3D Printing/Additive Manufacturing. 2018. Available online: https://3dprint.com/230626/3d-printing-an-improved-automotive-component/ (accessed on 2 March 2025).
52. *ASTM F2924-14*; Standard Specification for Additive Manufacturing Titanium-6 Aluminum-4 Vanadium with Powder Bed Fusion. ASTM International: West Conshohocken, PA, USA, 2014.
53. *ASTM F3001-14*; Standard Specification for Additive Manufacturing Nickel Alloy (UNS N07718) with Powder Bed Fusion. ASTM International: West Conshohocken, PA, USA, 2014.
54. Wohlers, T.; Caffrey, T. *Wohlers Report 2024: Analysis, Trends, Forecasts, 3D Printing and Additive Manufacturing State of the Industry*; Wohlers Associates: Fort Collins, CO, USA, 2024.

MDPI AG
Grosspeteranlage 5
4052 Basel
Switzerland
Tel.: +41 61 683 77 34

Vehicles Editorial Office
E-mail: vehicles@mdpi.com
www.mdpi.com/journal/vehicles

www.ingramcontent.com/pod-product-compliance
Lightning Source LLC
Chambersburg PA
CBHW041417210926
51850CB00013B/3300
* 9 7 8 3 7 2 5 8 3 8 7 9 0 *